Technological Utopianism in American Culture

Technological Utopianism in American Culture

Howard P. Segal

The University of Chicago Press • Chicago and London

For my parents
and
my twin brother, Robert

The University of Chicago Press, Chicago 60637
The University of Chicago Press, Ltd., London
© 1985 by the University of Chicago
All rights reserved. Published 1985
Printed in the United States of America

95 94 93 92 91 90 89 88 87 86   6 5 4 3 2

Library of Congress Cataloging in Publication Data

Segal, Howard P.
  Technological utopianism in American culture.

  Bibliography: p.
  Includes index.
  1. Technology—Social aspects—United States.
I. Title.
T14.5.S43   1985   303.4′83   84-8669
ISBN 0-226-74436-1
ISBN 0-226-74438-8 (pbk.)

Contents

A group of illustrations follows p. 32.

Preface

Technological utopianism is not the kind of conventional topic an aspiring graduate student in history ordinarily seeks out, or is directed toward, as he or she contemplates a doctoral dissertation and perhaps a first book. It plugs few scholarly holes, complements no earlier case studies and suggests few future ones, and offers only modest methodological innovation. In any event, a first-year graduate seminar at Princeton on antebellum America (with departing Professor Martin Duberman) aroused my interest in utopianism generally as a phenomenon worth further pursuit. The most important antebellum communities, I discovered, either had already been or were then being studied by others or else lacked enough materials to warrant sustained investigation. In addition, and despite their personal appeal in several cases, those communities were not, I concluded, as important to the overall development of American society as their organizers had hoped and as their historians might have wished. I decided that other expressions of American utopianism might offer better prospects. Gradually I saw a clear but as yet unexamined connection between American utopianism and the idea—and reality—of technological advance, especially in the late nineteenth and twentieth centuries. Of all the solutions then proposed for the nation's problems, solutions ranging from Christianity to taxation to revolution, technology appeared to be the most important, at least in retrospect.

But not necessarily the best solution: for to me, technology meant dull machines with largely negative consequences for society. Only after serious reading in the history of technology did I come to see technology as a social and cultural phenomenon as much as a material one, and as a phenomenon with profoundly mixed consequences for society. My book is hardly an uncritical defense of technology—or of technological utopianism—but it is, I hope, a balanced and fair assessment of both. I am indebted to my fellow former graduate student at Princeton, Arnold Pavlovsky, for suggesting early in my research that I think about technology in a less one-sided, more sophisticated manner and for suggesting basic readings in the field. I am no less indebted to David Billington, a member of Princeton's Department of Civil Engineering and a member of my dissertation committee, for broadening and sharpening my understanding of technology, particularly from the perspective of the engineer. I have repeatedly cited Professor Billington's conception of technology as both machines and structures because it is the most comprehensive, most accurate, and most persuasive conception of technology that I know. Professor Billington's well received writings on this and related aspects of technology indicate that others share this assessment of his arguments.

The history of technology is one of the few growing fields within the historical profession. Part of that growth may be due to increasing public interest in and concern about technology in the modern world; but part too is surely due to the receptivity to younger scholars from technical and nontechnical backgrounds alike of the field's principal organization, the Society for the History of Technology (SHOT), and of the Society's journal, *Technology and Culture*. Melvin Kranzberg of the Georgia Institute of Technology, who founded SHOT and edited *Technology and Culture* for many years, has been unfailingly supportive of my entry into the history of technology and of my active participation within SHOT. At various times over the past decade I have also received encouragement and assistance from other noted historians of technology: George Basalla of the University of Delaware, George Daniels of the University of South Alabama, Daryl Hafter of Eastern Michigan University, Raymond Merritt of the Minneapolis College of Art and Design, Carroll Pursell of the University of California at Santa Barbara, Mark Rose of Michigan Technological University, Bruce Sinclair of the University of Toronto, and, above all, Edwin Layton of the University of Minnesota. I remain grateful for their support.

Technology is indeed more than machines and structures, but the "hardware" cannot be ignored. My courses in the history of technology at the University of Michigan were greatly enhanced by field trips to the Edison Institute (Greenfield Village and the Henry Ford Museum) in nearby Dearborn. These visits to one of the world's foremost technological museums contributed significantly to my maturity as a historian of technology. I am indebted to John Wright, Director of Education; John Bowditch, Curator of Power and Shop Machinery; and Steven Hamp, my former student and present Curator of Archival and Library Collections, for their interest and assistance during repeated class and individual visits to the Institute.

Utopian studies is a smaller, less organized, and even younger field than the history of technology. I have received encouragement over the years from several scholars within the field, but I am most indebted to Warren Wagar of the State University of New York at Binghamton.

I am also indebted to several other historians. John Thomas of Brown University has enthusiastically supported my efforts since the day I sent him my dissertation prospectus. Dorothy Ross, now of the University of Virginia, was a member of my dissertation committee and was particularly helpful in placing technological utopianism in both intellectual and psychological perspective. Two University of Michigan professors specializing in other fields but sympathetic to my work, Arthur Mendel and John Broomfield, kindly read my entire manuscript and provided perceptive and helpful criticisms. So too did Arnold Pavlovsky. Finally, Solomon Wank of Franklin and Marshall College, my alma mater, has been a mentor since a freshman course in modern European history. More than anyone else, he has made me sensitive to the deeper meaning and significance of serious alternatives to existing values, institutions, and societies. If, apart from personal inclinations, I am today a more understanding student of utopias and other often ridiculed forms of social criticism, it is due in good measure to his stimulating lectures and seminars years ago and to his continued support of my work.

Two former colleagues in the now dissolved Humanities Department of the University of Michigan's College of Engineering, Robert Martin and Frederick Peters, each read portions of the manuscript and furnished excellent criticisms from the perspective of the literary scholar. No less important, each has in recent years provided moral and emotional support amid the painful upheavals of academic life in hard times.

My twin brother Robert, a scholar in religious studies specializing in the not unrelated area of myth and symbolism, read over every line of every draft of my dissertation and much of this book. His seemingly endless pleas for greater clarity and precision of language and argument often irritated but ultimately convinced me that those passages indeed required rewriting. Thanks to his criticisms, this book is far better written than would otherwise have been the case.

James Banner, who agreed to direct my dissertation, proved wonderfully attentive to both the style and the substance of my work and improved it markedly. Even after leaving Princeton to head the American Association for the Advancement of the Humanities, he read with great care and perception my entire book manuscript and enhanced it as much as he had my dissertation. For this, and for his persistent support, I am profoundly grateful.

I am forever thankful to the University of Cincinnati for a Taft Postdoctoral Fellowship and to Dalhousie University for a Killam Postdoctoral Fellowship. At Cincinnati, Zane Miller and Henry Shapiro, my sponsors, provided fresh and superb critical perspective on my dissertation and its pending revision, as did Alan Marcus, now of Iowa State University, then their graduate student. Without their individual and collective insistence on my viewing technological utopianism as a particular response to pervasive problems in late nineteenth- and twentieth-century America, and as a phenomenon with a vocabulary and a set of ideas reflecting a specific time and place in American culture, my book would have lacked historical roots. More than anyone else, they steered me away from a traditional, somewhat ahistorical approach to intellectual history—or the intellectual history of technology—and toward an approach more sensitive to historical discontinuities, to the fundamental differences between the past and the present frequently concealed by surface similarities. Alan Marcus' lengthy critique of my dissertation has proved invaluable during the years of revision into this book.

At Dalhousie, I enjoyed the pleasant company of historians from several countries and, as a temporary resident of Atlantic Canada, gained a new and very useful perspective on American history. I particularly want to thank Bala Pillay, then chairman of the History Department, for his many kindnesses toward me during my two years at Dalhousie.

In the absence of an organized movement of technological utopians, it has never been easy to try to locate all possible visionaries and their usually obscure works. I have been assisted by dozens of librarians and archivists throughout the country who responded to my many inquiries, even when, as was often the case, there was little or nothing to report. I am most indebted to the staffs of the New York Public Library, the Library of Congress, and various libraries at Columbia, Harvard, Rutgers, and Yale Universities, the University of Pennsylvania, and, above all, Princeton University for their assistance during my visits. A much appreciated research grant from the Penrose Fund of the American Philosophical Society enabled me to revisit several of these repositories as I revised my dissertation. Vickie Gordenier typed several drafts of my manuscript with care, speed, accuracy, and good humor.

Introduction

Technological Utopianism
as the Solution to
Problems in America

Between the appearance of John Macnie's *The Diothas; Or, A Far Look Ahead* in 1883 and Harold Loeb's *Life in a Technocracy: What It Might Be Like* in 1933, twenty-five individuals published fundamentally similar visions of the United States as a utopian society—visions that, they were certain, technological progress would eventually make real. More clearly, more systematically, and more intensely than any other group, these "technological utopians" espoused a position that a growing number, even a majority, of Americans during these fifty years were coming to take for granted: the belief in the inevitability of progress and in progress precisely as technological progress. But where their fellow Americans did not look far beyond the present, the technological utopians took these convictions to their logical finale: they equated advancing technology with utopia itself. This situation is my starting point, the spring from which my other observations and analyses all flow. Why these works appeared when they did, why no more such works seem to have appeared, what these works tell us about their authors and the real and ideal worlds in which their authors lived, and what they tell us about our own real and ideal realms—these are the questions that I will ask of those twenty-five utopian visions.

As separate phenomena, American technology and American utopianism have received considerable scholarly attention, especially in recent years.[1] What I will examine here is their historical connection, which, as far as I know, has never been systematically traced. Moreover, I will look at that connection before and after as well as during the years 1883–1933, the period of my particular concern. In dealing

1

with these ideas, I hope to dispel the customary association of *uto-pian* with *impossible*. For at least these twenty-five prophets and their followers actually believed that advancing technology would be the key to turning the impossible into the possible and even the probable.

To be sure, visions of the United States as a technological utopia antedate Macnie's book and postdate Loeb's. The earliest such work is John Adolphus Etzler's *The Paradise Within the Reach of All Men* (1833); among the latest is Buckminster Fuller's *Utopia or Oblivion* (1969). Moreover, the intellectual origins of technological utopianism in general may be traced back to European works such as Johann Andreae's *Christianopolis* (1619), Tommaso Campanella's *The City of the Sun* (1623), Francis Bacon's *The New Atlantis* (1627), Marquis de Condorcet's *Sketch for a Historical Picture of the Progress of the Human Mind* (1795), and the nineteenth-century writings of Henri de Saint-Simon, Auguste Comte, Robert Owen, Charles Fourier, and even Karl Marx and Friedrich Engels. But none of the Europeans made technological advance their panacea, as did all the American technological utopians.

The twenty-five technological utopians considered here thus share with their American predecessors and successors some degree of reliance on technological advance rather than other panaceas. Both Etzler and Fuller, for example, depend as thoroughly upon technolog-ical advance to bring about utopia as do Macnie and Loeb. What set apart these twenty-five futurists from the others are their *particular visions* of technological utopia. These differences in vision largely reflect differences in the general outlooks of the times in which they wrote. As I will show in more detail later, this set of prophets wrote in the heyday of that series of economic, social, and cultural transforma-tions called America's industrial revolution, where previous prophets like Etzler wrote in its initial stages and subsequent prophets like Fuller in its mature stages. Inevitably, all saw their visions through the lenses of their own times, and not just in general outline but in specific content: the precise technological (and non-technological) advances predicted and the forms they took. No less important, as I will also elaborate on later, those twenty-five technological utopians wrote as the dominant metaphor and model of American society was in transition from a mechanical to an organic social order. By contrast, Etzler and Fuller wrote while mechanical metaphors and models of American society held sway. The prevailing assumptions of their times affected all the prophets' visions. The historical significance of these utopians lies in the content of their visions, in their confidence in the accuracy of their visions, and in the relationship of their visions to the particular cultural context—and crises—of late nineteenth- and early twentieth-century America.

Except for *The Diothas* by Macnie and one article by Robert Thurston, all the works treated here appeared after Edward Bellamy's *Looking Backward* (1888). Several of the other twenty-three visionaries are known to have been inspired by the phenomenal success of *Looking Backward*. Several others may be suspected of having been inspired by it as well. Even *The Diothas* may owe something to Bellamy's work, for it gained modest popularity after 1888 as the alleged—and allegedly unacknowledged—inspiration for *Looking Backward*.[2] That only two other individuals besides Etzler and Macnie—Mary Griffith and Thomas Ewbank—published technological utopian works before 1888, and that so many published them comparatively soon thereafter, further suggests the impact of *Looking Backward* upon the development of this variety of American utopianism. So does the proliferation of American utopian works of all kinds in the fifteen or so years after 1888, critical though some were of Bellamy's scheme.[3]

With the grand exception of Bellamy, these technological utopians lived and wrote in obscurity.[4] A few of them knew one another personally, and a few more knew of one another's writings, but most of them apparently worked alone. Consequently, American technological utopianism never constituted a self-conscious movement and ideology, as did Populism, for example. Indeed, only in the 1930s, with the short-lived flourishing of the Technocracy crusade, did it become organized at all. What prominence was achieved by technological utopians came within their everyday callings as businessmen or professionals. Better known utopians included carriage-maker Chauncey Thomas, inventor and manufacturer King Camp Gillette, civil engineer George Morison, mechanical engineer Robert Thurston, professional writer Harold Loeb, and clergyman Solomon Schindler. Their lives, and those of their fellow technological utopians, will be examined in chapter 3.

First, however, I want to consider why Bellamy should have attained fame and influence while his fellow technological utopians endured obscurity. One possible explanation is simply that Bellamy, a seasoned writer, wrote better. A second explanation is that he wrote just at the onset of what has been termed the late nineteenth-century crisis of confidence in America, where all but one of the rest of them wrote during or after its peak. A third and complementary explanation is that imitations of *Looking Backward*, or unauthorized sequels to it, as were most of the works treated here, could hardly generate the enthusiasm of the original. A final and deeper explanation, which excludes none of the other three, is that the emphasis of *Looking Backward* on cooperation and community as well as on technological advance offered a more balanced and more appealing vision than the strictly materialist focus of nearly all the other works.

Thus *Looking Backward* was more than an attractive prediction of utopia to be brought about through social engineering. To argue, as Robert Wiebe does, that "Bellamy's book won its huge audience not as fiction but as a simple, logical essay combining so much that the discontented already accepted as gospel,"[5] does not sufficiently account for the presumed necessity of *utopian* versions of the "gospel" or the apathy that greeted similar utopian representations. Whatever the explanations, the works of those who followed Bellamy—as well as his own sequel, *Equality* (1897), a purer example of social engineering—never aroused the same enthusiasm as *Looking Backward*.

For too long, however, too many students of late nineteenth- and early twentieth-century America have relied upon *Looking Backward* as a guide to the fundamental nature of American culture at the turn of the century. They have substituted study of this one avowedly utopian work for comprehensive, systematic, and sustained investigation of the real world during this period.[6] Utopian works cannot themselves illuminate more than a portion of any real world culture, because, by their very design, they deviate from and often distort existing society, especially when their principal purpose is to change it. At most, they can identify particular values, trends, and problems in the culture that fostered them. They must therefore be employed cautiously, as means to full-scale historical inquiries, rather than as complete inquiries in themselves.

Similarly, the popularity of *Looking Backward* cannot alone account for the general popularity of the ideas of inevitable progress and of progress precisely as technological progress. For those ideas, as will be shown in chapters 4 and 5, preceded its appearance. *Looking Backward* may have popularized those ideas but it did not produce them. Given the existence of these ideas, we may wonder why there were not more works like Bellamy's, especially earlier or contemporary technological utopian works. *Looking Backward* may have lessened the popularity of later technological utopian works, but it cannot have inhibited earlier ones from appearing.

Without intending to account at this point for developments that preceded the publication of *Looking Backward*, I want to reconsider Wiebe's statement, "Bellamy's book won its huge audience not as fiction but as a simple, logical essay combining so much that the discontented already accepted as gospel." Although weak as an explanation of *Looking Backward*'s unique popularity, the statement points to the preexistence of a firm, even rigid set of beliefs about contemporary American society, if only among the millions of "discontented." The statement implies the preexistence of a coherent view of reality, which may properly be called an *ideology*.

Conceiving ideology in this positive sense, rather than in its more pejorative modern one, leads us to see ideology as an illumination rather than distortion of reality. This, in fact, was the original meaning of ideology when the term was first used in the late 1790s and early 1800s. The concept was devised by the savants of the Institut de France, and for positive purposes. To them *ideology* meant the understanding held by members of any social group about the way that group actually functioned. The size of the particular group to which the term applied was not critical, so long as it was cohesive enough to have such an understanding—or competing understandings. In other words, ideology was originally a normative concept. Though the name was new, the phenomenon, the savants contended, was not. Quite the opposite: ideology was a necessary condition of human existence, as modern students of culture have amply confirmed.[7] The particular forms of ideology naturally differed, but the concept in itself was value-free and not culturally relative in any way. Every social group needed an explanation of reality, and the explanation that developed and was then inculcated had enormous impact upon the thoughts, feelings, and behavior of its adherents.

As *utopia*, coined by Thomas More in 1516, was applied retroactively, so was *ideology* after its formulation around 1800.[8] The idea has remained current, although other terms, such as *worldview* (*weltanschauung*), have been substituted for it, especially to avoid the later negative connotation of *ideology*. More recently, *mentalité* and *archeology of knowledge* have been substituted as well, although not always with sufficient intellectual rigor.[9] The important point, however, is not the particular term used but the existence in every society of what will here be called *ideology*.

The appropriate functions of ideology will be detailed in chapter 9 of this study, in connection with the various functions of utopianism today. For now, I will observe only that societies often have one or more dissenting ideologies of this kind competing with and so criticizing the prevailing "official" ideology—as Wiebe implies in the case of *Looking Backward* and the readers attracted to its dissenting ideology. Indeed, utopianism is boldest as a rival ideology to the prevailing ideology of existing society. This kind of utopianism can be a potent vehicle of social criticism. It challenges the ingrained assumptions of existing society and offers significant alternatives to them. However, utopianism can also be a moderate or even conservative ideology, bridging rather than widening the gap between the real and the ideal worlds by demonstrating their relative proximity. This was exactly the achievement of *Looking Backward*.

Viewed as an ideologically conservative utopia, *Looking Backward* may be said to have appeared in a cultural context predisposed to favor its principal themes. Far from appealing to the discontented because it was ideologically radical, it appealed to them partly because it was ideologically conservative. And its conservatism gradually broadened its appeal beyond the already considerable ranks of the discontented and toward the American mainstream. If this does not explain the great numbers of Americans who were prompted to read this book, it does explain the empathy toward Bellamy of those who did read it.[10]

The ideological conservatism of *Looking Backward* may also help explain both the absence of more technological utopian works in turn-of-the-century America and the obscurity of those others that did appear. Put simply, there was only a limited need for additional declarations of the gospel of progress as technological progress. Once this notion was widely circulated and accepted —a process, to repeat, probably quickened rather than caused by *Looking Backward's* popularity—fewer Americans were intrigued by additional articulations of it.[11]

Technological utopias were not the only sort current at the turn of the century. Scores of other utopian works offering other panaceas— religion, taxation, socialism, and revolution among them—appeared in late nineteenth- and early twentieth-century America. And a considerable number of utopian communities, with equally varied beliefs and practices, sprang up during the period as well.[12] Yet none of these other expressions of utopianism, except perhaps Henry George's *Progress and Poverty* (1879), even approached *Looking Backward's* popularity and influence. Most utopian works and communities were as painfully obscure as the other technological utopian works.

Nevertheless, the mere appearance in late nineteenth- and early twentieth-century America of these many utopian communities and writings deserves our attention, because together they tell us something about that period of American history. Regardless of their particular values, forms, intellectual rigor, and popularity, all these works were intended as solutions to problems then confronting American culture and society. Moreover, all the genuine utopian writings at least were conceived as full-scale blueprints of their authors' version of utopia: that is, the nontechnical descriptions and drawings were intended to make clear the nature of utopian society— its physical appearance, its institutions, its values, and its inhabitants. Those sets of blueprints distinguished these utopians from the vastly larger number of Americans during this period who were mere rhetoricians of hope or of progress, or who were less visionary reformers seeking only piecemeal changes within existing society.

Utopian writers (and community builders) often agreed on the fundamental problems of the day even though they often disagreed on the specific solutions to them. Those problems included increasing poverty, unemployment, disease, rural and urban blight, immigration, political corruption, and centralization of economic power. Solutions included, besides technological advance, socialism, taxation, Christianity, education, and revolution. Equally important, large numbers of Americans of non-utopian bent likewise agreed that these were legitimate problems even though they sought more moderate solutions to them—primarily piecemeal reforms. Thus if most of the turn-of-the-century utopian messages were poorly received by the American public, it was not always because the public had a different perception of the problems at hand, a different ideology. Often it was the solutions proposed—an overall demand for utopianism rather than mere reform, or a particular scheme for reaching utopia—that drew this unfavorable response.

However received, the unprecedented outpouring of utopian writings in fin-de-siècle America reflected the hopes and the fears of far more Americans than ever read them. In this respect they were (partial) versions of what Wiebe aptly calls the "gospel," or what we may call the ideology of the discontented. As utopian solutions, however, they were both extensions and distortions of the gospel. Therefore, these writings do not inevitably drive us away from the mainstream of American culture and society of their times, as critics of utopianism contend; rather, these utopian writings, especially the technological utopian writings, lead us back into the mainstream—if by a circuitous route.

Technological utopianism, moreover, illuminates many larger and better known developments of late nineteenth- and early twentieth-century America. These developments range from conservation to corporate and government reorganization, from city planning to national planning, and from scientific management to Technocracy. I will examine technological utopianism not only in itself but also in relation to these broader developments.

To a lesser but no less important extent, I will also examine technological utopianism in relation to American culture as a whole from the late nineteenth century to the present. Although my study focuses on the decades immediately before and after the turn of the century, it includes contemporary American culture in its view. In particular, it investigates the impact of technology on our own culture and our culture's impact on technology. It thereby explores the nature of our contemporary "technological society."

In turning now to consider those twenty-five visions of technological utopia that I have introduced, I will emphasize the potential appeal, to a society already attracted to technological solutions for social problems, of unprecedented technological progress as the panacea for unprecedented social problems. I also will emphasize the potential appeal of technological progress as the means of at last making utopia real. And finally, I will stress the fears among at least some Americans that such unprecedented changes might become less the solutions to problems than primary problems in themselves. Consideration of the dream and the dreamers, however, will precede consideration of the possible nightmare.

Many persons today continue to equate advancing technology with utopia, and not just in America alone. At the same time, severe criticism increasingly is leveled in America and other technologically advanced societies against just this linkage. Not a few critics, in fact, have deemed technological progress and social progress to be outright antitheses. Part of this study will investigate the popularity and accuracy of both those notions in relation to the nineteenth and twentieth centuries alike. This investigation will primarily consider how far the technological utopians' predictions for the United States have been borne out by actual developments since they wrote; and secondarily, it will examine the hopes and fears that gave rise to those predictions in the first place. More bluntly, I will ask how technological progress, once hailed by millions as the panacea for virtually all of mankind's problems—and not merely its material problems— not only failed to solve a number of material and nonmaterial problems but, in the minds of thousands, became a principal problem in itself.

In criticizing technological utopianism, I do not mean to dismiss technology or utopianism in and of themselves. Every society has had and presumably always will have some degree of technology. To demand the virtual elimination of modern technology, a popular cry among some self-proclaimed humanists, is profoundly ahistorical and wholly unrealistic.[13]

The real issue here is the extent to which technological change has meant and may still mean genuine social improvement and the extent to which it has not and probably will not. To be sure, evaluations of "progress" are nearly as varied—and as value-laden—as schemes for perfection. Nevertheless, our society has great need of a more common agreement on the nature of technological progress. Toward that end, I suggest identifying some kind of plateau of technological progress beyond which technologically advanced societies like our own would proceed cautiously. It is essential that nontechnological progress—social, political, economic, and cultural— stays roughly even with technological progress. Such a proposal may itself appear utopian, but *utopian* does not always mean unfeasible. (This is discussed more fully in chapter 8.) Just such a conception lies behind recent demands for smaller scale, more decentralized, and more "appropriate" technology.

Like technology, utopianism ought not to be accepted or rejected altogether. Just as technology is intrinsic to every society, so utopianism, in the form of alternative or rival ideology, is intrinsic to many societies, although by no means all.[14] Utopianism is intrinsic to all those societies that allow—or fail to prevent—serious criticism of themselves and that in turn allow—or fail to eliminate—the possibility of serious social alteration. This is because every serious expression of utopianism—even as conservative a variety as technological utopianism—simultaneously reflects and criticizes the society that produces it.[15] Far from necessarily being escapist, genuine utopianism is intended to be played back on the real world in order to change it—that is, to make the real world more nearly perfect.[16]

At its boldest, utopianism represents thinking that is at once creative and critical about existing society. At its fullest, it offers an ideology to rival the prevailing ideology of society. While I personally advocate no particular expression of utopianism, including technological utopianism, I do advocate the conscious use of utopianism in general as a means of social criticism. This study, I hope, will prove to be not only a reliable account of technological utopianism but also a valid exercise in this brand of critical utopian thinking.

One

<hr>

The Vocabulary of
Technological Utopianism

<hr>

Before proceeding to describe the technological utopian visions and visionaries, I will define four principal terms. Three of these—*utopia, technology, culture*—will require extensive discussion. Once I have defined these, I will briefly discuss seven other terms—*evolution, equilibrium, efficiency, system, organization, planning, rationalization*—which are present or implicit in the description of technological utopia that follows and also in subsequent discussions of the various non-utopian reform crusades of the same period. As the present chapter will make clear, these widely used terms have more complex meanings than is commonly assumed. Further, their meanings have changed over time and must here be understood in their appropriate historical contexts.[1]

By *technological utopianism*, my first term, I mean a mode of thought and activity that vaunts technology as the means of bringing about utopia. There has been no previous book-length study in English of this subject, but a number of briefer publications have used the term, and with an interpretation identical to mine.[2]

By *utopia* I mean the perfect society, perfect at least in the eyes of its proponents.[3] Notions of what constitutes perfection obviously differ. If Carl Becker could call a book *Every Man His Own Historian*, perhaps in this book we could grant every person their own utopia. Nevertheless, it is useful to distinguish "genuine" utopias from "false" ones.

First, in a genuine utopia, perfection in both society and inhabitants usually entails a radical improvement of conditions as compared with pre-utopia—and, if other societies still exist, as compared with non-utopia. Utopia is—or should be—qualitatively different from pre-utopia and non-utopia. Radical change is necessary to achieve utopia, unless pre-utopia is already moving toward utopia—as is the case with American technological utopianism. Even here, however, considerable improvement in both society and its inhabitants is necessary. Both still must become perfect, ordinarily through the use of appropriate social arrangements: institutions, values, norms, activities, and so forth. Perfection does not come automatically: the inhabitants of utopia remain flawed by nature—save where, as in some utopian fiction, they are perfected through genetic engineering. Utopian society must maximize their virtues and strengths and minimize their vices and weaknesses. In discussing a utopia, the particular objectives and the means devised to obtain them define the particular perfection that is sought. *Perfection*, like *beauty*, is an empty word unless it is given specific contents.[4]

Second, not only the specific contents themselves but also their comprehensiveness further distinguishes genuine utopias from false ones. Genuine utopias seek changes in most if not all areas of pre-utopian society, where reform movements and other non-utopian crusades seek changes in only one or two of them, schools or prisons, for example. This is because the proponents of utopia are generally more dissatisfied with the basic structure and direction of their own, non-utopian, society than are the proponents of milder changes. The latter believe that perfection is utterly impossible or that their own society is already comfortably close to perfection. As Frank Manuel rightly observes, "neither specific reforms of a limited nature nor mere prognostications of the invention of new technological gadgetry need be admitted [to the ranks of genuine utopias]. Calendar reform as such would not qualify as utopian; but calendar reform that pretended to effect a basic transformation in the human condition might be."[5]

A third and final difference between genuine and false utopias is the seriousness of purpose found in the one but missing in the other. Whatever their particular forms and contents, all genuine utopias share the character described by George Kateb: "when we speak of a utopia, we generally mean an ideal society which is not an efflorescence of a diseased or playful or satirical imagination, nor a private or special dream-world, but rather one in which the welfare of all its inhabitants is the central concern, and in which the level of welfare is strikingly higher, and assumed to be more long-lasting, than that of the real world."[6] False utopias manifest only a portion of this attitude, if any. They frequently serve simply as means of escape from the real world; genuine utopias actively seek to make the real world a better place. If practicality of intention marks genuine utopian schemes, however, it is also true that the schemes proposed are often undermined by both impracticality and obscurity in their formulation.[7]

Utopia, then, is at once a more complex and more specific term than might be assumed. *Technology* and *culture*, on the other hand, are more complex and less specific terms and so are more difficult to define, as the abundance of competing definitions of each indicates. I employ *technology* to mean not only technological hardware itself but, more broadly, its use in establishing and maintaining an entire society. This conception goes beyond the older one of technology as hardware alone, inexplicably separate from the society which produces it—and which it affects in turn.[7] My conception brings technology closer not just to society but, I believe, to reality as well.

Nevertheless, technology as hardware alone must still be considered in order to understand technology's actual role in society. Even these conceptions differ enormously. The most sensible, to my mind, is that propounded by David Billington, a civil engineer. According to Billington, "technology, in the minds of many people, consists of machines. But machines provide only half a definition of technology: the other half consists of structures, those fixed buildings which form the physical foundations of society."[8] Structures include roads, bridges, dams, harbors, power plants, and skyscrapers; machines include cars, trucks, trains, airplanes, ships, pumps, motors, television sets, window air conditioners, and computers. Structures are designed to be static, permanent, large scale, and unique, and are custom made for a specific locale; machines are designed to be dynamic, temporary ("disposable"), small scale, and reproducible, and are mass produced for any number of sites. Thus, "if a bridge moves visibly, something is wrong," but "if a car does not move, there is trouble." Far from being autonomous, however, structures and machines are interdependent: "structures are built by machines, and machines have structures to hold them together." Structures, moreover, do not merely hold machines together but, as in buildings of various kinds, often house them.[9] Finally, both structures and machines require certain technological skills or techniques in order to be built. Such skills are an essential component of technology that deserves mention here. They enable structures and machines to be built in the first place and help determine their use in society.[10]

In addition to the distinction between structures and machines, another, less exact distinction can be applied to technology as hardware—its degree of complexity. Often *tool* is used to describe so-called primitive forms of technology and *machine* to describe advanced varieties. But these terms are imprecise and should be abandoned. As Billington has demonstrated, *machine* is only half of technology; and in any event the term could be applied to ancient and modern technology alike, as could *tool* and, for that matter, *structure*. Moreover, terms like *primitive* and *advanced* may prove no more illuminating, especially if value judgments are implied. Nevertheless, just as there are different forms of hardware, so there are ranges of complexity within *primitive* and *advanced* forms. Over time, especially in the West, technology has become more complex and more central to our everyday life. Recognition of this process provides an initial historical perspective on technological utopia.

The technological utopians recognized these changes, among others, in the real world in which they lived, and in part they based their predictions upon them. Since all of them wrote after the early stages of the industrial revolution in America, they all assumed that the tremendous changes already begun would eventually reshape American society. They were imaginative enough to incorporate in their works technological advances that at the time were merely ideas, not actual developments. None, however, anticipated the technological advances in electronics and in information processing that have come about since they wrote. The type of advances anticipated date the technological utopians' works and differentiate their visions from those of later technological utopians[11]—further proof of the validity and utility of distinguishing among stages of technological development.

Before discussing the (social) uses of technology as hardware, I will make one additional distinction, the one between *technology* and *applied science*. Although the terms are often used interchangeably, historically they are by no means synonymous. Only after the late eighteenth and early nineteenth centuries did the relationship between technology and science become sufficiently close for scientific principles to be applied to technology—technology as hardware. Before then, with few exceptions, science and technology were separate activities. They were pursued with different objectives—understanding the world (know-why) versus controlling it (know-how). And those who pursued each objective generally came from different social and educational backgrounds—well-to-do and formally educated scientists versus poorer and self-educated technicians. Acquiring knowledge for its own sake was sufficient for the one group; but the other sought to apply as well as acquire it. Any number of technological advances were accomplished by persons largely ignorant of the scientific laws that might have facilitated their invention; and many more scientists might have applied their findings than did, or cared to do so. If technology today is increasingly based upon scientific principles, and if scientific research today is as much applied as "pure," this interdependence is a relatively recent development.[12]

Technological utopianism, however, is more than hardware and applied knowledge. It is defined by its functions as well as its forms: the use of hardware (structures and machines alike) and, in addition, of knowledge (technical and scientific alike) to create and preserve an intendedly perfect society. Such a society may well be termed a "technological society"—without the negative connotation given the term by social critic Jacques Ellul[13]—because it ultimately patterns itself after the structures and machines and technical knowlege that at the outset helped to bring it about.

As we have observed, every society, however primitive, has some degree of technology, if only simple stick plows or bows and arrows. These elementary devices, and their more powerful and more complex successors, allow mankind to prevail within nature and survive the natural environment. Without such elementary devices, survival, and so society, would be problematic. Without their successors, increasing control over the natural environment, and in turn partial replacement of it by a man-made environment, would be inconceivable.[14]

What, however, distinguishes a technological society from previous societies is not only the extent, power, and complexity of its hardware and technical knowledge but also the extent to which technology shapes its values, its institutions, its techniques, and its way of life. Although Ellul's critique of technological society distorts the real world of even the most technologically advanced societies, it does describe many features of technological utopia—again, negatively rather than, as with the technological utopians, positively. Ellul's critique raises a question that I want to pursue later, whether there are values and orientations inherent in advanced technology. Put another way: Is the working of technology to some degree autonomous?[15]

If the term *technology* is difficult to define, so is the term *culture*. Like technology, culture has been defined variously, both very narrowly and very broadly.[16] As Raymond Williams has shown, the word originally meant " 'the tending of natural growth,' and then, by analogy, a process of human training." In the nineteenth century, the meaning changed from "a culture of something, . . . to *culture* as such, a thing in itself." It then acquired four specific meanings that reflected vigorous debates about its proper meaning and scope: (1) "a general state or habit of the mind"; (2) "the general state of intellectual development, in a society as a whole"; (3) "the general body of the arts"; and (4) "a whole way of life, material, intellectual, and spiritual."[17] These debates concerned first, whether culture was as narrow as (1)—or even (2) and (3)—or as broad as (4); second, whether culture was exclusively intellectual and artistic, as in (1), (2), and (3), or material as well, as in (4); and third, whether there actually were or should be an official, usually overt, "high" culture for society's elite and an unofficial, often covert, "low" culture for its masses. If nothing else, these debates made culture a public issue in America (and Europe) by the end of the nineteenth century. John Demos notes in an essay on colonial America, "virtually no one alive in 1670 could have made any sense of such a concept," but by 1940, Warren Susman observes in a companion essay, Americans had become acutely aware "not only of [American] culture but of the *idea* of [American] culture"[18]— so aware that they were nearly obsessed with discovering its "essence."

This concept of culture in its later forms—as a thing in *itself*, not merely a part of something else—was, as Williams demonstrates, a partial response first in Europe and then in America to certain changes brought about by the industrial revolution. And the discussions of the 1920s and 1930s on the nature of American culture, mentioned by Susman, represented a partial response to the recently acknowledged impact of technology upon American society.[19] Both these responses to technological change were as often negative as positive, but they were nevertheless responses to technological change. The extent to which advanced technology may shape a culture is a question that, like others raised in this study, must be asked twice: first of the imagined technological utopia, then of the real world America that was supposed to emulate it.[20]

For these reasons the material aspects of society as well as the nonmaterial must be included in any conception of culture purporting to reflect everyday life, whether real or ideal. In a companion essay to those of Demos and Susman, Neil Harris defines cultural history as the search for unities in society through linking artifacts (or objects) with ideas. The emphasis on unities may unfairly diminish society's disunities, but the attempt to connect specific objects and social generalizations is useful. It recognizes the need to consider material items along with ideas, values, and institutions when describing a distinctive American culture or, for that matter, any culture.[21]

However broadly defined, culture, like technology, does not encompass everything in society. Indeed, culture, again like technology, may shape society—and be shaped by society—without being identical with society. For culture, even in the widest sense, refers to only the texture of life, not to the strictly material processes of society—

such as industrialization and urbanization—which may make that texture possible.[22] If *material culture* often categorizes those material objects, *civilization* sometimes refers to those material processes, especially for technologically advanced societies like post–Civil War America. The term *technological civilization* (or *material civilization*) is not uncommon and would certainly apply to technological utopia.[23]

This brings us to the remaining seven terms. As used by the technological utopians and their reform-minded fellow Americans of the late nineteenth and early twentieth centuries, *evolution* means not simply gradual change but, more specifically, planned gradual change toward a particular goal: usually, an improved society of some kind growing out of the existing society, its complement rather than its antithesis. The Darwinian notion of evolutionary development, which at first glance seems directly relevant here, actually applies only peripherally. For Darwin applied his findings to human societies only tentatively and did not posit a specific, final goal like this one. His notion of random variation—unplanned, of course, by mankind— would produce a general improvement over time but included no concrete objective for any species or for the earth as a whole.[24]

Similarly, *equilibrium* means more than mere stability. It too refers to a condition of society and may be temporary or permanent or, in most cases, both: an alternating series of equilibriums and disequilibriums leading ultimately to an improved society of some kind that will constitute a permanent equilibrium. Before then, the process of change toward either equilibrium or disequilibrium or both may be gradual (i.e., evolutionary), as in biology, or spontaneous (i.e., thermodynamic), as in chemistry.[25]

Likewise, *efficiency* as used by the technological utopians and reformers does not refer simply to cheapness, that is, to saving the most money or time or energy or whatever. It refers, much more significantly in its advocates' eyes, to doing the most good for society, and only then to the greatest savings. Indeed, a somewhat costlier effort that produced more social good than a cheaper effort would be considered more efficient in this regard. Furthermore, the principal aim is less to promote efficiency than to avoid inefficiency, or waste, in all of these respects. Widespread inefficiency and waste are seen as socially deleterious, but losing some money or time or energy in themselves is not. The real "profit" of efficiency, therefore, lies in improving society as far as possible. Not surprisingly, *efficiency* in this sense was transformed during this period from a largely economic to a heavily moral crusade. Simultaneously, however, the moral crusade acquired a scientific cast.[26]

The very term *system* suggests a coherent, integrated order of some kind. The critical point, in the present context, is that *system* refers to a coherent, integrated, and at least partially planned social order in which the components operate regularly and harmoniously. The size and scope of such an order may vary, but *social system* usually refers to a setting no smaller than an entire city and no larger than the whole United States. The first application of *system* in this country was, appropriately enough, to a technological achievement: the so-called "American System of Manufacturing," which comprised precision instruments, interchangeable parts, mass production, and the like. Even there, as its historians have recently agreed, *system* encompassed more than machines and structures and economics alone. It included the relevant functions of government, of management, of labor, and even of culture. Indeed, these historians see the ability to conceive of a *system* as such as itself a notable achievement.[27] Certainly the broad general conception of *system* applies no less to *system* as used here. If anything, *system* here is far larger in size and scope and is avowedly, not implicitly, social and cultural in connotation. In addition, this *system* includes the integration of human activities into ever more complex patterns, many consciously designed by society's leaders and architects but others reflecting less formal daily choices by ordinary citizens. Yet *system* here is somewhat less stable than even a manufacturing system would likely be (development of the "American System of Manufacturing" was irregular and often disorderly). *System* as applied to a city, a state, a region, or the entire country presumes flux—a series of alternating, temporary equilibriums and disequilibriums—before the achievement of a more fixed and better planned system and so a more permanent equilibrium.[28]

The terms *organization* and *planning* will be considered in detail in chapter 6 in the contexts of occupational organizations and of city, regional, and national planning. At this point, I note merely that both terms presume an evolutionary scheme of some type with a variety of stages of development and with a hierarchy of components and stages. Both presume movement over time toward a largely fixed social system and a more permanent equilibrium.

Finally, *rationalization* does not refer merely to the spread of reason. Taken from the writings of Max Weber, it refers to the modern assumption that unprecedented scientific and technological advances now allow mankind to conquer nature and increasingly replace the wild environment with a controlled environment—as in the definition of technology itself. Moreover, the mysteries of nature are likely to be steadily deciphered by science and technology—by *calculation*, broadly conceived, above all. The world, in Weber's classic phrase, thereby grows ever more "disenchanted." The kind of largely planned system and more permanent equilibrium already noted would be a triumph of rationalization as used here. Ironically, Weber himself envisioned a gloomy future for a thoroughly rationalized Western world and saw the United States as potentially its most rationalized unit. Yet technological utopianism, as described next, and its contemporary reform movements, as discussed later, have sufficient faith in scientific and technological progress to avoid Weber's pessimism. Hence their (usually implicit) concept of *rationalization* is ordinarily positive.[29]

Two

American Visions of
Technological Utopia, 1883–1933

The visions of technological utopia treated here were found in forty articles, addresses, tracts, short stories, and novels. Several utopians wrote more than one work, and some of them wrote both fictional and nonfictional works. There is no qualitative distinction, however, between the fictional and the nonfictional technological utopian works. All envision a similar kind of utopia, and it is especially revealing that those utopians who wrote both fiction and nonfiction altered nothing of substance in moving from one genre to another. As Robert Wiebe observed about *Looking Backward*, and as Kenneth Roemer observes about many of the 160 utopian, semi-utopian, and anti-utopian works he collected and studied for the period 1888–1900 alone, "the fictional form was only a sugarcoating for the authors' realistic blueprints for the future."[1]

All but one of these twenty-five technological utopians provided fictional or nonfictional blueprints of their version of utopia, and Robert Thurston left sufficient piecemeal writings to allow reconstruction of his ideas. Because differences in form concealed no differences in content, and because the content for all twenty-five prophets, including Edward Bellamy, differed only in minor details, these separate visions of technological utopia may safely be treated as one—as, in effect, a Weberian ideal type. To treat these visions as one is not to flatten out history; rather, it illuminates the many common

Portions of this chapter first appeared in my article "American Visions of Technological Utopia, 1883–1933," *The Markham Review* 7 (July 1978): 65–76, and are reprinted by permission.

features that made these works a distinctive genre of utopian expression in this half-century.

Because technological utopianism has not been previously identified as a genre of utopian expression, the appropriate writings had not been brought together. I located many possible candidates through mentions of them in secondary studies, and I systematically searched the several major libraries cited at the outset. To my knowledge, I have been able to examine all conceivably appropriate works, and have chosen for study the forty that share a similar vision of technological utopia.

Except, of course, for *Looking Backward*, these writings are not familiar to most students of American literature and history. Specialists in American utopian literature, however, are acquainted with many of them: for example, John Bachelder's *A.D. 2050*, Herman Brinsmade's *Utopia Achieved*, Charles Caryl's *New Era*, Paul Devinne's *The Day of Prosperity*, Alvarado Fuller's *A.D. 2000*, Ludwig Geissler's *Looking Beyond*, King Camp Gillette's *The Human Drift*, Albert Howard's *The Milltillionaire*, John Macnie's *The Diothas*, Henry Olerich's *A Cityless and Countryless World*, Solomon Schindler's *Young West*, William Taylor's *Intermere*, Chauncey Thomas' *The Crystal Button*, and Charles Wooldridge's *Perfecting the Earth*—plus Bellamy's lesser-known *Equality*. All of these books, in fact, have been reprinted by various publishers within the past fifteen years. Similarly, specialists in American engineering history are aware of George Morison's *The New Epoch* and the major professional addresses by him and by Thurston. Likewise, specialists in American city planning history know Edgar Chambless' *Roadtown* and their counterparts in American aviation history know Albert Merrill's *The Great Awakening*. Collectors of rare books are likely to be acquainted with these works and also some of others not reprinted: for instance, Byron Brooks' *Earth Revisited*, Fred Clough's *The Golden Age*, Jeff Hayes' *Portland, Oregon, A.D. 1999*, Thomas Kirwan's *Reciprocity*, Harold Loeb's *Life in a Technocracy*, and D. L. Stump's *From World to World*.

The titles of these volumes are as intriguing as their contents (even though, as I have noted, the works generally lack literary grace). Several volumes, moreover, were illustrated, showing diverse aspects of life in technological utopia—among them the illustrations reproduced here. The titles, the illustrations, and even the contents of some of the works may give the impression of a lighthearted, even romantic, approach to the future. Several of these writings do indeed have such a tone. But all are fundamentally serious—some almost deadly so—about not just the shape of technological utopia but also the route leading to it. If anything beyond their common views of the future unifies these twenty-five technological utopians, it is their earnestness and their didacticism.

The technological utopians aimed at accurate prediction of the future, not at idle visions of a world someday somewhere. Moreover, the world they foresaw so specifically represented no break with the existing one—the world of 1883–1933—and many of the technological changes they predicted were, by 1883, already being discussed and, in some cases, developed. The difference between their utopias and the present was not qualitative but quantitative: they multiplied what they saw as the outstanding contemporary trend and predicted the greater and greater advance and spread of technology. This was not to be a sheer proliferation of machines and structures but an increasing use of technology in establishing and maintaining an entire society.

The utopians were not blind to the problems technological advance might cause, such as unemployment or boredom. They simply were confident that those problems were temporary and that advancing technology would solve mankind's major chronic problems, which they took to be material—scarcity, hunger, disease, war, and so forth. They assumed that technology would solve other, more recent and more psychological problems as well: nervousness, rudeness, aggression, crowding, and social disorder, in particular. The growth and expansion of technology would bring utopia; and utopia would be a completely technological society, one run by and, in a sense, for technology. Technology seemed to them a far more effective instrument of progress than the various panaceas proposed by other contemporary utopians.

The technological utopians' confidence in the imminence of utopia distinguished them from most other contemporary utopians, whose failed schemes often caused disillusionment and occasionally produced anti-utopians. Where most other utopians yearned forlornly for the ideal society, the technological utopians regarded it as a practical, quite sensible prospect.[2] A statement like the following, from the preface to Wooldridge's novel, is characteristic:

> This is a Utopian book, but its Utopia is not, as Utopias generally are said to be, in the clouds; on the contrary, it is worked out with much detail in accordance with a natural order of sequence from existing conditions, with every point definite in time and place, true in all fundamental physical features to the best maps, true also to the law of cause and effect and duly regarding the limitations of nature.[3]

Typical too is Thurston's description of the prophecy in one of his articles as "truly logical and scientific."[4] The utopians were even willing to specify utopia's time and place of arrival—usually within the next hundred years and within the existing boundaries of the United States.[5]

Yet, the means by which technological utopia was to come into existence were rarely specified. Virtually every technological utopia emerges as the final stage of some vague evolutionary process marked by few concrete events. "It took many, many years of strife and turmoil," recounts a character in Kirwan's novel in a typical explanation, "to bring about the present condition of affairs; but it came, in the natural course of social evolution." Slightly more specific, Bellamy's Dr. Leete (who guides Julian West through utopia) points to the "diffusion of knowledge among the masses . . . beginning with the introduction of printing." No technological utopian tract offers more precise information.[6]

The contrivances that transport protagonists from pre-utopia to utopia further illustrate the difficulties the utopian writers face making the transition from the present to the ideal realm. Some subjects simply "chance upon" utopia. Most arrive through mystical rather than practical means: dreams, hypnosis, death, prophecy, and time capsules. One moment the subject is in this world; a moment later he is in the next one, the barriers of time and space having been miraculously transcended. "On retiring that night," recalls Bachelder's narrator, "Morpheus soon took possession of my faculties, and with electric speed I was transferred to the Boston of A.D. 2000." Fuller's narrator explains that he had lain frozen in an ozone-filled container to await the establishment of utopia. "Life in death; death in life," says he mysteriously.[7]

Both their seeming inability to explain how utopia is to be established and their resort to largely magical devices for getting the protagonists there may indicate an underlying uncertainty among the technological utopians about the practicality of their schemes, an uncertainty which would belie their confidence in citing the exact location of utopia in time and space. This possibility will be examined in the next chapter. The utopians' vagueness, however, may reflect only a failure of imagination, not confidence. Put another way, their difficulty may be that of connecting history with the end of history.[8]

Clearly the coming of utopia is to initiate an age beyond history. Change, the bedrock of history, will cease, because perfection, the goal of history, will have been achieved. Proclaims a citizen of Wooldridge's utopia at the end of the novel: "Eternity is here. We are living in the midst of it." And proclaims Gillette at the conclusion of one of his tracts: "Heaven will be on earth."[9] History will become only ancient history, characterizing the past, not the present. As Morison puts it, "The lessons of history must be studied as showing the mistakes of the past, not as giving precedents to be followed now." In short, technological utopianism is uncompromisingly antihistorical.[10]

However tenuous the means presented for bridging the chasm between the pre-utopian age and the utopian age, the technological utopians never doubt that the gap can and will be bridged. The technological society will arrive and machines and structures will be its lifeblood. As Thurston declares:

> Man is turning his work in the world over to the more exact and more powerful machines that he has devised and built, and machinery here weaves his clothes, there gathers his grain, and yonder makes for mankind other machines. . . . Relieved from bodily toil, mankind is coming to that condition of social life in which mental activity absorbs its surplus energies, and the material and intellectual development of civilization are [sic] carried on together, and with continually accelerated progress.

For Gillette the world as a whole is a "mammoth factory," while for Chambless it is a giant "machine."[11]

Society is not, however, to be a mass of sooting smokestacks, clanging machines, and teeming streets. The dirt, noise, and chaos that accompanied industrialization in the real world are to give way to perfect cleanliness, efficiency, quiet, and harmony. Technology, like fire, will be domesticated. Boasts one inhabitant of Schindler's utopia:

> Our sanitary arrangements and lavatories are of the best, and easily accessible; our roads are well paved; smoke, cinders, and ashes are unknown because electricity is used now for all purposes for which formerly fires had to be built; our buildings and furniture, made of lacquered aluminum and glass, are cleansed by delicately constructed machinery that operates automatically. The very germs of unclean matter are removed by the most powerful of disinfectants, electrified water, that is sprayed over our walls, and penetrates into every crack and crevice.[12]

An inhabitant of Brooks' utopia proudly points to "The broad, shaded avenues, with pavements as smooth and clean as a floor; the absence of horses and heavy vehicles in the streets; the noiseless movement of the electric carriages; . . . the leisurely movements of the people . . . ; the absence of the rush and haste of business and of the fever of life and toil."[13]

Simultaneous with the taming of technology is to come the taming of nature. Wind, water, and other natural resources will be harnessed—into electricity above all, its supreme cleanliness and quiet befitting its supreme power. The mastery of nature is regarded as the fulfillment of man's destiny, the beginning of a new epoch for mankind, and the elevation of man to a status only slightly short of omnipotence. As a leading figure in Thomas' utopia says, "We give Nature and her vast forces an opportunity to work for us!" Similarly, Stump's novel is dedicated to "The proposition that the earth was made to live upon, and that all its elements were created to sustain life, and for no other purpose!"[14]

The envisioned domestication of both technology and nature will resolve the tension that Leo Marx, among others, has deemed irresolvable: the tension between the industrial and the agrarian orders, between the machine and the garden, a tension that Marx believes lies at the heart of the American experience.[15] The utopians would resolve the tension by the modernization rather than abandonment of the garden, by transporting it out of the wilderness and relocating it in the city—a city itself to be transformed from a lethal chaos into a healthy order. The new "industrialized garden" does not take the form of Marx's "middle landscape" or of Jefferson's ideal of an agrarian-based yet technologically proficient yeoman republic. Rather, it takes the form of a series of what have since been termed *megalopolises*: massive combinations of urban and suburban tracts covering vast areas. The title of Olerich's book describes it as *A Cityless and Countryless World*. Howard's vision is also representative:

> All cities are now circular in form the radii of which is [sic] one hundred miles, with an approximate circumference of seven hundred miles. . . . In all there are about twenty cities. . . . Indeed, by the all potent power of electricity, man is now able to convert an entire continent into a tropical garden at his pleasure.[16]

The megalopolises are tributes not only to philosophical ingenuity, which reconciles machine to garden, but also to the scientific planning of which the advocates of technological utopia are so proud. (Their pride would be justifiable if the planning success were actual, for scientific planning, whether social or physical or both, has had only limited success at any level in American history.)[17] In the utopian megalopolises, millions of persons live, learn, work, and play in perfect comfort, contentment, and happiness, ever free of dirt, noise, chaos, want, and insecurity. As a citizen of Devinne's utopia puts it, "It is because of this elaborate *planning* for our cities, that all men, in every city, have equal advantages and conveniences. . . . Today one city is just as good, just as beautiful, as another." For Gillette, such a giant city, like a machine, must "have no unnecessary parts to cause friction or demand unnecessary labor, and yet it must combine . . . all the necessary parts which will contribute to the happiness and comfort of all."[18]

Most immediately striking in each megalopolis are its buildings, numbering in the hundreds. Of various materials, shapes, and sizes, they fall into clear and neat patterns, with the tallest, sleekest structures concentrated in the center and the shortest and squattest along the periphery. Broad sidewalks and streets and wide plazas and parks border each building, even in the dense downtown areas. The cold, canyon-like atmosphere of many actual skyscraper cities is avoided. "As far as I could see," recounts Devinne's protagonist, "colossal, square, skillfully decorated marble palaces, built in various styles of architecture." "Can you imagine the endless beauty of a conception like this," asks Gillette, "a city with its thirty-six thousand buildings each a perfectly distinct and complete design, . . . each building and avenue surrounded and bordered by an everchanging beauty in flowers and foliage?"[19]

Also avoided is vehicular and pedestrian overcrowding, thanks to well-placed highways, walkways, bridges, and tunnels. Howard envisions the typical multi-layered design:

> All avenues and boulevards are required to be one thousand feet in width and interlined by a Triple Canopied Highway two hundred feet in width, with four hundred feet of lawned avenue on each side of the thoroughfare proper. . . . The . . . Highway is directly connected with all buildings on each side thereof as one continuous Hall or Public passage of three levels. . . . Directly beneath the Triple Highway runs a subway.[20]

Suburban areas resemble their urban counterparts, with which they are tightly integrated. As a visitor to Taylor's utopia describes it, "Roadways, or, perhaps more properly, boulevards, interlace the whole country. They are the perfection of road-building—smooth, even-crowned, and free from dust, water, or other offensive substance." And as a visitor to Macnie's utopia remembers, "We had left the train at what appeared to be a small village. Yet nowhere was to be seen any trace of that pervading lack of neatness and finish which, in our day, usually characterizes the country. . . . The buildings visible, though inferior in size to those of the city, were as solidly constructed, and of similar materials." An inhabitant of Thomas' utopia observes that "the cities are so numerous and so accessible that all the advantages they possess are easily obtained by those living in the country. Every country village has its pleasurehouse as well as its public library; and telephones and pneumatic tubes make these tributary to the city centres."[21]

What farm lands remain are also assiduously organized and linked to urban and suburban regions. Agriculture, states Gillette, "is an integral part of the producing machine, and the producing machine must be a unit." Moreover, "agriculture, like industry, when reduced to its lowest terms, is a problem of applied mathematics and engineering." Macnie's visitor recalls, "Not a waste corner, not a weed, was visible." And Thomas' inhabitant notes, "Horticulture has supplanted agriculture, and every acre is studied and stimulated to do what it can best do."[22]

Connecting all sectors of utopia are superbly efficient transportation and communication systems powered almost exclusively by electricity.[23] These systems enable the widely dispersed citizens of utopia to live and to work where they choose. As one of them puts it, "We have practically eliminated distances."[24] The means of transportation include automobiles, trains, subways, ships, airplanes, even moving sidewalks. Means of communication include pneumatic mail tubes, telephones, telegraphs, radios, and mechanically-composed newspapers. These devices usually only partially resemble those in use at the time of writing, if they existed at all, but rather accurately foreshadow those in use in later times.[25]

Technological advances pervade homes and offices as well. Electric clothes washers and dryers, dishwashers and dryers, refrigerators, ranges, vacuum cleaners, garbage disposals, air conditioners, phonographs, razors, and haircutters fill every home—again, devices sometimes unknown in that day but taken for granted in ours. "Domestic life," predicts Olerich, "will be attended with many comforts and conveniences." Underground pneumatic tubes from centralized distributors supply each home with all needed goods, from food to furniture. "It is," says Bellamy's West, "like a gigantic mill, into the hopper of which goods are being constantly poured by the trainload and shipload, to issue at the other end in packages of pounds and ounces, yards and inches, pints and gallons, corresponding to the infinitely complex personal needs of half a million people."[26] These various inventions, Gillette boasts, give virtual freedom from "all the annoyances of housekeeping" and "everything for comfort, economy, convenience, and freedom from care that a Corporate Intelligence could think of."[27]

Curiously, the technological devices that bring perfect *working* conditions are described in much less detail. Bellamy's rhapsody provides one example: "I need not tell my readers what the great mills are in these days—lofty, airy halls, walled with beautiful designs in tiles and metal, furnished like palaces, with every convenience, the machinery running almost noiselessly, and every incident of the work that might be offensive to any sense reduced by ingenious devices to the minimum." Clough's rhapsody provides another: "On the tour of inspection the sights they [visitors to utopia] saw were something wonderful to behold; acres of wonderful machinery running noiseless and doing perfect work."[28] Thus various kinds of automated machinery cut the working time and make the time spent working almost enjoyable.

Utopia's climate is as pleasant, and nearly as uniform, as everything else—again, the achievement of technology. Excessively hot regions have been cooled and excessively cool ones warmed; excessively wet ones have been made drier and excessively dry ones wetter. How? By the use of more vaguely described but enormously powerful machines and structures. With them, the utopians have been able to dredge, rechannel, and even create rivers and lakes; to irrigate deserts, heat the soil, flatten mountains, and clear the land; and to erect huge domes to capture and preserve sunlight. Declares Howard: "We have absolute control of the weather."[29]

As one would expect, the ethos of technology shapes the values as well as the physical dimensions of utopia, but the intensity of the influence is surprising. The inhabitants of utopia aim to be as efficient as their machinery. "The human machine," brags Brinsmade, "the greatest of all, has been . . . thoroughly understood and developed to its highest efficiency, something never previously done." Gillette sees workers and managers of his proposed international People's Corporation as literally—and happily—"cogs in the machine, acting in response to the will of a corporate mind as fingers move and write at the direction of the brain." He yearns to have "millions of individuals organized and moving like parts of a wonderful mechanism, from one field of production to another."[30] The very efficiency of technology inspires men and women to emulate the perfection of the working of the machine. Far from taking their ease because of labor-saving technology, they devote their leisure time to self-improvement, whether cultural, physical, or professional.[31]

In imitating the machine he has invented, utopian man finds fulfillment. Proclaims Olerich's protagonist: "We teach that labor is necessary and honorable, that idleness is robbery and a disgrace." Declares Thomas' leading citizen: "We value time as our first of all boons—it is our life—and we count every day another opportunity freighted with duties that we take pleasure in performing." Through their fictional figures, Thomas and Olerich equate work with play: "Willing work, in fields fitted to the capacity of the worker, is of itself one of the highest forms of pleasure," says Thomas; "work . . . gradually changes into play," says Olerich. Bellamy's Julian West actively seeks work:

> "It is a sensation which I never had anything like before," I said, "and never expected to have. I feel as if I wanted to go to work. Yes, Julian West, millionaire, loafer by profession, who never did anything useful in his life and never wanted to, finds himself seized with an overmastering desire to roll up his sleeves and do something toward rendering an equivalent for his living."

In short, as Brooks puts it, "This is the age of work."[32]

This cult of efficiency is reflected in every utopian activity and value judgment. To take the most stirring example, the virtue of cooperation is often preferred to individualism on the basis of its efficiency, and not its moral superiority. As Olerich insists, "Social and economic prosperity . . . can be attained only in a system which recognizes extensive . . . cooperation as its fundamental principle." Cooperation, says Gillette, "is the method which eliminates wasteful competition and duplication."[33] Thus holding that cooperation breeds efficiency, the utopians call for large, heterogeneous communities rather than small, homogeneous ones because they believe that large communities offer the widest opportunities for the kinds of contacts and friendships that spawn cooperation.

As well as fostering cooperation, efficiency shapes education, industry, and government. In education, technical subjects, especially the sciences and the vocations, constitute the bulk of the curriculum. Announces Gillette:

> The child under the People's Corporation will have such an opportunity for education as almost no child has today. . . . The Corporation will want its children to learn its business; the miracle of scientific production; the fairy tale of flour; the romance of rubber; the wonder of wool and silk. The child will get his education in the midst of production.[34]

Adds Chambless: "Instead of college . . . there will be an industrial university. . . . Pounding literature into the head of a natural born mechanic is both economic and mental waste. The universal query in Roadtown will not be what does he know, but what can he do."[35] Nontechnical subjects are not ignored, but they are included only insofar as they contribute to technical knowledge. As Macnie emphasizes, "The acquisition of the knowledge to be obtained from books, though by no means neglected, we regard as the least important branch of education." For "the intent of our present educational system," according to Gillette, "is to prepare the individual to take some position in the industrial machine."[36] Statements by two citizens of utopia summarize the role of education in society: "Knowledge has become power," says one; "The first and most important industry of this new civilization [is] . . . to cultivate brains," says the other.[37]

Upon completing their formal education, all inhabitants of utopia enter the "industrial army," whose very title bespeaks order, discipline, organization, and therefore efficiency. As Thurston envisions it, "All the world is falling into line, and the whole world-wide army is moving in concert if not under a single generalship. . . . [T]he captains of industry [command] . . . mighty armies overspreading all the fields of production of a whole country, even of many countries." A civilian army replaces the disbanded professional military. Every member, "no matter what station he fills," is, as Stump's narrator puts it, "a public servant."[38]

In this civilian army all citizens serve from twenty to forty years after completing their education. For this service, their education has prepared them, a further testimony to its practicality. All have technical skills, but they range from lowly manual labors to prestigious mental ones. As a resident of Kirwan's utopia acknowledges, "nearly every calling with us nowadays is industrial." Citizens with the greatest technical expertise head the industrial army, merit—and thereby efficiency—alone determining their status and that of every other utopian. Bellamy's Dr. Leete states, "The principle on which our industrial army is organized is that a man's natural endowments, mental and physical, determine what he can work at most profitably to the nation and most satisfactorily to himself." And Gillette predicts, "By following this rule, an industrial machine of highest efficiency will be secured."[39] Since in this meritocracy the highest merit accrues to those with the greatest technical knowledge,[40] the leaders of the industrial army might well be viewed as the technological equivalents of Plato's philosopher-kings.

Personal efficiency in utopia garners public and private honors but yields few tangible rewards. Hefty taxes rigorously prevent the accumulation of great wealth,[41] and indeed the availability of all needed goods and services at modest cost makes possession of riches pointless. To capitalists, the absence of material rewards would seem to rob people of any incentive to be efficient. But the utopians—who, in the characteristic fashion of American reformers, are not socialists[42]—do not perceive any such threat, which underscores the spell efficiency for its own sake casts upon them.

Efficiency governs government as thoroughly as it governs education and industry. The same technical experts who run the industrial army run the government, because expertise, not popularity, is what good government requires. "Administration, in a technocracy," Loeb explains, "has to do with material factors which are subject to measurement. Therefore, popular voting can be largely dispensed with. It is stupid deciding an issue by vote or opinion when a yardstick can be used."[43] Many utopias have no politicians at all. Proclaims Olerich: "We have no parties, no politicians, no election frauds, no political boodle, . . . no political congresses, parliaments, and legislatures." Officials are selected "always by mutual consent based on fitness."[44] Where politicians remain they are only figureheads. Loeb notes that the function of "Political government . . . would be showmanship. The routine of its executives would be made up of receiving distinguished guests, laying corner stones, making speeches about the rights of man, American initiative, justice."[45]

As technicians rather than politicians run the utopian government, the government is technical rather than political in nature. Since the basic laws and institutions of society have been fixed,[46] no legal, political, or ideological tasks remain, and therefore no lawyers, politicians, or "ideologues" are needed.[47] Only technical issues—that is, issues of efficiency—remain, and only technicians are needed to deal with them.[48] But for this very reason technical issues are more than "merely technical," and technicians more than mere technicians. In the words of Thurston, "The world owes all . . . to the inventor, to the mechanic, to the man of science. . . ." And in those of Morison, "Permanent success will depend not on commercial drummers, but on the civil engineer; not on the shrewd guesses of the so-called business man, but on the accurate knowledge of the manager." The manager alone, according to Gillette, "will sweep away all forms of waste."[49]

Not even the realms of culture and religion escape the ministry of technology and so of efficiency. The state supports culture, but culture supports it and its values. "In the past," rhapsodizes one technological highbrow, "the arts have led; in the future we shall see science leading and directing every development of the arts."[50] Some utopian writers go further. For them culture not merely sanctions efficiency but is efficiency itself. To Merrill "the greatest beauty" is that most "compatible with usefulness." Thurston equates the increasing output of "our mills, our factories, our workshops of every kind" with the conversion of "crudeness and barbarism into cultured civilization."[51] Finally, culture, in the form of lectures, concerts, and radio, is to sustain not just an elite but the entire population, its mass appeal attesting still more to the drive for efficiency.[52]

Religion promotes efficiency in the same ways as does culture.

First, religion upholds the value of efficiency—or more precisely, upholds the value on which efficiency rests: cooperation.[53] Utopian religion is little more than a demythologized belief in brotherhood and love of man. For Bellamy, religion is "dominated by an impassioned sense of the solidarity of humanity"; for Howard, "Pure religion we know to be pure love." A citizen of Kirwan's utopia says, "Our highest conception of a personal god is embodied in the perfect man. . . ." Gone are churches, rituals, creeds, ministers, and concerns for transcendent matters like the afterlife.[54] Second, religion, like culture, is efficiency itself. Technology and science are its gods, and their efficiency above all gives them their divinity. As Brooks says succinctly, "Real science is true religion." Thurston even labels the discoveries of science the modern version of "revelation and prophecy." And Morison unabashedly declares engineers the "new priests" of this new society.[55]

Efficiency is not, however, the only machinelike virtue that technological utopians seek to inculcate. Self-control is another, and equally important. Utopian man's control over himself mirrors the control that technology achieves over the environment. Technology tames both nature and itself, subduing the wind, water, fire, and climate, purifying its own wastes, staunching its own excesses, and replanting the gardens it has trampled underfoot. Utopian man aspires to similar control of human nature. In the words of Macnie, "If required to state the pervading characteristic of the manners of these people, I should say self-control. In proportion as man had become master of nature, it had become needful to become master of himself."[56]

Self-control is displayed in a variety of ways. Utopian citizens conform—in dress, in length of hair, in food, and in not smoking and drinking.[57] More important, family relations are uniform; the father rules every household, and the mother and children submit.[58] Only citizens judged healthy in body, mind, and morals can marry and bear children; the small minority of the unfit are not to be allowed to perpetuate undesirables.[59] Strict limits are imposed on the number of children permitted each couple, and divorce is forbidden except in cases of extreme incompatibility. Displays of emotion are rarely seen, and personal relationships are ordinarily formal and distant—in other words, controlled.[60]

Not even death excites emotion: mourning is kept brief and restrained, and bodies are cleanly and efficiently cremated.[61] Death is accepted as a natural phenomenon, the concern being for the improvement of this life. A teacher in Schindler's novel tells his young students, "Therefore, why, after all, should we trouble our minds in regard to a past that lies so far behind us. It is neither the past nor the future that should give us concern, it is the present. Here is this beautiful world, there is the span of life, granted to us to enjoy, and here is the work by which to make this life pleasant for ourselves and others."[62]

This, then, is the composite late nineteenth- and early twentieth-century vision of America as a technological utopia, a vision shared by at least twenty-five individuals. Except for minor details, their separate visions are fundamentally alike. Their visions can be treated as one vision of America's future, as a *de facto* ideal type. As we have seen, all twenty-five visions were intended by their originators as solutions to the principal problems of the day and as blueprints for eliminating virtually all problems forever. Whatever literary significance their works may retain, these visions deserve serious attention as unique sources of information about the times in which they were written. And because these works constituted predictions of our own times, it is intriguing to measure the gap between prophecy and fulfillment.

What these visions reveal about their originators will be considered in the next chapter. The historical background of these visions will be described in chapters 4 and 5. What the visions reveal about the society in which their originators lived will be examined in chapter 6. What they reveal about our own technological society and its possible future will be discussed in chapters 7 and 8. And what they reveal about the uses of utopian visions in general will be the subject of chapter 9.[63]

ROADTOWN

Cover illustration for Edgar Chambless, *Roadtown* (New York: Roadtown Press, 1910). Chambless foresees skyscrapers—laid on their sides rather than extending vertically—spanning the entire American countryside. No other technological utopian goes quite so far in this respect, but all carefully integrate urban, suburban, and farm land in their schemes.

"The Town Mansion," from Thomas Kirwan, *Reciprocity (Social and Economic) in the Thirtieth Century* (New York: Cochrane, 1909), n.p. Kirwan's town mansion is the more common suburban counterpart to the urban skyscraper. Superbly efficient transportation and communication systems enable the widely dispersed citizens of technological utopia to live and work where they choose.

Illustration from Herman Brinsmade, *Utopia Achieved: A Novel of the Future* (New York: Broadway, 1912), n.p. Brinsmade's typical airplane of the future reflects the technological utopians' ability to imagine devices then being discussed or even invented, but it also reveals the limitations of their imaginations.

Above: "New Utopia Schoolhouse," from Charles Wooldridge,
*Perfecting the Earth: A Piece of
Possible History* (Cleveland: Utopia, 1902), p. 251. Wooldridge's
schoolhouse is a tribute to the
vocational and scientific thrust of
education in technological utopia
and to the close connection between the school system and the
industrial army.

Right: Interior plan of the "New
Utopia Schoolhouse," from Wooldridge, *Perfecting the Earth*, p.
255.

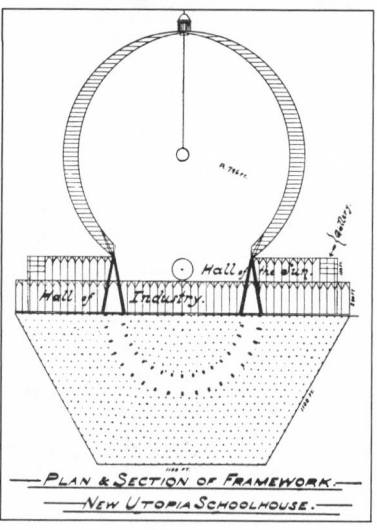

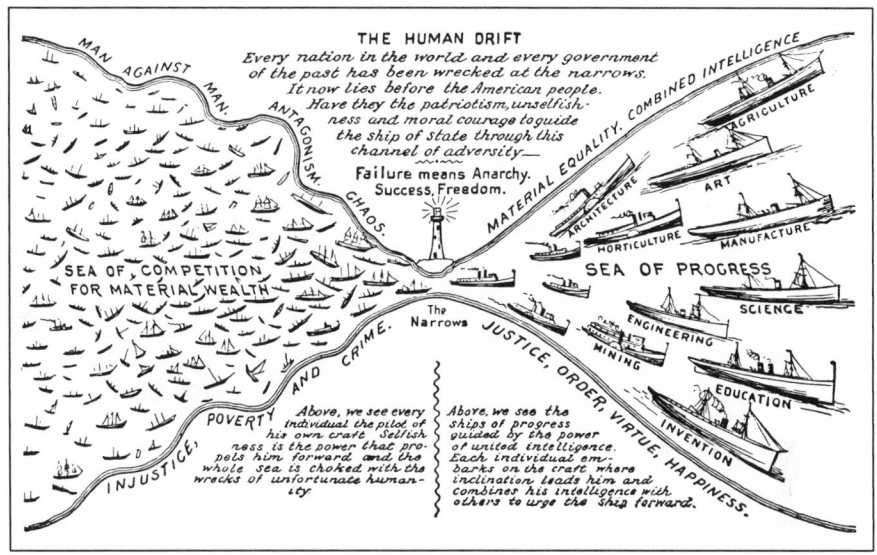

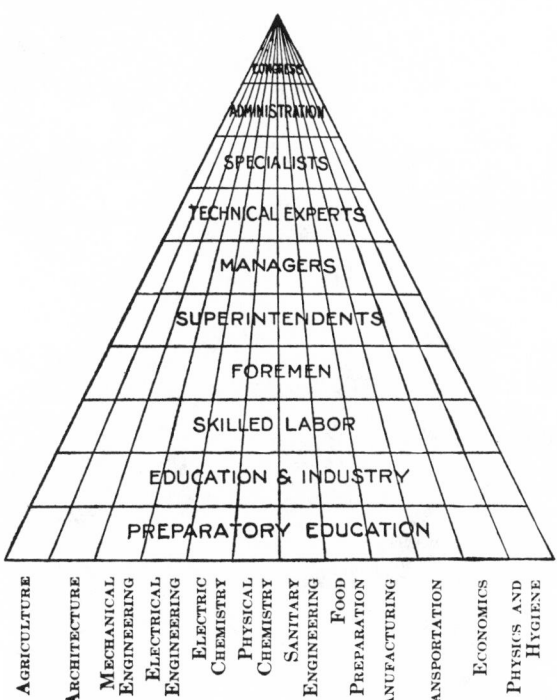

Above: "The Human Drift," from King Camp Gillette, *The Human Drift* (Boston: New Era, 1894), p. 18. Gillette's notion of progress that puts an end to human drift depends on order, organization, intelligence, and discipline, as well as on inventions.

Left: "Educational and Industrial Pyramid," from King Camp Gillette, *World Corporation* (Boston: New England News, 1910), p. 89. "The . . . pyramid is the symbol of the Educational and Industrial progress of the individual. The horizontal divisions represent the different planes upward until 'World Corporation Congress' is attained, whereas the divisions of the pyramid from base to apex represent the Grand Divisions of Industry—all of which finally merge into the 'World Corporate Congress.' Under this system the individual is free to choose his path of inclination, and his progress cannot be barred." Gillette's pyramid accentuates the ties between education and career advancement in technological utopia's avowed meritocracy.

Plan of the distribution of buildings, from Gillette, *The Human Drift*, p. 94. This typical two-square-mile section of a major city in Gillette's scheme integrates thirty-three apartment buildings with (A) five educational buildings, (B) five amusement buildings, and (C) five buildings for food storage and preparation. The outdoors similarly integrates lawns, avenues, and walkways.

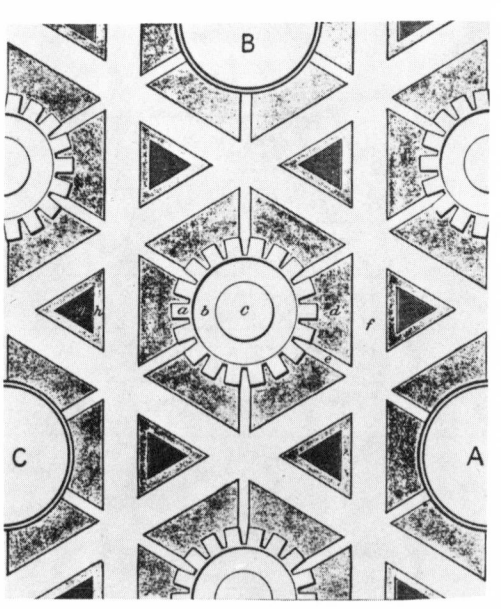

Plan of a single apartment building and its surroundings, from Gillette, *The Human Drift*, p. 96.

a Tiers of apartments.

b Inner court formed by the connecting of tiers of apartments, in form of a circle.

c Dining-room in center of court.

d Lawns surrounding buildings.

e Walks 50 ft. wide leading from buildings to avenues.

f Avenues 150 ft. wide.

g Glass domes, 200 ft. on each side, rising to a height of 100 ft. These domes give light to chamber below, and are directly over park in triangle form, that is 350 ft. on each side.

h Lawn surrounding dome of glass, thereby giving a continuous border of green to every avenue in the city.

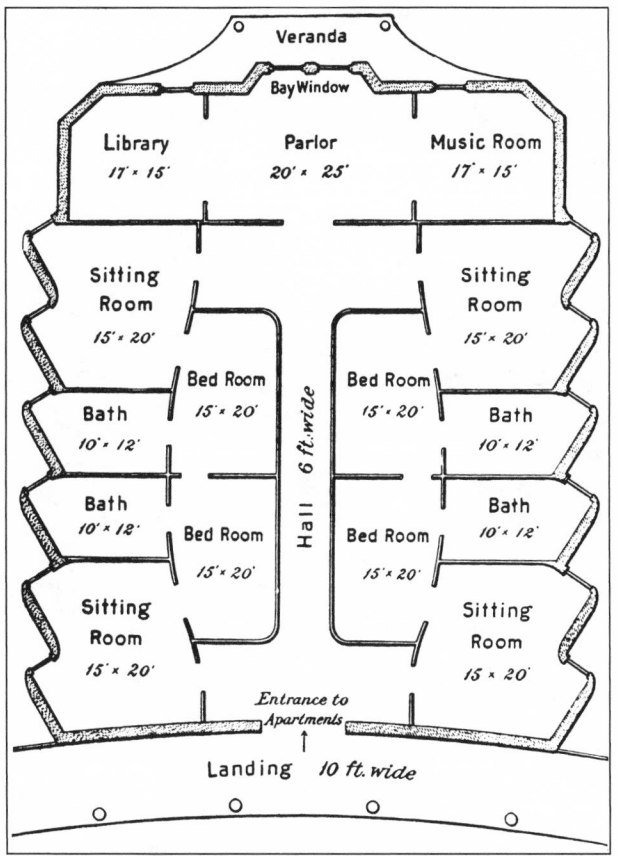

Above: Floor plan of a single apartment, from Gillette, *The Human Drift*, p. 100. This family apartment is designed to accommodate from four to eight persons.

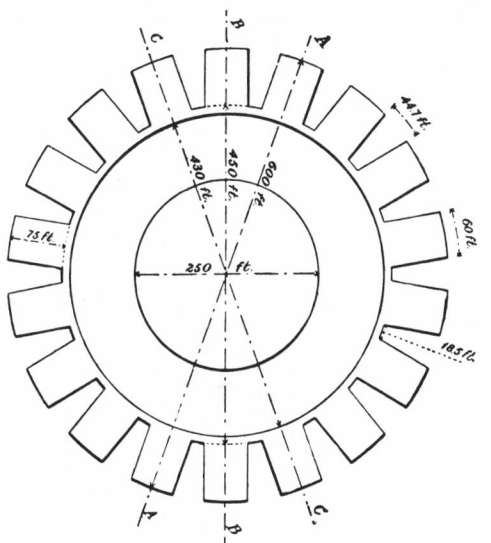

Left: Plan of a single apartment building, from Gillette, *The Human Drift*, p. 98. Here, as in the illustration above, there is some flexibility of design within individual units.

A Diameter of building over all, 600 ft.

B Diameter of court from apartment to apartment, 450 ft.

C Diameter of court from inner point of landing at each story, 430 ft.

Dining-room occupying central part of court, 250 ft.

Apartments, 60 × 75 ft.

Space between apartments on outer face, 44.7 ft.

Space between apartments where they join, 18.5 ft.

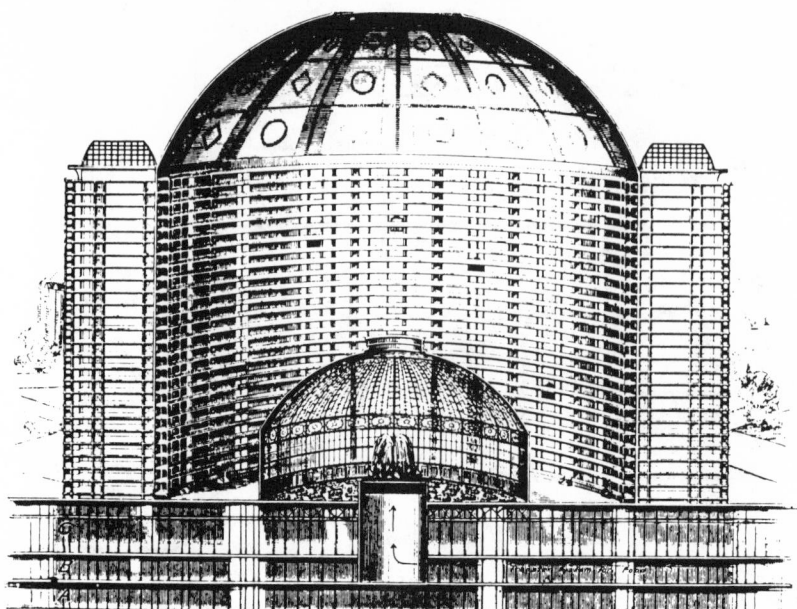

Above: Side elevation of apartment building, showing interior court, dining room, etc., from Gillette, *The Human Drift*, p. 104. Beneath the street are three additional levels: the first is for sewage, water, hot and cold air, and electrical systems; the second is for the transportation system; and the third is for pedestrians wishing to move about underground, particularly during inclement weather.

Right: Single tier of apartments in process of construction, from Gillette, *The Human Drift*, p. 106. According to Gillette, the building techniques and materials required for the structure would follow those then in use in large American cities.

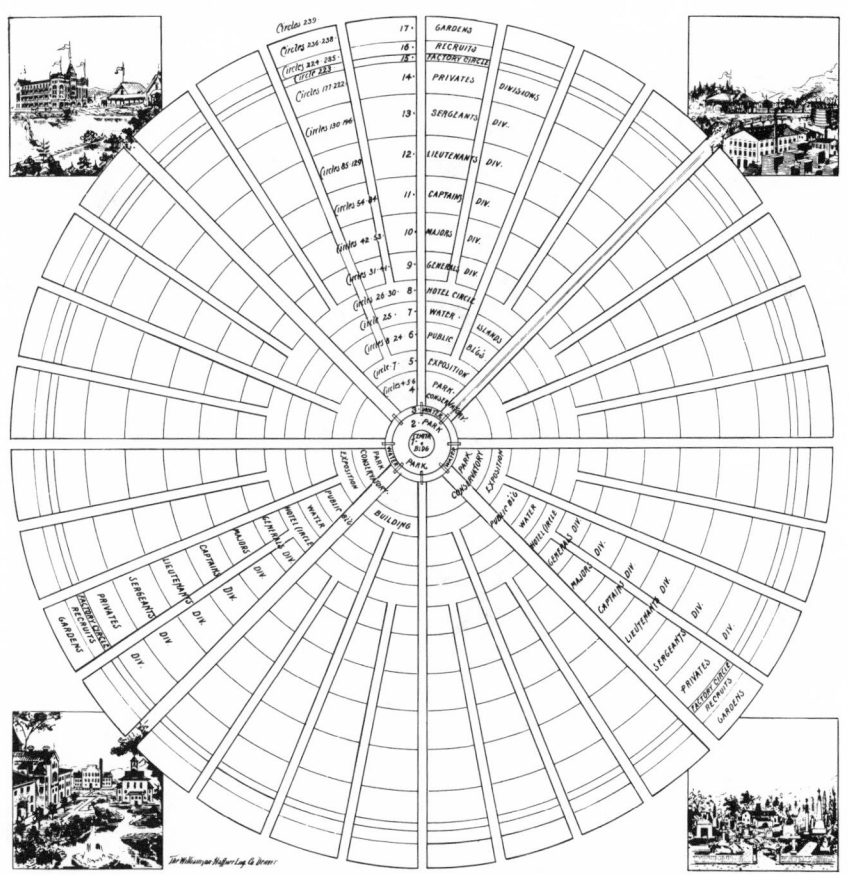

Outline of plan for new era model city, from Charles Caryl, *New Era* (Denver: n.p., 1897), n.p. Caryl's city assigns particular residential circles to members of the industrial army according to their rank. The extraordinary orderliness is common to all the technological utopians' schemes.

"Man Corporate," from Gillette, *World Corporation*, between p. 94 and p. 95. "He absorbs, enfolds, encompasses, and makes the world his own. He will do more; he will penetrate the confines of space, and make it deliver up its secrets and power, for Mind, the Child of the great Oversoul of Creation, is Infinite and Eternal." Gillette's corporate hero is the quintessential technocratic visionary.

Plan of a building, from Gillette, *The Human Drift*, n.p. Gillette's model apartment building, one of thousands he envisions in every major city, permits individual variation within each unit but requires uniformity without.

Sketch of center of new era model city, from Caryl, *New Era*, n.p. Caryl depicts one of technological utopia's metropolitan centers, the equivalent of the modern megalopolis.

Three

The American
Technological Utopians

Because the technological utopians were not members of an organized movement, it has not been easy to identify them. I have, I believe, scrutinized all conceivably appropriate figures, and each of the twenty-five on whom I focus wrote at least one work in the late nineteenth or early twentieth centuries that vaunts technology as the means of establishing an ideal society and offers a blueprint of that ideal society.[1]

What motivated these particular persons to write as they did has not been easy to determine. They shared the then pervasive American faith in progress through technology, but what inspired them to become technological utopians is not clear. The biographical information that I have been able to locate is often scant, and even when it is more plentiful, it provides at best a partial explanation. Psychology may shed some light on their motives; that approach will be explored at the end of this chapter.

Birth and death dates, first of all, indicate that the majority of technological utopians reached adulthood in the mid-nineteenth century. Of the nineteen utopians whose birth and death dates I learned, only one was born before 1820 and only two more before 1830; only three were born after 1870 and none after 1900. None of the nineteen died before 1890 and only two before 1900; four died after 1930. With one exception, all of these nineteen persons published their utopian works—several writing more than one—well past the age of thirty; some published them while in their sixties, seventies, or even eighties. To this extent their works represented their mature views of the present and the future, not the enthusiasms of youthful visionaries.

Geographical data are more difficult to obtain and less illuminating. They reveal that of the sixteen utopians whose birthplaces have been determined, eight were born in the Northeast, four in the Midwest, only one in the West, and none in the South. Three were born abroad.[2] Seven of these sixteen persons were born in cities, eight in small towns, and only one on a farm. Possible local and regional influences on these utopians are hard to determine. All of them moved at least three times during their lives. Few of them explain what, if anything, in their milieu predisposed them toward utopianism.[3] And most of their works disclose no trace of local or regional biases.

Occupational breakdowns are more helpful than geographical ones. That few of the technological utopians were professional writers is all too evident from even a cursory reading of their works. In fact, of the twenty-two utopians whose occupation is known, only eight were either journalists or novelists, and then not always in full-time positions. Three more were at least part-time writers of technical textbooks of various kinds.[4] Some of these writers, however, and others of the twenty-two as well, worked in fields tied directly to advancing technology. Three were engineers (one civil engineer, one mechanical engineer, and one aeronautical engineer), six were inventors (all but two of them businessmen or manufacturers as well), one was an architect, and one was a telegrapher. As for the rest, one was a carriage-maker, one a printing press foreman, one a mathematics professor, one a physician, and one a clergyman. Clearly, the majority of technological utopians were well-educated. Indeed, of the sixteen utopians whose schooling is known, fourteen attended high school, ten attended college, and four went on to graduate or professional school.

Finally, the technological utopians were nearly all white, Protestant, and male. There were no females and, as far as I can determine, two Jews and one Catholic and no nonwhites among them. Significantly, Kenneth Roemer's analysis of 154 late nineteenth-century American utopians of all kinds—the most comprehensive such analysis to date—correlates with my conclusions about the geographical, occupational, educational, racial, religious, and gender characteristics of technological utopians in particular. Although my group of visionaries spans a broader period than Roemer's, most of their utopian writings appeared before 1904, and only four appeared after 1915.[5]

What the data on occupation, education, race, religion, and gender suggest are that economically and socially, at least, the technological utopians were not marginal, alienated, disaffected figures but, on the contrary, successful, well-integrated Americans. Occupationally, the majority of technological utopians appear to have improved on their fathers' status,[6] in some measure because the majority were part of the ever-more prestigious field of technology. That the technological utopians appear to have been mainstream rather than peripheral members of society further indicates that technological utopianism was not a movement of revolt; it was a movement seeking to alter the speed but not the direction in which American society was moving.

A few bare facts about background or occupation cannot, of course, tell us very much about the personalities of any of the twenty-five technological utopians. Were they happy or unhappy, optimistic or pessimistic? In over half the cases, we cannot know; there is only the data assembled here and in the Appendix. In the remaining cases, I have found sufficient additional material to, in effect, bring those utopians to life.[7] The biographical materials that follow will offer a basis for some speculation about the psychology of utopianism in particular.

The only truly famous technological utopian, Edward Bellamy (1850–1898), hardly needs more than a sketch here. A barely known journalist, novelist, short story writer, and nonpracticing lawyer who had spent most of his life in his native Massachusetts, Bellamy was propelled overnight to fame by the extraordinary, and quite unexpected, success of *Looking Backward* (1888). Withdrawn, frail, often sickly, he was hardly the strong charismatic leader his millions of readers dearly sought. Yet he worked diligently, at the sacrifice of his remaining health, to try to effect the reforms he envisioned in his book and in other works. The Nationalist crusade he inspired eventually failed, and he devoted his last years to writing *Equality* (1897), his sequel to *Looking Backward*.

A sudden rise from obscurity to prominence—like Bellamy's—surely was the ambition of most of the other visionaries included in this study. Bellamy's success led other utopians to publish their own sequels to *Looking Backward*: Ludwig Geissler (no dates available) and Solomon Schindler (1842–1915). Chauncey Thomas (1822–1898), according to letters reprinted in his book, secured a publisher in 1889 for his manuscript of 1872–1878 only because of *Looking Backward*'s popularity. Certainly, John Macnie (1836–1909), who charged Bellamy with unacknowledged use of material from his neglected 1883 novel, would have enjoyed knowing comparable fame and fortune. Without repeating the tradition of Bellamy's alleged wholesale influence upon all subsequent utopian writers of his and successive generations—the view earlier criticized as simplistic—it can still be suggested that virtually all the other twenty technological utopians also sought to make their mark on their times.

Ten of the twenty-five technological utopians definitely had a youthful (as well as adult) fascination with technology, and that fascination may have helped steer them toward their utopian writings. It surely affected their career choices, for all worked at least part of their adult lives in jobs concerning some aspect of advancing technology. Several of these visionaries became prominent in their everyday callings. George Shattuck Morison (1842–1903), for example, a civil engineer, became the leading American bridge engineer of his day, and Robert Henry Thurston (1839–1903) became an equally distinguished mechanical engineer and engineering administrator, first at Stevens Institute of Technology and then at the Sibley College of Mechanical Arts of Cornell University.

Morison's success is particularly impressive in view of his being largely self-taught in engineering. At Harvard, his undergraduate work was in the liberal arts and he took his graduate degree in the law. But within a year of beginning law practice at a prestigious New York City firm, he decided to abandon that career for bridge engineering. Morison served as President in 1895 of the American Society of Civil Engineers, the highest professional organization of civil engineers in America.

Thurston contributed significantly to the growth and professionalization of engineering education in the United States. He served as first President of the American Society of Mechanical Engineers, an organization comparable in prestige to Morison's, from 1880 to 1882.

Although he did not rise so high in his profession, Albert Adams Merrill (1875–1952) became a pioneering aeronautical engineer and later a professor of aeronautical engineering at the California Institute of Technology. He was in love with aviation even before the Wright brothers taught him to fly a plane, and it is not surprising that his utopian novel appeared when he was only twenty-four.

As businessmen and manufacturers in private industry, John Bachelder (1817–1906), Byron Brooks (1845–1911), Chauncey Thomas, and King Camp Gillette (1855–1932) had different opportunities for achievement and recognition. Their youthful interest in technology translated into adult invention of various kinds, usually in conjunction with their businesses. Thus Bachelder helped refine the sewing machine, Brooks helped improve the typewriter, Thomas helped better the carriage, and Gillette, the only one of the four to obtain fame and fortune, invented the safety razor outright. Bachelder, Brooks, and Thomas were never destitute but they never acquired the riches and—thanks to his portrait on the wrapper of every Gillette blade—the public recognition of Gillette.

Gillette came from a fairly prominent family. His father was a successful inventor and businessman, and his mother was co-author of *The White House Cookbook*, one of the most famous and popular cookbooks of the nineteenth century. At the height of his entrepreneurial success, Gillette yearned to go on to reshape America, if not the world, in the mold of his utopian writings. Yet Gillette gradually lost his fortune, his influence on the company bearing his name, and, as advertising fashion changed, his portrait on the company's product. His last sad years coincided with the Great Depression; he died nearly penniless, forgotten as a man and as a thinker.

A no less unhappy tale is that of Edgar Chambless (1876–1936), who achieved neither fame nor fortune at any time in his life. Chambless was one of the ten visionaries who had a youthful fascination with technology. In his case it led to the adult occupations of architect and patent investigator. Chambless lost his savings in the stock market crash of 1893, and his "Roadtown" scheme reflected his strong desire at once to obtain new security and to serve society. After decades of vainly appealing to prominent Americans—from Woodrow Wilson to Henry Ford to Franklin Roosevelt—for support of his scheme, Chambless committed suicide in a lonely furnished room in New York City.

Suicide was also the chosen death of Henry Olerich (1851–1927), a jack-of-all-trades who was variously a writer, farmer, teacher, inventor (of a commercially unsuccessful patented tractor for cultivating corn), machinist, and lawyer and who also had a youthful interest in technology. In his suicide note Olerich claimed declining health as the principal reason for his act. But his lengthy autobiographical writings reveal a persistent desire for the public recognition that always eluded him. Indeed, Olerich tirelessly promoted his visionary views through newspaper ads and circulars; many quoted prominent contemporaries endorsing at least a few of his ideas. The overall impression, however, is one of desperation on Olerich's part. What fleeting fame he did acquire was due primarily to the talents of another: his adopted baby girl, whose experimental education at Olerich's hands made her a child prodigy and, for a few years, a Midwestern celebrity.

By contrast, Jeff W. Hayes (1858–1917) died as contentedly as he lived, despite the mysterious blindness that afflicted him after 1894. Hayes is the tenth and last of the technological utopians who had an early attraction to technology, and it is not surprising that he became a pioneering telegraph operator and builder and later telegraph company manager in the Midwest and the Far West. His cheerfulness in the face of adversity endeared him especially to the citizens of Portland, Oregon, his home from 1882 until his death and the subject of his technological utopian work.

Albert Waldo Howard (no dates available) may also have suffered from a serious permanent injury, although there is little biographical information about him. In 1901 Howard self-published two brief pamphlets on Beethoven and his deafness which implied that Howard was the reincarnation of that musical genius; the pamphlets spoke of a nonverbal "cosmic consciousness" that had been granted to gifted deaf people and would become a language of the future for talented hearing and deaf people alike. For Howard, however, as for Hayes, writing may have been a means of escape from the emotional solitude of incapacitating physical injury.

John Macnie suffered a different sort of tragedy in the prime of his life: the loss, for unspecified causes, of his young wife within a year of their marriage. An immigrant to the United States from Scotland, Macnie never remarried and apparently mourned her passing until his own. He joined the fledgling University of North Dakota in Grand Forks in 1885, where he became a highly respected and much beloved professor of several subjects. There he remained until his death. His many contributions to the institution were praised lavishly at the memorial service held for him at the University. Several of the memorial tributes noted his childlike simplicity and innocence, a trait rarely mentioned in connection with the other technological utopians—the vast majority of whom prided themselves on their tough-minded realism and lack of sentimentality.

For three of the remaining technological utopians, I have uncovered considerable biographical information. Solomon Schindler, Charles Willard Caryl (1858–1926), and Harold Albert Loeb (1891–1974) each led diverse, adventurous, and at times controversial lives. Schindler emigrated to the United States from Germany, where he had been ordained a traditional, or Orthodox, rabbi. His hopes of remaining in his native land as a secular teacher rather than as a rabbi were dashed by his outspoken opposition to Bismarck's policies. He eventually settled in Boston and became spiritual head of a new synagogue of aspiring businessmen. It developed into a bastion of liberal, or Reform, Judaism. Indeed, his synagogue was soon a popular meeting place for all varieties of area clergymen and laymen who, like Schindler, wished to make American religion more modern, more scientific, more intellectually respectable. This development increased Schindler's visibility but also made him the subject of wide debate. Ironically, Schindler ultimately lost his position for being insufficiently assimilated. His increasingly native-born congregants felt that he retained too many unrespectable mannerisms of the first-generation immigrant, not least a thick German accent. Schindler was as agitated over his congregants' unwillingness to accept the gospel of Bellamy Nationalism as he was over their casting him aside. An early New England advocate of the Nationalist creed, he had translated

Looking Backward into German four years before publishing his own sequel to it.

Caryl was an eighth-generation American from the Far West and at different times was a businessman, manufacturer, inventor (of a widely used commercial fire extinguisher), social worker, and founder of utopian communities. In the years immediately before and after publication of his technological utopian work, he was head of a prosperous gold mining company in Colorado. Unlike most of the other technological utopians, he had both the desire and the funds to test his utopian theories on a communal scale—not, as with most other utopian communities, as an end in itself but rather as a prelude to society-wide improvements. Unfortunately, all of his several communal experiments failed and all aroused controversy in the process. Caryl found himself charged successively—and not, it appears, unfairly—with the accidental deaths of a number of children, with the plagiarism of the ideas of other visionaries, and with the propositioning of young women through pornographic letters to them.

Loeb is the youngest of the technological utopians included in this study. His utopian work appeared nine years after the last previous one and, as will be seen in chapter 8, it was in a sense a transitional work. It was also a minor part of the Technocracy crusade of the 1930s, the first organized expression of technological utopianism. This crusade will be examined in chapter 6. Loeb was, for a time, a fairly prominent Technocrat but later broke with the principal body and formed his own. His interest in changing American society persisted long after the breakup of the latter organization. Loeb had other interests as well, for which he is probably better known. He was an active American expatriate in Paris in the 1920s and a compatriot of, among others, Ernest Hemingway, whose unflattering portrayal of Loeb as Robert Cohn in *The Sun Also Rises* caused Loeb much shock and pain. Because of family wealth, Loeb was able not only to travel and to write as he pleased (including three novels) but also to bankroll the expatriates' avant-garde magazine *Broom*. Curiously, as he confessed in his 1959 autobiography, *The Way It Was*, his concern for technology derived from his years abroad as an artist otherwise indifferent if not hostile to industrialized America. While on a visit with a fellow exile to the Burgundian countryside, he became intrigued with a seemingly out-of-place factory. Reflection, however, persuaded him of the aesthetic value and the potential social utility of technology. Loeb's desire to integrate art and technology makes his *Life in a Technocracy* a distinctive work even within the technological utopian genre.

Like several others among the technological utopians, Loeb led two or more comparatively separate lives. After his return to the United States in 1928, for example, he wrote primarily nonfiction, much of it technical. How much the inner tension of those diverse lives affected the utopian works produced by these individuals cannot be determined. The most that can be concluded is that at least some of the twenty-five visionaries had profoundly compartmentalized minds, but not necessarily unhappily so. Thurston, for example, envisioned a radically better tomorrow while arguing elsewhere that the depletion of the earth's coal and the sun's heat were leading to a cold and lifeless planet. Thomas was segmented in a different sense, divided between habitual rigid self-control and periodic anxieties that left him helpless. And Bellamy was a shy, withdrawn semi-invalid prompted by events first to write his stirring novel and then to lead a nationwide crusade to try to achieve its dream.[8]

Perhaps the most provocative example of such compartmentalization, however, is Gillette. As his recent biographer, Russell Adams, makes clear, Gillette somehow managed to keep separate throughout his adult life his business career and conventional beliefs in "free enterprise" from his writing career and unconventional beliefs in a superstate modeled on a supercorporation. He did not hide his writings from either his business associates or his family; his books were, after all, published under his own name, and his articles in prominent journals. Still, he almost never spoke to his associates or family about his social ideals and discussed them with only a handful of others. Moreover, he never permitted his social ideals to intrude upon his entrepreneurial practices, divergent though they were. To call Gillette hypocritical as well as compartmentalized would be unjust, especially if he himself saw no such contradictions. To call him a curious bundle of tensions would not, however, be unfair.[9] The same qualification applies to the other technological utopians whose personalities or sets of ideas contained apparent contradictions.

In any event, the twenty-five technological utopians were a diverse lot. Despite roughly similar economic and social backgrounds in all cases where I have been able to find information (for over half, no substantial information exists), they were an eclectic group which arrived at a common ideology from many different beginnings. Their eclecticism—and eccentricity—may not, however, set them completely apart from their fellow Americans. In publishing utopian works—whatever their motives or the importance of those works in their lives—they were not, of course, average Americans, or even average American reformers. Only a relative handful of their fellow citizens became utopians of any variety. Yet the technological utopians' works were clearly intended to improve, not to undermine, the mainstream. They were eccentric in their methods, not their goals.[10]

Further, their methods were not entirely unshared among their peers. Several prominent engineers besides Merrill, Morison, and Thurston did publish essays and articles similar in tone if not in detail to theirs; and so did a number of equally prominent architects, city planners, and industrial designers. The stigma often attached to what is today termed "futurism" may well have discouraged engineers and others professionally involved in advancing technology from greater public involvement. The fear of being labeled a "dreamer" may have stopped many professionals from writing anything speculative at all.[11]

It is, of course, possible that professionals in the field of advancing technology simply took for granted that a better society would come about, without the help of visionary schemes. Whatever the reasons for their scant numbers, the twenty-five visionaries were conservative reformers not completely alone in their approaches.

In short, the technological utopians in this study ought not to be dismissed as mere crackpots. On the other hand, they ought to be accorded modest recognition for their relative unconventionality. They were not simply pale reflections of more conventional contemporary reformers; they were significant precisely as utopians. To miss this is to miss technological utopianism's relationship to the mainstream of late nineteenth- and early twentieth-century American culture. Just as technological utopianism is at once a reflection and a distortion of that culture, so the technological utopians are at once a reflection and a distortion of their contemporary fellow Americans— and in their psychology as in all else.[12]

It is a truism that any kind of utopianism reflects the visionary's hopes or fears or both. Sometimes these feelings lie below the level of consciousness. This holds for utopian writings no less than for utopian communities and organizations. Often hopes and fears appear in combination, because the fears are what give rise to the visions which constitute the hopes. Sometimes the hopes are weaker than the fears and the visions temporary and fragile defenses against greater fears.

Roemer characterizes the outlook of most of his 154 late nineteenth-century visionaries as "pessimistic optimism."[13] The technological utopians were not free of that mixed outlook either, despite their confidence not only in the accuracy of their visions but also in the practicality of their panacea: technology. They might not have been entirely delighted with their technological utopias if they had actually achieved them.[14] I want to consider this point in some depth, to avoid the risk of reducing the technological utopians' individual or collective vision to their individual or collective personalities, thus distorting or ignoring its content and its relationship to the real world in which they lived and wrote—and which they hoped to perfect.[15]

At least in part, the technological utopians were responding to widely acknowledged problems of their day. Their writings were intended as comprehensive solutions to those problems. Like countless other Americans, they were concerned about seemingly unprecedented social disorder, including crowding, competition, aggression, selfishness, and rudeness; the ever-present threat of disease, whose source in microorganisms was just beginning to be understood; the unsanitary habits and habitats that spread the agents of epidemic and death; growing secularism and decline of belief in God and an afterlife; and the pervasive sense that individuals had lost control over the self, family, and community, and even advancing technology.[16] The influence of these commonly perceived problems was surely stronger than any differences in the specific backgrounds and personalities of the visionaries; they reacted to the same conditions as did millions of other Americans. The utopians' anxieties were thus rooted in concrete developments of particular historical circumstances.

I want to suggest also that the technological utopians' anxieties may have been partly the universal anxieties of the kind described by Freud. As I have argued elsewhere,[17] the obsession in the technological utopians' writings with order, cleanliness, neatness, efficiency, and control sounds very like Freud's "anal character." Likewise, the obsession in those same writings with security, dependency, and the relatively passive satisfaction of needs and wants is congruent with Freud's "oral character." The fictional inhabitants of technological utopia exhibit an extreme mixture of these traits—to the extent of seeming neurotic in these ways. Certainly the presence of these neurotic tendencies in the characters in their fictional works may suggest something about the prophets of technological utopianism themselves.[18] I do not want to press such extrapolations too far: the evidence is too scant to justify labeling the prophets themselves as real life anal and oral neurotics.[19] And in any event, human beings are far too complex to warrant such reductionist analysis. Nevertheless, it is reasonable to see reflections of the prophets' psyches and personalities in the utopias that they envisioned; indeed, their own characters may have predisposed them to become utopians in the first place—just as historical conditions shaped the kinds of structures they proposed.

To return to the hopes and fears of the technological utopians, their normal human anxieties about the present and future took nothing away from the sincerity of their beliefs in technology as the solution to society's problems—including any problems created by technology itself. In fact, the principal problem they foresaw with technology was that it might not remain the servant of mankind but someday might become the master. We must not overemphasize their fears while legitimately qualifying their hopes.[20]

Even if we see the technological utopians' confidence in the future as mixed with anxieties about the troublesome present, we need not suppose that the achievement of technological utopia would have necessarily disappointed them. The capacity for suspending or reconciling ambivalences or outright contradictions may be what made technological utopia, for its exponents, the perfect society. Simply, technological utopianism actually may have been the most practical way of dealing with certain otherwise overwhelming contemporary problems, whether personal or social. This seems to have been the view of several other American utopians of their day.[21] Surely this was true of technological utopians such as Bellamy, Gillette, Loeb, Thomas, and Thurston. Utopianism need not be viewed as necessarily either deviant or naive.

Similarly, and equally important, to examine the technological utopians' lives and writings for indications of their hopes and fears is not to equate their lives with their writings. Undoubtedly their lives and their writings are related. But they are not identical—just as the sleeping dreams of individuals, which Frank Manuel has identified with utopianism,[22] are clearly related to but still different from their waking lives (and dreams of ideals). The sleeping dreams of most individuals do not become their own, much less their society's, conscious goals. Whether the utopians' lives were happy or unhappy or a mixture, their writings finally stand apart, as things in and of themselves.

In this study, psychology can, as history does elsewhere, serve to connect the technological utopians and their visions with the mainstream of American society; and the utopians' hopes and fears can in turn illuminate the mainstream and its ideology. I have tried to make such connections in this chapter. Manuel correctly contends that "the utopia may well be a sensitive indicator of where the sharpest anguish of an age lies"; utopia may equally well indicate the fondest hopes of an era.[23] As I will show in the next chapter, the American technological utopians had ample historical precedent, dating back to sixteenth-century Europe, for at least qualified hope that they could make their utopia real.

Four

The European Origins

Technological utopianism emerged in America only in the nineteenth century, but it is rooted in two earlier developments in Europe: rapid technological advances, which began in the eighteenth century and have continued, and a gradual transformation of the concept of utopia from the impractical to the practical, a transformation that began in the sixteenth century and has continued into the twentieth. Although the second of these two developments will be stressed here, the two are intertwined. In large part it was those enormous technological advances that made utopia look like an achievable goal. Only when, in the eighteenth century, technology advanced sufficiently to offer the prospect of an "affluent society" for many did most utopian schemes begin to appear at all realistic. That is because most utopian schemes have presupposed the availability of adequate food, clothing, and shelter, and only in modern times could their availability be taken for granted. Even those schemes which, like the first four to be discussed in this chapter, were anything but materialistic and hedonistic, still presupposed those basic necessities and could not have been realized in their absence. Only in the nineteenth century, however, did these two developments coalesce—and they did so first in the United States.[1] By tracing that coalescence back to its European origins we can best understand its final integration in nineteenth-century America. The history of utopian thought as a whole has been surveyed many times before, but the history of technological utopianism has not been.

Thomas More's *Utopia*, which appeared in 1516, marks a crucial shift in the image of utopia.[2] It is true that the term *utopia*, which More

(1478–1535) coined from Greek roots, means "nowhere" (although it also means "good place"). And it is true that More considered human nature depraved. Nevertheless, there is evidence that he was the first to hold out the prospect, albeit dim, of actually establishing a perfect society and thereby altering man's nature. That evidence remains despite his failure to explain how human nature is to be sufficiently tamed to enable the perfect society to be established in the first place.[3]

The evidence is threefold. First, More locates utopia within contemporary semifeudal society rather than in, say, an agrarian paradise of long ago or far tomorrow.[4] His thus implicitly striving to perfect his own society rather than forsake it surely bespeaks a grain of optimism. Second, he expects man, not God, to establish utopia. Dependence on God would mean a vote of no confidence in man and, in the wake of Christ's delayed return, the consignment of utopia to a very distant future. Dependence on man rather than God can be said to distinguish utopianism from millenarianism.[5] Third, More provides a detailed description of utopia, not merely a set of abstract principles, and his description reflects a close evaluation of his own society, not unanchored speculation. According to J. H. Hexter, his originality lay here—"not in the bare idea of a community of property and goods" but "in the exactness, the precision, and the meticulous detail with which he implemented his underlying social conceptions."[6] Certainly it is here that More foreshadows the systematic social planning that characterizes technological utopianism.

The planning envisioned by More, however, is far less dependent on actual technological advances than is the planning in technological utopianism. Indeed, More provides his utopians with only three technological improvements, none of them decisive: the hatching of eggs by artificial heat, the charting of planetary movements by various instruments, and the invention of unspecified "clever . . . war machines."[7] To achieve economic subsistence, his utopians rely on the efficient use of existing agrarian and craft devices. To achieve improvements in human nature, they rely on the introduction of various social and cultural arrangements—above all, the abolition of money and of private property. In contrast, technological utopians rely on the prosperity wrought by technological advances. Like the technological utopians, More's utopians are hungry for new, scientific knowledge of the world, but their purpose is as much to serve God as to improve society: "They think that the investigation of nature, with the praise arising from it, is an act of worship acceptable to God."[8]

The Pansophists, a small number of late sixteenth- and early seventeenth-century visionaries, forge the first real connections between utopianism and technology. In their greater optimism, they stand

closer than More to technological utopianism. Their religious orientation, however, sharply differentiates them from secular-minded technological utopianism, for they seek a civilization, called Pansophia,
which harmoniously joins Christianity and science.[9]

The three Pansophists whose works are most akin to technological
utopianism are Tommaso Campanella (1568–1639), an Italian Domin-
ican friar; Johann Valentin Andreae (1586–1650), a German Lutheran
minister and teacher; and Francis Bacon (1561–1626), the scientist,
philosopher, man of letters, and, like More, Lord Chancellor of England. Campanella's *The City of the Sun* appeared in 1623, Andreae's
Christianopolis in 1619, and Bacon's *The New Atlantis* in 1627.[10] Like
More's *Utopia*, these three works recount the adventures of travelers
who have discovered civilizations unknown to Europeans and have
returned home to announce their finds. Those civilizations prove not
merely to have equalled the cultural and material achievements of the
West but to have surpassed them.

From the outside, each of the three utopias, like More's, resembles
a medieval European town—modest in size, with high towers and
thick walls. Within those walls, however, life is considerably more
pleasant than life in any medieval European town ever was. The
inhabitants of all three utopias enjoy comfortable living and working
quarters; handsome public buildings and gardens; academic and
vocational schooling; limited working hours; cultural and recreational activities; a rough equality of opportunity; a relatively equal
distribution of property; an honest, efficient, apolitical government;
and above all the practice as well as the preaching of Christianity.
Except for the last, these features reappear in technological utopianism, in different form.

The historical significance of these utopias is the attention they
give to technology. For example, Campanella's utopia boasts an
efficient water and sewage system, labor-saving agricultural
machines, and mechanically propelled ships. Similarly, Andreae's
utopia prides itself on its water and sewage system, its fireproof stone
buildings, and its lighting system. Bacon's ideal society vaunts the
most impressive achievements of all: the conservation of water and
the harnessing of its power, a system of weather prediction, the
invention of clocks, the production of artificial metals, underwater
travel, flying, the preservation of bodies after death, the curing of
diseases, and the prolongation of life—all of these achieved through
technology. Finally, the utopias of Andreae and Bacon harbor full-
fledged research institutions, which aim at achieving further technological advances.[11]

To say that their technological advances are what make these
utopias utopia would, however, be too bold. Clearly, these advances

improve life. They make life pleasanter, easier, healthier, and safer—in short, happier. It is just as clear that the creators of these utopias value technology. If nothing else, the prominence they give their research institutions attests to this. Indeed, Andreae terms his citizens' think tank the "innermost shrine of the city" and Bacon his the "very eye of this kingdom."[12]

Nevertheless, technological progress in these utopias is only a means to an end and not, as in technological utopianism, virtually the end in itself. The end in Pansophism is the service of God; the attainment of peace and brotherhood and happiness, which technology helps achieve, is viewed as such service. Knowledge of the world, on which technology rests, and mastery of the world, which technology yields, are also "in the greater service of God," for with knowledge and mastery of the world comes a keener appreciation of God's handiwork.[13]

Utopia, The City of the Sun, Christianopolis, The New Atlantis, and all other tracts composed between 1516, when *Utopia* appeared, and the outbreak of the French Revolution comprise what Frank Manuel calls "Utopias of Calm Felicity."[14] Although these utopias are based on confidence in the possibility of improving human nature, their authors remain sufficiently wary of mankind to propose establishing limits within utopia. To keep human nature in check, they envision a fixed, unchanging society without further technological progress. Their rejection of technological progress, which for the technological utopians is the foundation not just of society but of happiness as well, is part of their ultimately religious rather than materialistic aims. They seek surcease from toil not simply to enjoy leisure but even more to contemplate and worship God. Release from ceaseless labor, provided by technology, is for them a means and not an end.[15]

Two works by the Marquis de Condorcet (1743–1794), the Enlightenment philosopher, break radically with this notion of utopia and with many of the other notions that characterize previous utopias. The works, both published posthumously, are (1) *Sketch for a Historical Picture of the Progress of the Human Mind*, which traces the remarkable economic, political, and social progress made by mankind from ancient times to the 1790s and which predicts even greater progress in the future; and (2) a commentary on *The New Atlantis*, an updating of Bacon's vision, predicting scientific and technological advances surpassing those imagined by Bacon.[16] In both works Condorcet departs fundamentally from the "calm felicity" of previous utopias—first, by envisioning endless progress rather than endless tranquillity, and second, by denigrating rather than embracing organized religion.

Condorcet's break with other aspects of earlier utopias is no less sharp. He evinces an unprecedented optimism about the prospects for realizing utopia: its realization, he believes, is virtually at hand. He transcends local, regional, and even national boundaries to envision a worldwide utopia. And he grants technology an unprecedented role in establishing utopia. In other ways Condorcet does not so much break with previous conceptions of utopias as enrich them. Where previous visionaries tended to ignore history, Condorcet finds meaning in it: he sees mankind as moving firmly toward utopia. Where previous visionaries tended to concern themselves only with mankind's physical and spiritual well-being, Condorcet is concerned with its emotional well-being also. Finally, where previous visionaries tended to seek a blanket equality of opportunity for all, Condorcet seeks to develop what he sees as the inevitably differing talents and abilities of individuals.

In all the aspects in which Condorcet differs from his predecessors, he is closer to technological utopianism. Yet he is no technological utopian himself, for he credits mankind's advances to things other than technology—to increasing secularization, education, and equality, for example. And he envisions eternal progress rather than the eventual culmination of all past progress in a specific kind of society. Thus the technological advances he so carefully and lovingly delineates are only indications of the way society is moving generally, not blueprints for a specific future society.[17]

The same changes that bring Condorcet closer to technological utopianism mark, according to Manuel, the beginning of "open-ended utopianism," the successor to the "utopias of calm felicity."[18] Where the latter are local, religious, static, and ahistorical, the former are universal, secular, dynamic, and historical. Although Manuel has only a limited interest in its increased emphasis on technology, open-ended utopianism is nevertheless the forerunner of technological utopianism—and not only substantively but chronologically as well. Open-ended utopianism flourished in the late eighteenth and early nineteenth centuries, the period just preceding the emergence in America of technological utopianism. Indeed, Manuel's chronological boundaries for open-ended utopianism might properly be extended to encompass technological utopianism itself throughout American history. For apart from its ahistorical quality, technological utopian writing certainly qualifies as universal, secular, and dynamic in both nature and orientation.

Of all the open-ended utopias, those of Henri de Saint-Simon (1760–1825) and Auguste Comte (1798–1857), his one-time protégé, most nearly approximate the vision of technological utopianism. Notwithstanding considerable similarities between them, their visions are distinct from each other and require separate treatment.

Among the most popular utopias of the nineteenth century, Saint-Simon's vision has continued to attract disciples throughout the West, making him among the most famous and influential of utopian theorists.[19] Saint-Simon's vision rests on his analysis of the state of Europe in his time. He argues that the intellectual, social, political, and cultural unity that Europe once enjoyed has collapsed under assault by Protestantism, Deism, empiricism, nationalism, and commercialism. A new unity must be forged, and its basis must be ideological. The ideology that is to forge this unity is science, which will replace the divisive and shaky world views currently presented by religion. Science is to be applied in the practical form of "industry," which includes both manufacture and distribution and which amounts to technology.[20] Priests and politicians, the old rulers of Europe, are to be supplanted by scientists and technicians.[21]

Far more the philosopher of utopia than its designer, Saint-Simon offers no blueprint and so falls short of being a full-fledged technological utopian.[22] Moreover, near the end of his life he backed away from his exclusively technological vision and urged a religious as well as technological panacea. For he came to feel that man had spiritual as well as material needs. Religion must therefore be allotted a place alongside science and technology, but religion in the form of brotherly love rather than as theological dogmas. In place of the reign of scientists and technicians, there is now to be a triumvirate of scientists and technicians; industrialists and managers; and artists, teachers, and philosophers—but no clerics.[23]

Comte became Saint-Simon's disciple in 1817 and remained with him until 1824, when he broke with the master over both personal and intellectual matters. The exact causes of their disagreement, and the precise contribution of each to the several works they had earlier published together, have been the subjects of prolonged debate. Of importance here, however, is their agreement, even after their parting, on six fundamental aspects of what I have termed the European origins of technological utopianism: (1) the need for a new social order in Europe and perhaps elsewhere in the wake of the disorder brought about by the industrial revolution in England and the political revolution in France; (2) the need for science and technology to solve major social as well as technical problems; (3) the need for technical experts to solve both those problems and, in time, to run society; (4) the need to control the unenlightened masses in order to effect these changes; (5) the need to establish a new European hierarchy based on not social origins but natural talent and society's requirements; and (6) the need to abandon mass democracy and, in turn, politics.[24]

Comte was a more systematic thinker than Saint-Simon. Not content with Saint-Simon's comparatively modest philosophical formulations, he created a full-fledged philosophical system that tried to encompass every aspect of human existence. This "science" of mankind and society—or "sociology," as he began characterizing it in 1839—in his view represented the discovery and application of the natural laws of the universe. Variously called "Positivism" and the "New Social System," it provided an intellectual foundation for the modern discipline of sociology which grew out of it. Its principal expressions were his multivolume *Cours de philosophie positive* (1830–1842) and *Système de politique positive* (1851–1854).[25]

Not only did Comte think more systematically than Saint-Simon, he also worried more than his former master about the implementation of his theories. Dividing all of history into stages, Comte envisioned the gradual realization of his philosophical system, beginning with the imminent Positive Revolution and ending with the more distant Positive State. To persuade the ignorant masses of the virtue of these impending changes and to dissuade any from challenging them, Comte envisioned a revolutionary dictatorship. Although such a body, he cautioned, would be transitional, "liberty" and "equality," the ideals of the French Revolution, would be replaced by "order" and "progress." The revolutionary ideals Comte denounced as antiprogressive, their replacements he praised as fully compatible with each other. Comte thereby proved less tolerant of individualism than did Saint-Simon, whose writings are less authoritarian than Comte's and, as indicated, than the technological utopians as well.[26]

More zealously than Saint-Simon, Comte attempted to spread his ideas throughout the West. If anything, however, he had less success than Saint-Simon, whose remaining disciples celebrated his name even as they often reinterpreted his ideas.[27] Yet Comte's ideas, more than his former mentor's, influenced the development of the sociological discipline.

Finally, Comte followed Saint-Simon in expanding his vision to accommodate a spiritual side along with the material side. Although their specific conceptions of this dimension differed, its presence in their work showed an ultimate reluctance on the part of either to embrace a purely technological utopian vision. Indeed, Comte went as far as to predict that the complete administration of the Positive State by technical experts would allow others in society—unspecified "thinkers"—free reign for their imaginations. As Manuel puts it, "The temporal order of the future did not occupy a central place in Comte's considerations because in the end this was the lesser order."[28] This may account for the absence in Comte's writings, as in Saint-Simon's, of the kind of detailed blueprints found in the writings of the purer technological utopians.

In short, all European prophets of technological progress—from More to the Pansophists to Condorcet to Saint-Simon and Comte—either refrained or retreated from endorsing unadulterated technological advance. Their reasons differed, but none was finally a technological utopian. To each, other aspects of life either were or became no less important than technological advance.

The reservations about technological advance, articulated especially by Saint-Simon and Comte, were echoed in very different quarters in late eighteenth- and nineteenth-century Western Europe: in conservative circles of particularly England and France. The intense desire of Thomas Carlyle, John Ruskin, and William Morris, among other prominent English social critics, for a renewed organic society amid the upheavals of the industrial revolution that began in England in the 1750s, are familiar enough not to require detailing here.[29] Similar pleas among somewhat less familiar French social critics—most notably the Vicomte de Bonald (1754–1840) and Pierre Edouard Lemontey (1762–1826)—originated as reactions to the French Revolution but were later applied to the English industrial revolution as well.[30] Worries concerning mechanization and specialization of work and leisure, decline of communal institutions like the family and the church, unprecedented materialism and utilitarianism, and the breakdown of social and political order thus gave radical and conservative social critics alike some grounds for agreement about restricting technological development.

What Samuel P. Huntington has written about the absence of a conservative utopia applies to all these conservative social critics:

> No one is born to conservatism in the way in which a Mill is born to utilitarianism. The impulse to conservatism comes from the social challenge before the theorist, not the intellectual tradition behind him. Men are driven to conservatism by the shock of events, by the horrible feeling that a society or institution which they have approved or taken for granted and with which they have been intimately connected may suddenly cease to exist. The conservative thinkers of one age, consequently, have little influence on those of the next.[31]

The real problem, Huntington suggests, is *what* to conserve. That problem challenged more radical social critics and nearly made the terms *conservative* and *radical* as applied here irrelevant.[32] Robert

Owen (1771–1858), Charles Fourier (1772–1837), and Karl Marx (1818–1883) and Friedrich Engels (1820-1895) were among the few radical critics who offered serious schemes for making beneficial use of various technological achievements without letting technological advance—or, for that matter, the threat of technological advance— become virtually an end in itself. More than either the other "radical" European visionaries or the "conservative" critics of industrialization, they proposed specific means of regulating and accommodating—as opposed to merely stopping or severely limiting—technological advance. Their various schemes deserve elaboration here.

The son of an English ironmonger and saddler, Robert Owen had only an elementary education before being apprenticed to a draper. After several years' work in that and other occupations, he was, at age twenty, chosen manager of a cotton mill. Under his direction, the mill was reputed to produce Great Britain's finest yarn. At twenty-four he became managing partner of another prominent concern that four years later purchased Scotland's New Lanark Mills. For the next three decades Owen headed this largest of Britain's cotton-spinning enterprises. While there he established a model factory community that greatly improved working and living conditions, reduced working hours, and educated workers' children in innovative ways. Gradually, Owen acquired considerable fame and fortune.

Dissatisfied, however, with what seemed a slow rate of social improvement not only at New Lanark but also throughout Britain, Owen in 1824 journeyed to America, using his personal savings to found a model utopian community at New Harmony, Indiana. Unlike New Lanark, New Harmony failed completely, and five years later Owen returned to Britain, though not to his business, from which he retired. For the rest of his life he worked tirelessly on behalf of his nation's laboring classes, advocating in speech and in print a combination of trade unionism, socialism, secularism, and spiritualism. Meanwhile his disciples founded other Owenite communities in Great Britain and the United States. Although most of his and his disciples' efforts were unsuccessful, Owen died confident of the ultimate vindication of his ideals.[33]

Owen was not a profound thinker, or, despite his entrepreneurial skills, a very practical communitarian. His ideas and practices left much to be desired. Yet unlike the European prophets of technological progress discussed earlier, Owen actually lived—and worked— amid profound technological change, clearly benefiting enormously from the English industrial revolution. But Owen gradually perceived certain drawbacks of mass industrialization, which he hoped might be remedied: he did not advocate wholesale retreat from technological change but restrictions on its development; and, no less important, he showed as much concern for improving the social, cultural, and moral lot of his industrial workers and their families as for improving the machinery.

Owen's solutions to problems were at once colored and undermined by his unapologetic paternalism, a trait he acquired as a successful manager, and by his naive belief in the power of human reason to discern, treat, and eliminate all social ills and the power of an appropriately structured environment to shape human character in desired directions. Owen's intended beneficiaries readily recognized his paternalism and often resented it—just as did the residents of later model company towns such as Pullman, Illinois, and Gary, Indiana.[34] But he held fast to the reins of control and to his seemingly contradictory faith in both reason and the environment. Owen envisioned a rational, enlightened, and compassionate business and government elite determining the environment, and so the character, of the industrial masses. Moreover, he adopted outright his friend Jeremy Bentham's utilitarian idea of the greatest good for the greatest number, as defined by that same elite. As one student of Owen has aptly put it, "the relationships, close, deferential, and interdependent, which were thought peculiar to pastoral Britain, were capable of reconstitution within industrial society."[35] These convictions were expressed in many of Owen's writings, often with unnecessary abstraction and repetition, but were most clearly spelled out in two of his earliest and most famous works: *A New View of Society* (1813) and *Report to the County of Lanark* (1821).

A New View of Society treats Owen's achievements at New Lanark. From the outset of his efforts there, Owen understood, as few other industrialists did, that the discipline required for the efficient operation of any large-scale factory-based system must be undergirded by as pleasant a factory life as possible under the circumstances and also by a comfortable domestic environment. Thus Owen furnished decent food, clothing, and shelter for all his workers, and at reasonable costs. He reduced their working hours as well. He provided them and their families with opportunities for educational, religious, and recreational activities. There was mandatory free schooling for all children between five and ten, who learned reading, writing, and arithmetic, plus good manners and good morals. Here as in other Owenite communities, the educational activities were the most praised by outsiders. Repeatedly, and in the workers' presence, Owen and his associates linked high moral standards to high profits. They claimed to have eliminated the uncleanliness, idleness, drunkenness, lewdness, and other social disorders that had plagued their predecessor, David Dale, the mills' founder. Consequently, announced Owen,

"Those employed became industrious, temperate, healthy; faithful to their employers, and kind to each other; while the proprietors were deriving services [and profits] . . . far beyond those which could be obtained by any other means. . . ." Here was the justification for Owen's dictum: "Train any population rationally, and they will be rational."[36]

A New View of Society contains only cryptic remarks concerning Owen's hopes to educate similarly the whole of the British working population. Not so his Report to the County of Lanark: here is a detailed plan to transform that segment of British society, not by renovating existing communities but by creating new ones. Having failed to secure passage through Parliament of even modest factory reform legislation, having witnessed increasing poverty and discontent after the end of the Napoleonic wars in 1815, and, above all, having uncovered the physical and psychological enslavement of countless factory workers to the machines and structures allegedly invented to improve their lot, Owen came to advocate a return to the soil on a mass scale—not to abandon technology, but to redeem and purify it.

By means of well designed and tightly organized cooperative communities to be set up throughout the British countryside, Owen hoped to restore that oft-lauded sense of community supposedly lost in the transition from agrarian to industrial society. Such communities would engage primarily in farming and only secondarily in manufacturing, but they would utilize the latest machinery and agricultural science. Moreover, they would be able to grow enough food to supply an expanding national population which, according to some well-respected economists of the time, was otherwise doomed to starvation. The affluence necessary for nearly all serious utopian schemes would thus be available.[37]

As Owen envisioned them, these communities would consist of multistoried parallelograms, each housing between three hundred and two thousand men, women, and children "in their natural proportions." If Owen described New Lanark in his earlier work as "living machinery," then these structures might well be deemed pioneering middle landscapes.[38] More than either New Lanark or New Harmony, both of which he acquired rather than constructed, these communities would be scientifically planned. So, too, would be the work and leisure spaces and activities of their inhabitants. More than in New Lanark, diversity of daily routines would be institutionalized and, ideally, exhaustion and boredom would be avoided. Ideally, too, these communities would be self-sufficient in virtually all respects and would attract citizens from all classes. Their inhabitants would eventually be part of one large communal family replacing smaller separate families.

Alas, Owen's parallelograms were never established to his specifications, and those that were set up failed. For all their scientific planning and incorporation of technical advances—such as special valves in each room to regulate heating, cooling, and ventilation—the parallelograms were nevertheless more conservative than New Lanark. Like New Harmony, they were so predominantly agrarian that they were more retreats from industrializing society than accommodations to it. Still, they were intended by Owen as devices that would balance and integrate technological progress with social progress, and, despite their limitations, their beneficial intent ought to be recognized.

Charles Fourier was the son of a moderately successful cloth merchant. Like Owen, he had a modest formal education and early exposure to the marketplace. But where Owen's youthful departure from home was voluntary, Fourier's was not. Owen enjoyed his apprenticeship to a draper, but Fourier despised his less elevated position as a merchant's assistant. The experience, in fact, was one of several that eventually led to Fourier's obsession with ridding the world of greed and hypocrisy. Fourier's contempt for existing society was deepened by his traumatic experiences as an unwilling participant in the French Revolution. He lost his paternal inheritance—and nearly his life—because of various false accusations against him, and he suffered temporary imprisonment and years of forced service in the army as an aftermath of those charges. In 1795 Fourier managed to become a clerk in a cloth concern and, a few years later, a traveling salesman. The latter was a slightly livelier job; more important, it allowed Fourier time to think and to write. In his later years Fourier lived in several French cities and towns before finally settling in 1826 in his favorite locale, Paris—again, as a lowly clerk in a textile importing firm.

After several provocative preview essays, Fourier's first book appeared at last in 1808; his second, not until 1822. These and subsequent works, however, received little attention, despite Fourier's strenuous efforts to publicize them. Their turgid prose, annoying repetitions, and inscrutable neologisms—all infinitely greater than in Owen's writings—contributed to their obscurity. Only in his last years did Fourier gain any notice and any following, but then he won admirers throughout the world and enjoyed greater success than did Owen both in spreading his gospel and in establishing communities in Europe and America. Even then, however,

Fourier remained a reclusive bachelor whose theories of passionate attraction were cosmically distant from his personal practices.[39]

Fourier persistently preached that his version of utopia could come about only in small communities whose inhabitants knew one another well, in varying degrees, not in big cities with anonymous masses. He asserted the economic and moral superiority of agriculture over manufacturing. Yet Fourier recognized that communal living in itself was no panacea for the problems he diagnosed. For true happiness, the passions must be released completely, he maintained. Conventional community and family life tragically repressed and perverted the passions. Hence new societies must be created to replace existing ones, just as both Owen and the nineteenth-century American communitarians insisted. Like their communities, his phalansteries, as he called them, also would be models for the reconstruction of his own nation and, eventually, of all Western Civilization. But Fourier's communities also failed to be realized according to his specifications and so failed to serve the grander purpose he intended.

Fourier's blueprints for utopia were too extraordinarily detailed and complex for any comprehensive description here. I can only list his ardent commitments. (1) He would replace separate families with a single large one for each phalanstery, a goal akin to Owen's; (2) he would institutionalize variety of work (no more than two hours per day at any given task), of leisure (similarly diversified), and of pleasure (no monogamous marriages and ever-changing liasons); (3) he would encourage and entice rather than coerce children to become socialized and educated, and he would have children perform tasks unpleasant for adults but not for them; (4) finally, he would promote and reward creativity within every citizen, including scientists and technicians as much as artists and other humanists. Fourier determined, with characteristic mathematical precision, that each human being had a complicated mixture of twelve passions (five sensuous, four group, and three distributive, as he categorized them); he then sought in each of his phalansteries male and female representatives of every one of the 810 possible combinations of psychological character—or approximately 1,600 inhabitants. The rigidly symmetrical multistory rectangles he envisioned for each phalanstery would be designed to contribute to this remarkable variety of activities and of associations. In effect, Fourier would establish communities that, through intricate devices, would foster group spirit and self-fulfillment simultaneously. If he was a far more meticulous calculator and quantifier of values than was Owen, he was also far more tolerant of both passion and diversity.[40]

Fourier's schemes have been ridiculed as often as praised, and their rediscovery in the 1960s prompted both reactions. In his defense, Fourier lived in a largely pre-industrial society and so, unlike Owen, did not write in direct reaction to the excesses of the English industrial revolution. His proposed phalansteries were predominantly agrarian and rural, making only modest provision for crafts and light industries, with little place for cities. He missed the significance of the profound economic and social changes that Saint-Simon, twelve years his elder, as well as Owen, his contemporary, readily grasped. That his own society had not yet undergone industrialization may explain, but not entirely excuse, this crucial omission.

Nevertheless, Fourier's concern—or obsession—with fulfilling mankind's psychological as well as material needs was not just provocative but prophetic. The means he devised for institutionalizing diversity and for avoiding exhaustion and boredom in his utopian communities may not themselves be appealing—certainly the rationalizations he offered for them were labored. But just such problems have arisen in our technological society, one as thoroughly organized as Fourier's imaginary universe.

It is reasonable to assume that Fourier, being neither reactionary nor antitechnological, would have been cautious but not rigid in introducing further technological developments into his phalansteries, had they been readily available in France. More than even Owen's envisioned parallelograms, Fourier's proposed communities were social machines and social structures in themselves; and they were akin in comprehensiveness, although not in design, to the technological utopians' own ideal worlds. Also like them, they would be basically stagnant once achieved. Unlike them, they would not consciously treat their inhabitants merely as components of the social mechanism, as miniature machines and structures. Fourier did not view the proliferation of technology in any form, including the human one, as essential for the achievement of utopia.[41]

Karl Marx and Friedrich Engels were less specific about the future than either Owen or Fourier, but they were more sensitive to the liberating as well as enslaving potential of modern technology. Although they deliberately avoided the kind of detailed blueprints of the future that Owen and Fourier provided,[42] they repeatedly hinted at a society radically superior to the existing capitalist one, which would utilize modern, especially automated, technology as a principal means of freeing the proletariat. The proletariat would be liberated not simply from their long-standing alienated labor but also for other, more varied and fulfilling activities. More than Owen and even Fourier, they readily perceived the psychological as well as the physical consequences of work in industrializing Western society. As with Owen and Fourier, one need not accept all of their gospel to appreciate that insight.

Whether Marx and Engels were technological determinists is the subject of considerable controversy. Nevertheless, it appears that they were not, certainly not in the crude form represented by the familiar lines from *The Poverty of Philosophy* (1847): "The handmill gives you society with the feudal lord; the steam-mill, society with the industrial capitalist."[43] Rather, they primarily emphasized the mode of production—that is, the prevailing mode of labor or productive activity—as conditioned by the existing state of technology. The mode of production was primarily social, and the mode of production under capitalism was wage labor.

Man, Marx and Engels maintained, had always been a tool maker until, during the maturation of the industrial revolution that began in England in the 1750s, his tools were transformed into machines and removed from his control. (Like many students of technology, Marx and Engels equated *tools* with primitive forms of technology and *machines* with advanced varieties; this imprecise distinction ought not, however, confuse the larger issues they raised.)[44] They described a transition from early capitalist manufacture, which still relied on human skills despite a high degree of worker specialization, to the "machinofacture" of "Modern Industry," which did not. Machine-centered industry brought into being a set of material productive powers that, in turn, would eventually destroy capitalism from within and create a new mode of production and so a new society. The revolutionary changes that they anticipated were predominantly economic and social but followed from the inability of the existing mode of production—wage labor—to accommodate new technological advances.

The economic and social ramifications of the transition to wage labor, as described by Marx and Engels, are sufficiently well known to need no detailing here. As important but less familiar are their visions of advanced capitalist machines as protohuman monsters, regenerating themselves, even constantly improving themselves, amid the inexorable demand for ever more products. As they wrote in *Grundrisse* (1857–1858):

> ... once absorbed into the production process of capital, the means of labour undergoes various metamorphoses, of which the last is the *machine*, or rather, *an automatic system of machinery* ("automatic" meaning that this is

only the most perfected and most fitting form of the machine, and is what transforms the machinery into a system). This is set in motion by an automaton, a motive force that moves of its own accord. The automaton consists of a number of mechanical and intellectual organs, so that the workers themselves can be no more than the conscious limbs of the automaton.[45]

However evocative this and additional passages may be of earlier and later visions of "autonomous technology," they were not intended to portray that "animated monster"[46] as beyond the control of either the bourgeois capitalists at the outset or the proletariat in the end. Marx and Engels persistently affirmed their belief that human beings make and remake their own history and are not the passive objects of cosmic forces.[47] In the case of powerful technological instruments, the proletariat had to gain control over both the bourgeois capitalists who built them and the machines themselves. Doing so would entail not just the familiar triumph of one class over another but also the less familiar technical education of the proletariat. The very size and complexity of automated equipment had denied the proletariat not only authority over technology but also their knowledge of its inner workings. The proletariat had to be provided with technical knowledge to accompany and enhance their technical skills.

Marx and Engels did not, of course, envision the education and triumph of the proletariat as an end in itself. Rather, they expected to redirect modern technology to lessen the burdens of work for all citizens and to produce sufficient goods and services for all citizens to create a universally affluent society. If more concerned than Owen with fulfilling workers' noneconomic desires, they were less concerned than Fourier with making work enjoyable. They argued in *Grundrisse*, "This does not mean that labour can be made merely a joke, or amusement, as Fourier naively expressed it. . . . Really free labour, the composing of music for example, is at the same time damned serious and demands the greatest effort."[48] Work of whatever type was part of the realm of necessity, as they called it, and so part of the human condition.[49]

Consequently, the following famous lines from *The German Ideology* (1845–1846), ironically drawn from Fourier's writings, must be read as reflecting the spirit rather than the letter of Fourier's yearning for diversifying both work and leisure:

> . . . in communist [i.e., utopian] society, where nobody has one exclusive sphere of activity but each can become

accomplished in any branch he wishes, society regulates
the general production and thus makes it possible for me to
do one thing today and another tomorrow, to hunt in the
morning, fish in the afternoon, rear cattle in the evening,
criticise after dinner, just as I have a mind, without ever
becoming hunter, fisherman, shepherd or critic.[50]

Moreover, Marx and Engels saw technical activities, such as inventing and improving machinery, as no less creative than the artistic
ones usually so labeled. They saw creative technicians as mental
more than physical laborers, recognizing their intellectual along with
their manual abilities.[51] Fourier, in contrast, left little room for technical work of any kind apart from agriculture (as horticulture).

For Marx and Engels, then, the same technology that would help
destroy capitalism would thereafter help build a genuinely good, if
not utopian, society. And if the automated equipment at the heart of
the new technology could be enslaving, it could also be liberating. As
Marx pronounced in an 1856 speech on the dialectical thrust of
technology under capitalism:

On the one hand, there have started into life industrial and
scientific forces, which no epoch of the former human
history had ever suspected. On the other hand, there exist
symptoms of decay, far surpassing the horrors recorded of
the latter times of the Roman Empire. In our days everything seems pregnant with its contrary. Machinery, gifted
with the wonderful power of shortening and fructifying
human labour, we behold starving and overworking it.[52]

In effect, Owen, Fourier, and Marx and Engels offered different
versions of what I have called that plateau of technological progress
beyond which technologically advanced societies would proceed
cautiously, in order to effect a roughly equivalent measure of nontechnological—of social, political, economic, and cultural—progress.
None approached the concept explicitly, but all provided a glimpse of
such a stage of technological and social development. That kind of
plateau, as I will discuss in chapter 8, may offer a real life alternative
to technological utopianism.

The visionaries considered in this chapter were not, of course, the
only European utopians between the time of More and of Marx and
Engels. Nor were they the only European utopians who treated technology in their works. Most notably, Etienne Cabet (1788–1856),
author of the popular *Voyage to Icaria* (1840) and leader of an ill-fated
1848 expedition to the United States to found a real utopia, was more

technocratic in orientation than any of the visionaries discussed here. His novel, written under both Owenite and Fourierist influences, was much more rigid, centralized, and homogeneous in character and content than any of his mentors' schemes. In addition, it imagined more technological advances than did their model societies: huge factories manufacturing standardized goods evenly distributed to the citizens, uniforms denoting everyone's age and occupation, daily food allocations delivered to everyone's door, glass-covered sidewalks, dust collectors, omnibuses, and so forth. These innovations, plus an "industrial army" so regimented that all were made to work, give the novel the look and feel of an early technological utopia—and an especially dull, authoritarian one at that. Cabet is a minor figure within European utopianism, if a key link to the American technological utopians.[53] In any case, he and his fellow visionaries were practically the only European utopians in these four centuries who gave technology a moderately if not fully central role in their schemes to bring utopia about.[54] Although only Cabet was—or wished to be—a full-fledged technological utopian, all the visionaries contributed to a tradition of utopian thought in the West that led to technological utopianism. They made utopianism a legitimate mode of thought by making utopia seem possible, even probable, rather than impossible.[55]

Five

The American Origins

In light of the tradition of utopian thought in the West established by More's *Utopia* and later European writings, it is natural that American technological utopianism should be partly European in origin. Indeed, the most basic notions underlying the vision of America as a technological utopia derive from prior European notions about America: America as a potential utopia; America as a potentially advanced society rather than as a permanent primitive paradise; and America as an advanced society technologically as well as politically, economically, socially, and culturally. These notions represent applications to America of the originally European concepts of utopianism and of progress per se.[1]

Taken together, these ideas about the future in general and about America's future in particular contributed significantly to what I have described as the prevailing ideology of American society of the late nineteenth and early twentieth centuries. These portions of that ideology (which of course included many additional ideas as well) influenced Americans long before and long after the fifty-year period central to this study. Yet the technological utopians, like innumerable other Americans of their and other generations, reassessed and revised these basic ideas in light of American experiences unforeseeable by those original European exponents.[2] Specifically, the technological utopians envisioned America as a probable, not merely a potential, utopia; and this utopia was to be brought about primarily through technological changes rather than through a combination of political, economic, social, cultural, and technological changes. They made technological progress equivalent to progress itself rather than merely a means to progress, and they modeled their utopia after the machines and structures that made such technological progress probable.[3]

Since its discovery and first settlement by Europeans, America had been the object of utopian hopes abroad, and those hopes fed America's own. What made America a potential utopia was its status as a blank slate on which a new society could be written and its possession of enough natural resources to provide material plenty for all. Coincidentally, these were among the chief advantages sought by the European utopians, although they did not necessarily seek them in America.[4]

Nevertheless, the *potentiality* rather than reality of America as a utopia must be emphasized. Too often it has been assumed that abundant natural resources alone guaranteed that America would become an advanced society, even a utopia, and that virtually all Americans from the seventeenth century on agreed with that proposition. But there is a considerable gap between the possession of abundant resources on the one hand and the proclamation of them as utopia on the other.[5] In the American case at least four factors stood in the way of a simple conversion of resources into utopia: (1) the inheritance of a European tradition of utopianism; (2) the need to convert natural resources into finished products; (3) the existence of another civilization—that of the Indian—that viewed European civilization as an invader and as a rival for territory; and (4) the consequent existence of a partly settled territory rather than a virgin land.

That European civilization, or remnants of it, displaced and in large part destroyed Indian civilization can merely be noted here, however profound the consequences for both sides. Here I will mention only three points: first, the clash was between two civilizations rather than between civilized Europeans and savage Indians; second, despite this clash, a unique new civilization resulted from the more peaceful contacts between those two existing civilizations; and third, the ironic outcome of that clash was the creation of several areas of wilderness out of partly settled territory, because of the disease and destruction the Europeans wreaked upon the Indians and their lands, although not always deliberately.[6]

Even when the physical impact of European settlement was more benign, the threat to the new arrivals and their offspring of permanent scarcity frequently loomed far greater than the promise of eventual abundance.[7] Until at least the mid-nineteenth century, considerable numbers of European immigrants and their native-born descendants could not take abundance for granted. Instead, they conceived of even America's natural resources as finite, not infinite. The finished products developed from those resources and the wealth realizable from their manufacture, purchase, and distribution were also seen as finite—not only because the raw materials were limited but also because other sources of wealth could apparently not be created. Hence these Americans saw their economic realm as ultimately a closed system; moreover, as a closed social system, as we have seen. Consequently they conceived of America as a *potential* utopia but hardly a probable, much less an existing, one.[8]

What, beginning in the mid-nineteenth century, changed these early American assumptions about scarcity was not the discovery of ever greater natural resources far beyond the initial areas of European settlement. Rather, it was the invention of various devices that not only converted raw materials into finished products at an infinitely greater pace but also provided opportunities at last to create new wealth. New wealth could be created in three ways: (1) by the use of machines and structures to locate, extract, and transport previously unknown or unmovable natural resources; (2) by the invention and manufacture of synthetic raw materials and in turn finished products; and (3) especially by the use of machinery to build canals, steamboats, and railroads, and in due course farms, towns, and cities, all of which transformed much of America's vast lands into fluid—and infinitely expandable—capital. Not only did technological advances make possible all of these developments but technology itself, as I have defined it, includes the first two of them and part of the third. Technology, you will recall, includes the adaptation to and conquest of the natural world and, eventually, the creation of a man-made world as an addition to or partial replacement of it.

One of the earliest, most penetrating, and most publicized reports on industrialization in America made these very connections among natural resources, technological advance, and abundance. Interestingly, *The Report of the Committee on the Machinery of the United States* (1854–1855) was the work of foreigners: a group of prominent British citizens sent by their government to examine the nature and uses of American technology in what was beginning to be called "The American System of Manufacturing." The group came to attend the New York Industrial Exhibition of 1853, the American counterpart to the British Crystal Palace Exhibition of 1851, but delays in the opening gave the visitors an unexpected opportunity to investigate American industry outside of the exhibition itself.

In a special report of his own attached to that of the group as a whole, committee member Joseph Whitworth, an eminent engineer and manufacturer of machine tools, observed, "It is not for a moment denied that the natural resources of the United States are immense, that the products of the soil seem capable of being multiplied and varied to almost any extent, and that the supplies of minerals appear to be nearly unlimited. The material welfare of the country, however," he added, "is largely dependent upon the means adopted for turning its natural resources to the best account, at the same time that the calls made upon human labour are reduced as far as practicable."[9] The means Whitworth cited were technological: the construction of modern machines and structures, the use of interchangeable parts, and the organization of scarce labor. The complete *Report* raised disturbing questions about the previously assumed technological superiority of Britain over the United States.

In any event, these technological advances resulted in the reconceptualization of America as an open system rather than as a closed system; and again as a social as well as an economic system. It is hardly accidental that the first American technological utopian writings—those of Etzler, Ewbank, and Griffith—appeared just as those advances appeared in pre–Civil War America. Those and later technological advances transformed realizing utopia from something impossible to something possible and even probable. As the idea of America as man-made rather than natural utopia became a distinct possibility, the original Puritan notion of America as the site of God's millennial kingdom on earth faded in popularity. Dependence on man rather than God, as we have seen, distinguishes utopianism from millenarianism.

The unprecedented abundance brought about by technological advances did not, however, have uniformly beneficial consequences. Not only was the distribution of raw materials, finished products, and the wealth accruing from them grossly unequal, but concerns arose about the impact of wealth on its very recipients, the newly rich. Puritan concerns that abundance might produce moral corruption resurfaced, made more acute by unparalleled abundance and the absence of the kind of rigorous ethical restraints that the Puritans imposed on themselves.[10]

By the time the first of the post–Civil War technological utopias appeared in 1883, prosperity was widely viewed as a problem in itself as well as a solution to other problems. In *Progress and Poverty*, social critic Henry George had argued four years earlier that poverty accompanied material progress, and material progress produced its own malaise among its beneficiaries. Like his highly popular work, the technological utopian works examined here were intended as solutions to, among other problems, the unexpected and unprecedented problem of abundance. Like George, the technological utopians were confident that the problem could be solved. To use technological advances to solve a dilemma created in some measure by technological advances (or their misuse) struck them as neither paradoxical nor problematic. Again, they did not see technology as in any respect a problem in itself.[11]

Whether most other Americans of the late nineteenth and early twentieth centuries had a similarly uncritical view of technology is uncertain and, since there were no reliable public opinion polls in that period, impossible to establish.[12] What can be stated with assurance is that American attitudes toward technology have shifted over time and have occasionally been quite critical. It is equally certain that an overwhelming majority of Americans from colonial times until the present have accommodated themselves and their communities to technological change, if not always enthusiastically. American attitudes toward technology, like the origins of American technological utopianism, have been complex. We need, therefore, to examine the evolving conceptions of technology itself in both the real and ideal worlds of nineteenth- and early twentieth-century America.

One cannot, of course, fully reconstruct American attitudes toward technology for that period or any other portion of Americans' ideologies. Materials are too few and the past is too different, too discontinuous, from the present for that. But we can probe more deeply the conceptions of technology found in the writings of earlier generations in order to achieve a partial reconstruction. This goes beyond the question of whether Americans generally liked or disliked technology. Such a question, characteristic of contemporary survey research, hardly touches how Americans *conceived* of technology. Their conceptions of technology surely affected their likes and dislikes and, in not a few instances, may have shaped them.

If, as Thomas P. Hughes convincingly argues, American attitudes toward technology have altered at least in part in response to actual technological developments, so too have American conceptions of technology.[13] In fact, the very introduction and popularization of the term *technology* came about shortly after America's industrial revolution began. Hugo Meier conducted an extensive search through late eighteenth- and early nineteenth-century American writings and concluded, " 'Technology' was a word probably quite unknown to Americans of the late eighteenth century."[14] Only in 1829, through the publication of Jacob Bigelow's *Elements of Technology*, did the term begin to be commonly used. That the first technological utopian work in America, Etzler's *The Paradise Within the Reach of All Men*, appeared four years later is hardly accidental. With *Elements of Technology* as with *Looking Backward*, however, the influence of even a popular book cannot alone account for the popularity of its ideas.

Bigelow (1786–1879) enjoyed a long and multifaceted career. He was simultaneously a prominent botanist, physician, professor of "materia medica" at Harvard Medical School, and, from 1816 to 1827, Harvard's first Rumford Professor "of the Physical and Mathematical Sciences as Applied to the Useful Arts." In addition, from 1847 to 1863 he was President of the American Academy of Arts and Sciences. Clearly, he was an influential citizen.[15] It was from his lectures as Rumford Professor that *Elements of Technology* derived. *Elements of Technology* does not deserve the obscurity that has been its fate.[16] If the book did not coin the term *technology*, it did help popularize it and no doubt introduced it to many readers in America.[17]

Bigelow's conception of technology is therefore important. His conception is at once strikingly modern in its comprehensiveness and, to contemporary readers, strangely unfamiliar in some of its language. The subtitle of *Elements of Technology* called technology "the application of the sciences to the useful arts." By this Bigelow meant not only that technology was science applied for utilitarian purposes but also that the sciences and the arts were closely related. He wrote, "Whenever we attempt to draw a dividing line between the *sciences*, usually so called, and the *arts*, it results in distinctions, which are comparative, rather than absolute. In many branches of human knowledge, the two are so blended together, that it is impossible to make their separation complete."[18] If today's readers are familiar with technology as applied science (however historically dubious this equation may be), they are much less likely to think of connections between the sciences and the arts.

Bigelow's conceptions of applied science and useful arts reflect the period in which he wrote, which was the beginning of America's industrial revolution. During this period, the early and mid-nineteenth century, the actual relationship between technology and science became sufficiently close for scientific principles to be applied to technology and for technology genuinely to be applied science. Before then, science and technology were, with few exceptions, separate phenomena. What became known as the "industrial revolution" in part was the widespread application of science for utilitarian purposes. At the beginnings of technology, dating back to the seventeenth century, it was described as an "art" or "arts" with a "scientific" focus and language.[19] These original formulations persisted until Bigelow's day and beyond, and his incorporation of them in his book was natural.

Bigelow, however, broadened the conception of technology. Although he provided no fuller definition than the one just quoted, the components of his book revealed technology's true scope. The

"elements" of technology included numerous materials, both natural
and man-made, ranging from minerals to glass and from wood to
paper; equally numerous machines and structures, from bridges to
printing presses and from canals to railroads; the processes of discovering or inventing and refining and producing all of those materials and machines and structures; the technical knowledge, skills,
and equipment needed to carry out all of those processes; and the
history of the development of all of those processes and techniques. In
its comprehensiveness, especially in its implicit recognition of technology as a social and cultural as well as material phenomenon,
Bigelow's conception of technology is strikingly modern. It is comparatively modern as well in several of the specific elements it discusses.

Yet *Elements of Technology* does not pretend to be an original
analysis of those many elements, much less a pathbreaking history of
any of them; it is an up-to-date catalog of them "now published," in
the words of its cover, "for the use of seminaries and students." In this
respect the book is merely one in a line of such catalogs dating back to
the earliest systematic writings on the history of technology.[20] *Elements of Technology*, therefore, is as much a reflection of its own and
previous times as a preview of the future.

Nevertheless, *Elements of Technology* would not have been as
popular as it apparently was in the early and mid-nineteenth
century,[21] if it had been nothing more than a catalog, even an up-to-
date catalog. Surely its preview of the future, however modest in
length and restrained in tone, accounts in some part for the influence
it achieved. Indeed, in demonstrating the magnitude of many and
varied technological achievements, Bigelow intended to kindle both
pride in mankind's progress so far and confidence in future progress.
Bigelow asserted:

> The ancients, who were but recently descended from
> barbarians, were obliged to make the most of small means,
> because the stock of previous or common information,
> from which they could draw, was extremely limited. The
> moderns have the accumulated learning of ages before
> them, and have only to select and apply their agent from
> among a multitude of means already discovered.[22]

Moreover, where the ancients "possessed the quick eye, the expert
hand, acute taste and unwearied industry," the moderns "substitute
preparatory science, economical computation, and mechanical
power."[23] The result, Bigelow concluded, is that "with less bodily
strength, and probably with not more vigorous intellects [than the

ancients], we have acquired a dominion over the physical and moral world. . . . We convert natural agents into ministers of our pleasure and power. . . ."[24] Such progress to date explained Bigelow's belief in further progress in the future. As he proclaimed in his inaugural lecture as Rumford Professor in 1816, "every thing [technological] is permanent and progressive."[25]

Thirty-six years after *Elements of Technology* appeared, in 1865, Bigelow delivered a major address at the recently founded Massachusetts Institute of Technology. He observed that in 1829 the term *technology* "was not then in use nor was it generally understood. . . ."[26] By 1865, however, the situation had changed:

> It has happened in regard to technology that in the present century and almost under our own eyes, it has advanced with greater strides than any other agent of civilization, and has done more than any science to enlarge the boundaries of profitable knowledge, to extend the dominion of mankind over nature, to economize and utilize both labor and time, and thus to add indefinitely to the effective and available length of human existence. And next to the influence of Christianity on our moral nature, it has had a leading sway in promoting the progress and happiness of our race.[27]

Technology had obviously become a more familiar and more widely used term, just as the physical manifestations of technology had become more familiar and more widely used in the course of industrialization. Equally important, technology had become distinguished from science, as a thing in itself and as an enterprise requiring as much formal education as science (even as its relationships with science had become closer).[28] Finally, technology had begun to achieve the goals for Americans predicted by Bigelow in his 1816 inaugural address as well as in his book: control over the natural environment and the creation, as a partial replacement, of a manmade environment; and, in the process, savings of both labor and time. By 1865, Bigelow's prognosis concerning America's future was more expansive: "Our country, with its vast territory, its inviting regions, its various populations, its untrammelled freedom, looks forward now to a future which hitherto it has hardly dared to anticipate."[29]

Two years after *Elements of Technology* appeared, and two years before Etzler published *The Paradise Within the Reach of All Men*, Timothy Walker offered perhaps the foremost defense of technology in the antebellum period. Entitled a "Defense of Mechanical Philosophy," it was a response to Thomas Carlyle's pessimistic 1829 essay, "Signs of the Times." Walker (1802–1856), a graduate of Harvard College and Harvard Law School, was, at the time, a promising lawyer in Cincinnati.[30] Carlyle (1795–1881), of course, was the prominent English man of letters and social critic.

"Were we required to characterise this age of ours by any single epithet," Carlyle had announced, "we should be tempted to call it, not an Heroical, Devotional, Philosophical, or Moral Age, but, above all others, the Mechanical Age. It is the Age of Machinery, in every outward and inward sense of that word. . . ."[31]

Carlyle's critique of the Mechanical Age was subtle and complex. He opposed the "inward" effects of industrialization more than he did the "outward" effects. It was not the proliferation of machines and structures but their misuse that troubled him: most notably, the exploitation of factory workers and their families, the preoccupation with material means instead of spiritual ends, and the overorganization and dehumanization of society. He lamented, "There is no end to machinery." He opposed the growth of mechanistic thought in education, in art, in literature, in philosophy, in religion, in politics, and in other spheres. As he sighed, "Men are grown mechanical in head and heart, as well as in hand."[32] Carlyle believed that a relatively homogeneous and placid organic society had, through industrialization, been replaced by a markedly heterogeneous and agitated mechanistic society. Yet without apparent contradiction he looked to machinery, properly reemployed, as both instrument and embodiment of not just material improvement but also spiritual uplift. Only later in his life did he despair of this dream.[33]

Walker's reply to Carlyle, like Carlyle's critique itself, is as significant for the conception of technology that it embodies and of the society technology was bringing about as it is for its defense of technological and social progress. Yet the contents of both essays have frequently been reduced to their respective optimism or pessimism regarding the future—an intellectual strategy which is not merely inadequate, as already observed, but, in these two instances, misleading. For Carlyle was by no means wholly pessimistic about the future of British society in 1829, and Walker was not wholly optimistic about the future of American society in 1831. Walker wrote, "But let us not be misunderstood. The condition we speak of, is not one of perfection. This we neither believe in, nor hope for. Supposing it possible in the nature of things, it would be any thing but desirable. For with nothing left to achieve nor gain, existence would become empty and vapid."[34]

Equally important, Carlyle and Walker agreed that a new age of some kind either had come about or was in the making and that a return to pre-industrial society—a dream of several other critics of industrialization, notably John Ruskin and William Morris—was neither practical nor desirable (a point with which Bigelow was in complete agreement). No less important, Carlyle and Walker recognized that the triumph of "Mechanism" represented as much intellectual and psychological changes as material and social ones. Indeed, both would likely have agreed with Bigelow's statement in *Elements of Technology* that "The application of [mechanical] philosophy to the arts," more than any physical changes, "may be said to have made the world what it is at the present day."[35] Finally, and most important, Carlyle and Walker conceived of technology in similar ways even though they differed in large part over its consequences. Walker's alleged failure to "find an American adversary" for his views may be less significant than his unintended kinship with the foreign adversary he did find.[36]

Certainly Walker disagreed sharply with Carlyle over the meaning of the intellectual and psychological and, to a somewhat lesser extent, material and social changes that were taking or had taken place. Walker characterized Carlyle's objections to the triumph of mechanism as sheer mysticism. "In plain words," Walker contended, "we deny the evil tendencies of Mechanism, and we doubt the good influence of his Mysticism. We cannot perceive that Mechanism, as such, has yet been the occasion of any injury to man."[37] Quite the opposite: mechanism, Walker sought to demonstrate, had produced and would continue to produce unprecedented progress in all of the spheres in which Carlyle was alarmed. Americans, British, and eventually other advanced populations would live in unparalleled physical comfort, because various machines and structures would redesign or replace the natural environment and provide reliable substitutes for human or animal labor. The considerable leisure thus obtained would in turn generate unparalleled intellectual and social advances, and "there would be nothing to hinder all mankind from becoming philosophers, poets, and votaries of art." For that matter, "The whole time and thought of the whole human race could be given to inward culture, to spiritual advancement"—thereby alleviating Carlyle's foremost concerns. Walker thus refuted Carlyle's apprehension that "in our rage for machinery, we shall ourselves become machines."[38]

Their differences, however, should not obscure Carlyle's and Walker's agreement that a "mechanical" social order was replacing an "organic" social order and that technology was not simply the principal means of this transformation but its very model and so its end. This change underlay all of the other changes they listed. It represented change of the most fundamental kind: alteration in the prevailing American and British ideology, in the way that increasing numbers of persons in those two societies viewed and, more precisely, divided up and pieced together their world.

Technology in the form of "the machine" became the principal metaphor for conceiving of both society and the universe. The two were seen as inextricably linked, God having created a fully mechanized universe and having delegated to mankind responsibility for creating a suitably mechanized earth. Such conceptions of a mechanical social order—and of an alternative organic one—are not, of course, peculiar to the early nineteenth century. They occur, or recur, throughout at least Western history, including the history of utopianism.[39] These conceptions have not simply alternated in regular cycles and remained the same in content; their alternations have been irregular and their contents have changed as well.[40] Where, for example, the clock was a favorite metaphor for the conceptions of a mechanical social order during the Middle Ages and early modern periods,[41] the stationary steam engine and spinning machinery of the textile mill (and later the factory in general) became the metaphors of the early nineteenth century.

The invention in their respective periods of the clock and of the textile mill was not necessarily the cause of these particular conceptions of the mechanical social order; the precise relationships between actual technological changes and changing conceptions of technology are complex and varied. Nevertheless, the proliferation of technological changes—as in the cases of first the clock and then the steam engine and the spinning machinery—surely contributed to the widespread adoption of new conceptions of technology and, beyond that, of new conceptions of society itself. In Carlyle's words:

> We term it indeed, in ordinary language, the Machine of Society, and talk of it as the grand working wheel from which all private machines must derive, or to which they must adapt, their movements. Considered merely as a metaphor, all this is well enough; but here, as in so many other cases, the "foam hardens itself into a shell," and the shadow we have wantonly evoked stands terrible before us and will not depart at our bidding.[42]

To Carlyle's dismay, the metaphor became a working model of British society. Walker, however, was pleased with the image: "We employ the mechanic lever to lift weights, which our unassisted strength could not lift. Why not employ the *social* lever in the same way?"[43]

This analysis can be pursued further still. Carlyle and Walker alike conceived of the new mechanical social order as one in which, like the gears and pulleys of the machines and structures that they described, the parts of society were interlocked, relatively unspecialized, roughly equal in their importance, and fairly interdependent. The gears and pulleys included all members of society, all their spheres of activity, and all their institutions. Some gears and pulleys naturally did more work than others; but if one minor gear or pulley failed, so did the entire Machine of Society. This illuminates Carlyle's statement that:

> With individuals, in like manner [to institutions], natural strength avails little. No individual now hopes to accomplish the poorest enterprise single-handed and without mechanical aids; he must make interest with some existing corporation, and till his field with their oxen. In these days, more emphatically than ever, "to live, signifies to unite with a party, or to make one." . . . [Citizens] have lost faith in individual endeavour, and in natural force, of any kind.[44]

It likewise illuminates Walker's dreams for the "inward" cultivation of "all mankind" and the "whole human race." Or, still better, Walker's expectations of national if not international harmony:

> From the ineffable harmony and regularity, which pervade the whole vast system, we deduce the infinite power and intelligence of the Creating Mind [i.e., God]. Now we can perceive no reason, why a similar course should not be pursued, if we would form correct conceptions of the dignity and glory of man. . . . Examine the endless varieties of machinery which man has created. Mark how all the complicated movements cooperate, in beautiful concert, to produce the desired result.[45]

Walker bluntly advised the individual citizen, "Let him unite with those whose opinions agree with his, and he adds another unit to the sum."[46] The mathematical language was not coincidental.

"From the effect," Walker continued, "turn your attention to the cause. Examine the endless varieties of machinery which man has created."[47] "Cause and effect" represented yet another dimension of the mechanical social order, for it meant that touching one portion of the social mechanism would move another portion—just as in the textile mill. Machines and structures on a national or international scale would produce desired alterations in the social, intellectual, and psychic as well as material fabric. Walker provided several examples of the last:

> Where she [Nature] denied us rivers, Mechanism has sup-
> plied them. Where she left our planet uncomfortably
> rough, Mechanism has applied the roller. Where her
> mountains have been found in the way, Mechanism has
> boldly levelled or cut through them. Even the ocean, by
> which she thought to have parted her quarrelsome chil-
> dren, Mechanism has encouraged them to step across. As if
> her earth were not good enough for wheels, Mechanism
> travels it upon iron pathways.[48]

The correspondence between technology and society, as between one
part of the machinery and another, was therefore one-to-one. Carlyle
lamented, "Thus it [allegedly] is by the mere condition of the
machine, by preserving it untouched [i.e., unimproved, not literally
untouched], or else by reconstructing it, and oiling it anew, that man's
salvation as a social being is [allegedly] to be ensured and indefinitely
promoted."[49]

In all of these respects, the new mechanical social order, like the
textile mill that was its specific model as well as its specific metaphor,
constituted what both Carlyle and Walker called a "system" or a
"social system."[50] Such a system was not necessarily static, and the
two agreed that further changes in British and American society were
inevitable even as they disagreed on the desirability of those changes.
Carlyle clearly preferred an organic social order with "spontaneous
growth" reflecting the "Dynamical" rather than "Mechanical" nature
of man.[51] We may note that a yearning for an organic social order is
characteristic of many disaffected individuals and societies in Western history. But just as the mechanical social order has had various
specific configurations, so has the organic; again, the two have not
simply alternated in regular cycles. The label is not an explanation; it
is necessary to examine the particular organic social order in question. Nevertheless, Carlyle's yearning is again less significant than his
reluctant agreement with Walker on the triumph of the mechanical
social order and, with it, of technology.

Were Carlyle's and especially Walker's conceptions of technology
and of the society technology was bringing about merely their own,
their own essays would not warrant such extensive analysis. Their
essays are important because they are representative of the changing
ideology of countless other Americans (and Britons) of the early and
particularly middle nineteenth century. Without attempting to reconstruct the way in which all Americans at that time viewed and divided up their world, I believe that we can take the particular mechanical social order described by Carlyle and Walker as both the principal
metaphor and the principal model for a majority of Americans. It did

not, to be sure, constitute the whole of their ideology; nor did most Americans necessarily think so "systematically" about their society and themselves. Yet other historians who have studied the first six decades of the nineteenth century—historians of wide-ranging phenomena, including culture, government, political parties, reform movements, education, religion, and medicine—confirm at least indirectly not only the existence of this new view of America and Americans but also its pervasiveness.[52] Moreover, the outlines of this view are discernible in Bigelow's *Elements of Technology*, not just in its hopes for America's future but also in its comprehensive conception of technology. What Carlyle and especially Walker contributed were explicit, forceful expressions of that new view. They did not create it. For rhetorical purposes, they may well have exaggerated their application of the new mechanical model to the living social order. If so, they imitated–or foreshadowed–none other than the technological utopians, who consciously took to their logical finale the positions of many of their contemporaries. In both instances, however, the intention was not to distort truth but to reveal it.

Neither Walker nor, for that matter, Bigelow was a utopian. Neither furnished the kind of fictional or nonfictional blueprints that marked the technological utopians, beginning with Etzler in 1833. More important, neither believed in the possibility of perfection, or virtual perfection, through technology or any other means. Still, Bigelow and Walker believed firmly in progress and in progress through technology. Walker wrote, "They who feel this divine impulse, know that the labors of kindred spirits in past ages have not been in vain. They see Atlantis, Utopia, and the Isles of the Blest, nearer than those who first descried them. . . . if, with this explanation"—the above-mentioned lack of belief in perfection—"our views should pass for visionary, we cannot help it."[53]

There is not, of course, any necessary contradiction between a belief in progress and a lack of belief in perfection (or ultimate progress). Bigelow, Walker, and the other Americans whose views they represented combined the two stances without difficulty. They appreciated another aspect of the mechanical model of social order that we are discussing: namely, the equilibrium among the parts of a single machine or among the machines of an entire textile mill or factory. The social equilibrium was relative rather than absolute because it allowed for change, or what Bigelow, Walker, and others called "improvements." Such improvements led to progress, but progress simultaneously meant the preservation or restoration or achievement of some degree of stability. The technological utopians strongly believed in stability, but stability following the attainment of perfection; theirs was thus a more rigid equilibrium and, we may assume, a more mechanical social order.

Yet just as we must distinguish among changing conceptions of technology and of the society technology was bringing about for Americans generally, so we must examine these conceptions as held by the American technological utopians. Like their fellow Americans, the technological utopians of the early and mid-nineteenth century— Etzler, Ewbank, and Griffith—did not view, divide up, and piece together their world in the same way that the twenty-five late nineteenth- and early twentieth-century technological utopians did. The latter visionaries, like the other Americans of their period, had experienced changes of almost infinite number and complexity during the preceding several decades, changes that inevitably affected— even if they did not necessarily determine—those visionaries' utopianism.

Moreover, and more specifically, just as a significant strain of European utopianism in the early and mid-nineteenth century adapted a less mechanical and more organic view of both the real and the ideal worlds,[54] so did the late nineteenth- and early twentieth-century American technological utopians. What Frank and Fritzie Manuel wrote about French utopianism of the former period applies in large part to American technological utopianism of the latter period:

> While passion for equality is marked among the Frenchmen of the eighteenth century, it is condemned by the "classical utopians" of the nineteeth. The eighteenth-century plans are often mechanically egalitarian—all men eat the same food and perform virtually the same amount of labor with a minimal differentiation of tasks. Men are like interchangeable counters. The nineteenth-century organizations were more complex: the egalitarian ideal took the shape of an opportunity for equal self-actualization, and because desires and capacities varied, the whole fabric of the utopia tended to become more intricate. With the introduction of imagery from the biological sciences, men were regarded as parts of a social organism and they assumed distinctive characters as well as functions. . . .[55]

The parallels between the two varieties of utopianism are not exact, and the factors that gave rise to these alterations in French utopianism may not have given rise to their American counterparts. Nevertheless, it is evident that the technological utopianism of the earlier American visionaries is substantially different in those general respects from the technological utopianism of the later ones.

The utopian writings of Etzler, Ewbank, and Griffith have been examined in detail elsewhere[56] and need only be summarized here for purposes of comparison; they are listed in the Appendix. Their authors' lives have been examined in somewhat less detail because not much information about them has survived.[57] Consequently, a brief sketch of what is known about each author may be useful here.

Etzler was a German-born engineer whose exact dates are unknown. He became a close friend of bridge builder and Hegelian visionary John Roebling and, following a stay in the United States in the 1820s, returned to America with Roebling and other Germans in 1831. For various reasons Etzler split off from Roebling and spent much of his later life traveling throughout the Americas and Europe vainly seeking financial backing for a variety of schemes involving the use of technology to improve mankind's lot. Besides his *Paradise* of 1833, he published several essays and tracts in the 1840s and is familiar mainly as the object of a sarcastic essay by Thoreau in the *United States Magazine and Democratic Review* of November 1843. Thoreau faulted Etzler's *Paradise* for excessive optimism, materialism, and authoritarianism. His life after 1846 is completely unknown.

Ewbank was born in England in 1792 of humble parentage and came to America in 1819, like Etzler, hoping to make his fortune. He had a much more successful career than Etzler, working successively as an inventor, a manufacturer, and from 1849 to 1852, as United States Commissioner of Patents. He published several essays and tracts; among other places, he published some of them in the annual reports of his high office—an action for which he was criticized by some Congressmen. He died in 1870.

Less is known about Mary Griffith than about Etzler and Ewbank. She was apparently born in America about 1800 and, when her one utopian work appeared in 1836, lived in New Jersey. That work is a novella, "Three Hundred Years Hence," published as the longest part of a collection of her writings entitled *Camperdown; or, News from Our Neighbourhood.* Her other works were primarily about horticulture, an activity she undertook commercially after her husband died. She died in 1877. Given the modest number of persons involved, any statistical comparisons between the lives of these visionaries and those of their successors, described in chapter 3 above, would not be illuminating.

But comparisons between their respective writings are illuminating. To begin with, there were obviously far fewer earlier technological utopian writings than later ones. Second, the earlier writings were, with the exception of Griffith's novella, nonfictional essays and tracts, where over half of the later writings were fiction—novels and short stories. Third, the earlier writings were considerably less detailed than the later ones. Not only were Etzler and, to a lesser extent, Griffith reluctant (or perhaps unable) to disclose the particulars of their technological advances, but they and Ewbank, too, were not very specific about the nontechnological dimensions of their utopias. Fourth, and not unrelated to point three, the earlier writings were less precise and less modern in whatever descriptions they did provide of both the technological and the nontechnological aspects of utopia than were the later writings. Because of the differences in the development of American society between the early and late nineteenth century, Etzler, Ewbank, and Griffith could barely imagine what their successors could readily conceive. As we have seen, the very term *technology* came into common use in America shortly before the first technological utopian work appeared. And as I have suggested, the relationship between actual technological changes and changing conceptions of technology and of society is close if imprecise.

For these reasons, it is not surprising that technology in the earlier technological utopian writings was, as an instrument and embodiment of progress, treated as a new, even mysterious phenomenon, where in the later writings it was treated more as a fact of life—if a most important fact of life. The "thrill of the technological transformation" that Hughes found throughout the early and mid-nineteenth century writings he examined (by no means all of them utopian) became what he called the more restrained "realization of technological power" in their late nineteenth-century counterparts, including the technological utopian writings described here (few of which he studied). In the works of Etzler, Ewbank, and Griffith, technology gave man mastery over nature, and this was the first step toward the permanent affluence required for any serious utopian society. To these writers, *technology* meant both the application of natural forces and the invention and production of various machines and structures to ease physical labor and save time for leisure. The later technological utopian works did not discount these activities and achievements, but they envisioned other, complementary ones: the creation of a man-made environment to partially replace the now-tamed natural environment, and the reduction of waste throughout society. The growing power of man in the later writings invariably meant the shrinking power of God, whose role in the earlier writings was considerable. Indeed, Etzler, Ewbank, and Griffith saw technology and so society advancing according to an at least implicit divine plan; but their successors saw man becoming God and the engineer his priest. The danger that technology might get out of human control—might, that is, become autonomous—did not generally trouble the later visionaries.[58]

The basic difference, however, between the earlier and later technological utopian writings was the increasing orientation toward an organic social order. The transformation was far from complete, as the many references in chapter 2 to the machine and to the machine as metaphor and model for society make clear. The emphasis on uniformity of citizens and on cause-and-effect change also reflects the mechanical model of social order. Yet the material discussed in chapter 2 does show a shift in outlook from the earlier visionary writings: (1) the use of organic metaphors and models; (2) the allowance of substantial differences among the components of society, from citizens to institutions to machines and structures, and the consequent hierarchy within each component; (3) the greater specialization of these components and the greater complexity and interdependency of their relationships; (4) the organization of individual citizens into groups at every stage of their lives; (5) the notion of change as either more evolutionary or more spontaneous than cause and effect; (6) the use of thermodynamic metaphors taken from chemistry (along with the organic ones taken from biology) and the additional notion of change as producing a temporary instability, which, of course, ultimately achieves equilibrium; and (7) the broadening of *system* from implicitly social (as well as material) to explicitly social.

Despite being transitional in this regard, the technological utopianism described in chapter 2 takes certain of these developments to extremes. The rigid meritocracy which characterizes technological utopia is perhaps the best example of its social order reflecting an organic model. But other examples, ranging from its systematically arranged buildings to its systematically arranged industrial army departments, come readily to mind. As the mechanical social order of Etzler, Ewbank, and Griffith reflected the real world tendencies of their time, the societies imagined by the twenty-five later visionaries may exaggerate but do not distort tendencies found in the America of their time. As I will show in the following chapter, a less mechanical and more organic social order was emerging as both a metaphor and a model in late nineteenth- and early twentieth-century America. In both technological utopianism and the real world, the change was not simply part of an eternal cycle but rather reflected developments peculiar to the time. Finally, the partial abandonment of the mechanical in both the real and ideal social orders did not mean the abandonment of technology as the principal instrument and embodiment of progress. Instead, the particular organic social order that arose in both realms reflected the continuing influence of technology, but in other, more complex forms than earlier in the nineteenth century. Even so skeptical an observer of progress as Henry Adams acknowledged the continuing significance of technology in a superb chapter of *The Education of Henry Adams* (1907) entitled "The Dynamo and the Virgin." Despite his reservations about the course of Western society and culture, Adams recognized the unprecedented, perhaps supra-human, power of twentieth-century technology as represented by the dynamo that he saw during his visit to the 1900 Paris Exposition.[59]

In light of the ideological transitions just described, Meier's statement about the assimilation of what he calls the "technological concept" is at once perceptive and incomplete:

> The assimilation of the technological concept was a gradual process, uneven in degree and haphazardly directed by the self-interest and the enthusiasms of inventors, engineers, mechanicians, interested politicians, publishers, and writers. It was a very real thing, however, and the admittedly "technological frame of reference" of the present day American is in large degree the result.[60]

The statement is perceptive because it recognizes the complex and unsystematic manner in which ideological change of this fundamental kind usually occurs; no one person or event or piece of writing can bring it about. It is perceptive as well because it recognizes that such change—as, for example, from a particular mechanical to a particular organic conception of the social order—is not a mere abstraction but a real thing; and is real even if the majority of persons involved do not view the change as consciously or as systematically as a Carlyle or a Walker. But the statement is incomplete because it does not recognize the further alterations in the "technological concept" since 1850, when Meier's study ends. According to Meier, "By 1850 much of the spade work was done. . . ."[61] But the "assimilation" has continued to the present day.

The assimilation of technology, either as a concept—in whatever form—or as a social development patterned on such a concept, did not mean automatic, nationwide acceptance of either. It is impossible to date precisely the initial consciousness among Americans of the concept of technology, much less their initial consciousness of the transformations that were increasingly incorporating technology into daily life. For there was not, in fact, any specific date.[62] Yet there is evidence that to at least some Americans in the early and mid-nineteenth century, as to more Americans later in the century, technology appeared to be as much a problem as a solution to problems. Here again, however, conceptions of technology and of the society technology was bringing about were in transition.

To the majority of earlier critics and defenders of technology,

technology on so large a scale was a new phenomenon of unprecedented power and potential—whether positive or negative. Most of
the principal arguments in favor of technological advance were also
employed by its opponents—simply turned around and seen negatively. These debates centered upon whether technological advance
was excessively materialistic and anti-intellectual, antireligious, and
anti-aesthetic; whether it was antidemocratic and antirepublican; and
whether it was exploitative, impersonal, and dehumanizing. The very
machines and structures that introduced the industrial revolution in
America—steam engines, steamships, steam locomotives, textile
mills, and factories—and that were defended by some as instruments
and embodiments of progress were attacked by others as the antithesis
of progress. A number of historians have recently shown that the
colonial American ideology of "republicanism" was caught up in
these debates; they thus had great potential for determining the nation's fate.[63] No less important for America's future were the related
debates over whether industrialization here would entail the kind of
urban blight then plaguing England, the scene of an earlier industrial
revolution; and whether machinery would consume the American
countryside and trample over farms and villages, as it already had in
much of England. This concern was voiced by many European visitors to the United States between 1830 and 1860.[64] For foreign and
native critics of technology alike, the prospect of America as a potential utopia had become the prospect of, in our contemporary parlance,
a potential dystopia.

Criticism of technological change in the early and mid-nineteenth
century took the form not only of speeches and writings but also of
protests, strikes, and other organized expressions of bodily resistance
to certain "improvements."[65] Where the speeches and especially the
writings were frequently the products of intellectuals and social critics physically and psychologically removed from the industrial revolution, the protests and strikes involved the industrial workers
themselves. Their perspective was obviously unique. There were few
American equivalents of the English Luddites, or "machine breakers"
(whose legendary destructiveness has been downplayed by revisionist historians).[66] The majority of Americans, including workers,
accommodated themselves and their communities to technological
change, if not always enthusiastically.[67] The utopian view of America's destiny no doubt affected many of them. As Americans in the
early and mid-nineteenth century often asserted, America's special
nature would somehow allow the nation to preserve its pre-industrial
character while absorbing the latest technological developments—a
growing number by now of American origin.[68]

Ironically, one of the contributors to the same 1854–1855 British *Report of the Committee on the Machinery of the United States* that generally praised the "American System of Manufacturing" criticized the Americans he met for being *too* optimistic, for assuming that utopia was just around the corner. George Wallis, Headmaster of the Government School of Art and Design in Birmingham, observed in a separate special report complementing Whitworth's:

> The extent to which the people of the United States have as yet succeeded in manufactures may be attributed to indomitable energy and an educated intelligence, as also to the ready welcome accorded to the skilled workman of Europe. . . . Only one obstacle of any importance stands in the way of constant advance towards greater perfection, and that is the conviction that perfection is already attained. This opinion, which prevails to a larger extent than it would be worth noting here, is unworthy of that intelligence which has overcome so many difficulties, and which can only be prevented from achieving all it aspires to, by a vain-glorious conviction that it has nothing more to do.[69]

Like the mechanical social order itself, these criticisms of it did not disappear entirely by the late nineteenth century. Such concerns remained or became the concerns of many other Americans and, in many instances, remain the concerns of contemporary Americans. Still, as a newer conception of the social order gradually replaced the older one, different criticisms of technological change gradually arose in response to newer technological developments. Those criticisms primarily concerned not the prospect but the reality of a seemingly affluent technological society. They reflected anxieties over the gap between utopian expectations and actual achievements. The changes to which they were responses will be discussed in the next two chapters.

Having examined the role of technology in the construction of American ideologies, let us now investigate the exact forms of America's accommodation to technological change. By *accommodation* I do not necessarily mean wholesale or universal acceptance, for the industrial revolution in America had its critics as well as defenders. Rather, I mean a preponderantly positive response to technological changes and a general equation of such changes with "progress." Still, the sheer fact of accommodation cannot be overlooked. Americans *did* make their peace with vast changes in their ways of life. Far too often historians both teach and are taught that accommodation between the technology of the industrial revolution and pre-industrial society was neither widely sought nor widely achieved. Instead, the two are seen as utterly antagonistic. Such antagonism is expressed as machine *versus* nature, or the city *versus* the country, or civilization *versus* wilderness. This distortion of American history leads to the kind of pseudo-romantic and vain quest for a pretechnological past that was popular in the 1960s.

Of particular importance here are the diverse efforts throughout the nineteenth and early twentieth centuries (and beyond, to the present) to create what Leo Marx aptly calls a "middle landscape": a reconciliation of some sort between technology and the pastoral (or domesticated nature). In *The Machine in the Garden*, Marx describes several such attempted reconciliations in the period between the American Revolution and the Civil War. These included efforts by St. John de Crèvecoeur, Thomas Jefferson, Tench Coxe, Timothy Walker, and Ralph Waldo Emerson.[70] It is Marx's intellectual achievement to have offered a more accurate and a more attractive account than previously available of the relationship between technology and the pastoral in eighteenth- and early nineteenth-century America. As the notion of America as a potential utopia must be revised, so must the notion of America as either wholly wild or wholly industrialized. And so, in turn, must the conception of technology in that period.

Marx nevertheless contends that technology and the pastoral proved irreconcilable after the Civil War, thanks to unrestrained and unregulated technological "advance." The fears of antebellum critics of technology, he argues, were amply confirmed by the various social ills clearly associated with industrialization: exploitation, crowding, disease, and so forth. After the Civil War, he declares, the middle landscape was no longer a realistic social and cultural ideal but rather became a cheap rhetorical device to mask a painfully different reality. In effect, one conception of technology—or of the society technology was allegedly bringing about—had given way to another.[71]

As I have suggested elsewhere,[72] Marx's lament is at once historically incorrect and premature. There continued to be many attempts to create new versions of the middle landscape, and some of them were successful. Without simplifying the content and variety of these attempts, three new versions can be identified: what I have called the urban, the suburban, and the regional. In brief, the urban version includes carefully landscaped cemeteries, walkways, playgrounds, and above all, parks. It is epitomized by New York City's Central Park. The suburban version embraces similarly well-designed communities of various kinds within walking or riding distance of cities. They may be subdivided into commuter, industrial, and garden suburbs and are exemplified by, respectively, Shaker Heights, Ohio, and the

Country Club District of Kansas City, Missouri; Pullman, Illinois; and the "garden cities" inspired by the English reformer Ebenezer Howard. The regional version includes a handful of efforts to integrate cities, suburbs, towns, and farms into a comprehensive whole. The foremost example of this middle landscape is the Tennessee Valley Authority. The existence of these three types of middle landscapes does not, of course, mean that most American cities, suburbs, and regions were necessarily either well planned or planned in similar fashion.[73]

We can identify a consciousness among the majority of the conservationists, landscape architects, city planners, and social critics involved in these undertakings that new versions of the middle landscape were both possible and desirable. Moreover, these new versions were not idle exercises in nostalgia for antebellum America but were instead avowedly modern and *scientific* improvements of both the natural and the man-made environment. The leading example of such scientific planning of those two environments was the work of Frederick Law Olmsted (1822–1903), whose multifaceted career has been rediscovered and reappreciated in the past decade.[74] Once again, the conception of the society being brought about by technology had changed, but in less dramatic fashion than sketched by Marx.

These accommodations between technology and the pastoral in the real world bear directly on the vision of technological utopia described in chapter 2 and in turn on the relationship between those two realms. Technological utopianism itself represents an example of the middle landscape: specifically, an extreme version of the regional variety, or what I earlier called megalopolis. This America of the future is to be composed of several massive combinations of urban and suburban tracts embracing practically the whole territory of the United States. Moreover, those megalopolises are to be just as scientifically planned as their real world counterparts, probably more so. As the next chapter will show, efforts in the real world to achieve urban and suburban, much less regional, planning on a systematic scale have fallen far short of this technological utopian ideal.

Yet the actual attempts to plan natural and man-made environments bring us back to the idea of America as a potential utopia. This is because many of the efforts to bring about utopia in fact were prompted by that very notion. Not only were technology and the pastoral to be reconciled, but so were existing society and utopian society. As Cecelia Tichi has recently demonstrated, a number of influential Americans—from the Puritans through Walt Whitman— sought to improve and then to blend the natural and the man-made environments exactly to expedite the creation of a utopian society (or, in her term, a millennial kingdom).[75] Damming streams, clearing forests, draining swamps, erecting towns and cities, and inventing and manufacturing the machines and structures necessary to accomplish those tasks were consequently acts of national—and, in the millenarian tradition, spiritual—fulfillment. Those activities at once reflected and generated an overall, though of course not unqualified, optimism about America's destiny.[76]

Still more optimistic in this regard were the dozens of avowedly utopian communities that sprung up throughout nineteenth- and early twentieth-century America. Although the majority of these communities were established outside of urban and industrial areas— usually to achieve a measure of autonomy— many of them engaged in small crafts, in other light industries, and even in manufacturing, as well as in agriculture. They did so to enhance their prospects not only for economic and social survival but also for attaining a genuine utopia once such survival had been assured. Although not themselves technological utopias, technological advance being for them only a means to other ends and not necessarily the principal means at that, these communities represent small-scale models of the middle landscape. Their persistence into the heyday of technological utopian writings is further evidence of the persistence of various forms of mediation between technology and the pastoral. It also further confirms the connection between such mediation and the utopian potential of American society.[77]

In short, the American origins of technological utopianism reflect rather clearly the complexity of technological change in nineteenth-century America and the impact of technological change on existing notions of America as a potential utopia. This is not, however, to reduce technological utopianism to a mere reflection of developments in the real world, for technological utopianism invariably takes actual developments to extremes. No less important, we will see next, developments in real world America between roughly 1883 and 1933 in turn reflected tendencies toward technological utopianism.

Six

■

Technological Utopianism and the Development of Modern American Society

Technological utopianism is a strain of American culture significant in its own right, as another variety of American utopianism, and also illuminative of many larger and better known developments of late nineteenth- and early twentieth-century America. These developments range from conservation to corporate and government reorganization, from city planning to national planning, and from scientific management to Technocracy. As we will see later in this chapter, technological utopianism takes all of these actual developments to extremes, and these developments in turn reflect tendencies toward technological utopianism in the real world America of the same period.

In addition to such connections between the content of technological utopianism and these other phenomena, there is a definite, related connection between the form of technological utopianism and these other phenomena. This connection derives from a change in form within American utopianism in general during the nineteenth century. The change was from communities as the principal expression of American utopianism to writings. It took place late in the nineteenth century, with the outpouring of utopian writings of vari-

Portions of this chapter first appeared in my articles "The Uniquely American Faith in Utopia through Technology," *World's Fair* 2 (Fall 1982): 11–15; "From Utopian Communities to Utopian Writings: A Change in Form and Purpose," *Communal Societies: Journal of the National Historic Communal Societies Association* 3 (Fall 1983), 93–100; and "Utopian Fairs," *Chicago History*:12 (Fall 1983): 7–9, copyright 1983, The Chicago Historical Society; they are reprinted by permission.

ous kinds, not just technological utopian writings. The change is therefore not directly attributable to the Civil War, long the supposed leading cause of social alteration in nineteenth-century America. The war may, however, have played an indirect role, just as the revolutions of 1848 in Europe played a similar role for a comparable transition there from utopian communities to writings.[1]

Surprisingly, the significance of this change in form has barely been recognized. Yet this change in form represented a considerable change in content within American utopianism overall. It marked the passing of the confidence of serious utopians that American society could be effectively reformed by the continued creation and duplication of small-scale communities, the original principal expression of American utopianism. Instead, serious utopians began to turn to other means of articulating their ideals and of trying to realize them. Utopian writings then became their principal form.

Writings, they determined, whether fiction or nonfiction, provided richer possibilities than communities both for conceiving alternatives to existing society and for attracting popular support to the alternatives. Utopian speculation was less risky when placed on paper than when placed on land and thus was frequently more inviting to potential converts. Moreover, where a community could solve problems, if at all, on only a local scale, a book could do so on a much larger scale, if just in theory. The very ability to treat problems somewhat abstractly yet also comprehensively in a book no doubt contributed to the popularity of that form. Not having to worry about the everyday concerns of a real life utopia, no matter how small in size, surely allowed for—if did not necessarily produce—bolder speculations about the future.

No less important, however, the turn to utopian writings acknowledged, if only covertly, the greater complexity of late nineteenth-century America, compared with the rest of the century. The kind of social experiments possible in small-scale communities could not readily encompass those emerging problems in the real world that utopian writings of whatever variety proposed to solve. And utopian writers often agreed on the fundamental problems of the day even though they often disagreed on the solutions. Indeed, the problems created by the unprecedented complexities variously described as "industrialization," "urbanization," "immigration," "labor unrest," and the like were on the minds of many, including those in the non-utopian reform movements, from conservation to Technocracy, outlined above and filled in below. Hence the connections both in form and in content between technological utopianism and those other contemporary developments.

Edward Bellamy himself concluded that the times demanded utopian writings rather than communities before he began to write *Looking Backward*. As he observed retrospectively in 1892, "In a broad sense of the word the Nationalist movement did arise fifty years ago" in the form of "the Brook Farm Colony and a score of phalansteries for communistic [communal] experiments." But then the overriding concern for ending slavery redirected the energies of "these humane enthusiasts." Now the time for such small-scale enterprises had passed. He wrote in the following year, "We nationalists are not trying to work out our individual salvation, but the weal of all, and no man is a true nationalist who even wishes to be saved unless all the rest are." Consequently, "A slight amendment in the condition of the mass of men is preferable to elysium attained by the few."[2]

I do not suggest either that American utopian writings appeared no earlier than the late nineteenth century or that American utopian communities had disappeared by then. True, this was commonly assumed until quite recently. But several scholars have now shown that there were utopian writings as far back as the late eighteenth century;[3] and Robert Fogarty has demonstrated that utopian communities have persisted from the late nineteenth century until the present.[4] Nevertheless, the issue here is not the sheer persistence of communities but the shift to writings as the principal means of reform. As Charles LeWarne concludes about the late nineteenth- and early twentieth-century utopias on Puget Sound in western Washington, "Communitarianism in the 1890s was but a detour on the path of reform. . . . it was futile because it was an anachronism from an earlier and presumably simpler time."[5]

Similarly, I do not suggest that the greater complexity of late nineteenth-century America compared with earlier periods means that antebellum America was as devoid of institutions and as socially and culturally chaotic as some in the 1960s and early 1970s claimed. Those who described the antebellum utopian communities as "anti-institutional institutions" went too far in interpreting them as nearly unique responses to disorder and fragmentation.[6] Not only was American society at the time more cohesive than they assumed, but the postbellum communities Fogarty lists were no less critical of their day's predominant institutions and values than were their antebellum predecessors. The key point, again, is that the critical function assumed by utopian communities (and other crusades for change) earlier in the nineteenth century was taken over in large measure by utopian writings (and by other crusades) later in the century.[7]

Finally, I do not suggest that the utopian writings in general were eventually any more successful than the utopian communities in general in providing either practical or popular solutions to the problems of American society at any juncture in the nineteenth and early

twentieth centuries. Both forms of utopian expression had few followers and little influence throughout that period—with the exception of a handful of communities like the Shakers and a handful of
writings like *Looking Backward*.[8] And both forms proved unable to
translate their models of utopian society, however feasible in
themselves,[9] into vehicles for widespread social change. True, the
communities' members often sought withdrawal from the real world
in order to create legitimate and attractive alternatives to it. And true,
their language concerning change—such as we either know about or
can reconstruct—was usually less scientific and less predictive than
was that in the writings. But the ultimate purpose of most of the
communities' members was the same as that of most of the authors: to
perfect the real world through their respective good works.[10] By contrast, as many students of American utopian communities of the
1960s and 1970s have noted, most recent communities have functioned as at least attempted escapes from the real world. This has
occurred even when the ultimate goal was to reenter the world in
order to improve it; and even when the communities have been
located inside cities rather than, as with most of their predecessors,
outside of them. It is not surprising that scholars have found a greater
pessimism about the future in the most recent communitarians than
in their predecessors. And many of the utopian communities of the
1960s and 1970s have been much more critical of the technology of
their day than were their predecessors of the technology of theirs.[11]

The persistence of some antebellum utopian communities into the
late nineteenth and early twentieth centuries and the establishment of
numerous new ones in the latter period does not undermine the point
that none of these groups developed mechanisms for effecting
genuine change. If anything, the newer communities achieved less
popularity and less influence than the older ones. Like their predecessors, they were not intended as escapes from the real world, but
often—more often than their predecessors—they functioned as such.
Few of these communities, old or new, had any effective plans for
growing or spreading. At most, the original utopians or their direct
disciples formed one or more similar communities elsewhere. Similarly, most of the utopian writings of the nineteenth and early twentieth centuries were intended to be blueprints for actual change, but
they remained only literary blueprints.

Contrary to popular stereotypes, as we have seen, many of the
utopian communities, particularly the longest lasting ones, were
quite favorable toward technology. But this did not greatly enhance
their appeal. Likewise, that many of the utopian writings, including
the technological utopian writings, were quite conservative in their
critique of existing society did not markedly improve theirs. Nor were
these writings able to take advantage of contemporary technological

advances that would have permitted their being printed in ever larger numbers at ever lower costs. There was not sufficient demand for them. At most, the new availability of mass publication facilities and techniques may have spurred some writers to seek to publish their visions rather than leave them unwritten or circulate them privately. Only when a utopian work like *Looking Backward* attracted an initial audience did the availability of those facilities and techniques make a substantial difference by allowing for mass reproduction.

Yet, of course, the technological utopian writings, unlike the utopian communities or other utopian writings, were successful collectively in one fundamental sense: their message was widely and warmly received, even though it was spread by other means and they benefited little from its appeal. Their collective vision appeared in a cultural context predisposed for various reasons to accept its principal themes. But as we have seen, there was limited need for multiple declarations of the gospel of progress and of progress as technological progress.

The non-utopian reform movements also fit comfortably within this gospel of technological progress. If anything, they contributed more effectively to its propagation that did most of the utopian writings (and communities). Yet they too were responses to widely perceived contemporary problems that they hoped to solve. Their crucial difference from technological utopianism was their lack of the kind of comprehensive scheme for social change found in the latter. These reform movements sought only limited, piecemeal improvements in American society and to this extent were even more conservative than technological utopianism, whose alterations were really conservative extrapolations from trends in existing society. Unlike the technological utopians, the conservationists, the corporate and government reorganizers, the city and national planners, and the scientific managers and Technocrats saw no need for wholesale alterations in the fabric of American society. Nor did they see the need to predict the future as did the technological utopians—however near the easier future may have seemed. Yet these reformers were as confident as the technological utopians that their particular panaceas would gradually solve America's fundamental problems, without endangering its basic institutions and values.

These piecemeal reformers very likely had individual and collective anxieties about the present and the future of American society akin to those of the technological utopians. People do not lead reform crusades unless they are dissatisfied with some aspects of their society. Many Americans, especially in the late nineteenth century, shared those concerns, whether or not they became reformers. In any event, the precise balances of hopes and fears for either the utopians or the reformers cannot be determined and, if it could, would not provide an appropriate yardstick for judging the merit of their respective schemes.

We know that there were reformers, or perhaps more accurately would-be reformers, of the period who were openly anxious, indeed pessimistic, about America's future. They were proponents of what Frederic Jaher has called "cataclysmic thought." Their ranks included notables such as Ignatius Donnelly, Homer Lea, Mary Lease, Jack London, and Brooks and Henry Adams.[12] Like the technological utopians, they stressed the momentousness of the social, political, and economic upheavals taking place. Unlike the technological utopians, they despaired of any solution to them. All would culminate in a cataclysm, they predicted. Meanwhile, they looked forlornly to the past, to a more settled, more homogeneous, more agrarian Golden Age that could never be restored—if it had ever existed.[13] By contrast, the technological utopians looked confidently to the future, to salvation through technological development.

But technological development, broadly conceived, encompasses all the non-utopian, piecemeal reform movements that we will now examine. More precisely, it is technology in the comprehensive sense used here that embraces them all; technology, not as machines and structures alone, but as employed in the organization and evolution of society and also as technical skills and knowledge. For all of those movements sought to use technology in this inclusive sense to reorder American society on the model of a giant machine (with equivalent structure)—just as, much more explicitly, technological utopianism did. That these reformers assumed no need for the kind of full-scale change throughout American society sought by the technological utopians is certainly significant. But no less significant is the way these reformers shared with the technological utopians not simply a metaphor but a model of American society for the present and future: a de facto technological society, with technological advance as much the end of change as the means.

The specific kind of technological society sought by the technological utopians was that largely organic model which, late in the nineteenth and early in the twentieth centuries, was in the real world steadily replacing a largely mechanical model dating back to the early and mid-nineteenth century. As we saw earlier, this change from a mechanical to an organic model of society was not limited to an intellectual elite but rather was occurring as a fundamental, "ideological" change throughout much of American society—a change that embraced the contented and the discontented alike. Just as the debate between Thomas Carlyle and Timothy Walker symbolized (and perhaps exaggerated) the emergence of the earlier mechanical social order, so the differences that we have noted between the three initial technological utopians and their twenty-five successors symbolized (and likewise perhaps exaggerated) the emergence of the later organic social order. In both cases actual technological developments had a definite impact on these ideological developments but did not necessarily cause them. A grasp of these ideological changes is essential to an understanding of the various reform movements that we will now discuss in connection with technological utopianism. Without such a grasp, not only the individual movements but also their relationship to general trends recently traced by other historians may not be readily comprehensible.

As we have seen, the organic model of society that emerged in late nineteenth- and early twentieth-century America involved more than a shift to organic metaphors. First, it brought a new emphasis on differentiation, specialization, hierarchy, complexity, organization, and explicit systematization among the components of present and future society: that is, its individuals, its institutions, its machines and structures, and its groupings of each. Second, it saw change as either more evolutionary or more spontaneous than "cause and effect" and as producing a temporary instability, if leading ultimately to equilibrium. These points surely illuminate the notions of a "search for order," a "quest for community," and a desire for an "organizational society" that, as several historians have suggested, characterized American society during this period.[14] The kind of society that, these historians persuasively argue, America was coming to be certainly presumes differentiation of parts, and the like, and change as other than cause and effect. Moreover, it presumes features like "technical systems," "functional associations," "rules with impersonal sanctions," and "mechanisms of continuous management" that these same historians show to be fundamental to the development of modern, organizational America as well.

Technological utopianism itself did not, of course, cause the "search for order," the "quest for community," or the desire for an "organizational society." Nor, of course, did it alone spur any of the specific reform movements from conservation through Technocracy. Rather, the profound ideological change reflected in the emerging organic model of society—a change in how Americans viewed, divided up, and pieced together their world—accounts in large degree for the general trends toward an organizational society and in turn for those specific reform movements. Technological utopianism and all the specific movements alike embody that model of society. Technological utopianism, however, represents an extreme version both of the movements and of the general trends. Certainly the vision of technological utopia is of an extraordinarily, if not excessively, well ordered society, with every aspect assiduously planned, prescribed, integrated, and fixed in most cases for all time. All this organization within technological utopia, however, is supposed to produce not a sense of faceless impersonality but, on the contrary, a sense of community; for society is to be composed of countless interlocking and hierarchical local, regional, and national units. These facets of utopia alone would seem to justify characterizing it as a decidedly "organizational society." But there are abundant examples, as well, of "technical systems," "functional associations," "rules with impersonal sanctions," and "mechanisms of continuous management." Hence my original claims that technological utopianism takes actual developments in real world modern America to extremes and that these developments reflect tendencies in that same society toward technological utopianism.

The examples that follow will further support these claims. I will not offer every possible example, because doing so would mean providing a capsule history of much of modern America. For instance, I have left out eugenics, which wielded its greatest influence between 1905 and 1930, during which time many scientists and laymen pressured the federal government to restrict the immigration of supposedly inferior southern and eastern Europeans.[15] Campaigns to prevent defective individuals from marrying, having children, and even living among healthy fellow citizens proved less successful.[16] Eugenics began as a reform crusade but ended, by 1930, as a reactionary one. Technological utopianism has little affinity with eugenics, since utopianism assumes the regular good health and character of the overwhelming majority of its citizens. Yet it, too, envisions some selective breeding; the few citizens somehow unfit mentally, physically, or morally are to be barred from marrying and bearing children.[17]

As pervasive as were these tendencies toward technological utopianism, real world America of this period obviously fell far short of becoming technological utopia. Nor did trends toward an organizational society capture anything near the whole of American culture.[18] Consequently, the examples of related reform movements discussed here illuminate the accuracy—and the limitations—of the technological utopians' understanding of their own time and also the contemporary world developments that surely influence their prophecies.[19]

At first thought, conservation may seem to be a most inappropriate reform movement with which to begin. Conservation may seem as far removed a reform crusade from technological utopianism as one could imagine, given not only the traditional romantic stereotype of conservationists but also the popular American view of nature and technology as opposites. Nevertheless, the very notion of the deliberate preservation, much less the orderly use, of natural resources suggests organization, planning, systematization, evolutionary change, and efficiency, and in the senses already spelled out. Moreover, it was in the late nineteenth and early twentieth centuries that the growing concern for conservation finally led to the creation of a formal, national conservation movement.[20]

The leaders of the movement were not simply romantic nature lovers but, as Samuel Hays has shown, were often hardnosed scientists, technicians, and other experts in resource management. If the more romantic conservationists wanted to preserve the wilderness whole and unspoiled, their scientifically minded allies wanted to put it to use, but to do so with care and planning and with newly developed techniques. During this period, the opponents of conservation were not greedy developers, at least not always, but an indifferent public. Thus "corporations often supported conservation policies, while the 'people' just as frequently opposed them."[21]

Hays concerns himself with only the conservation movement, but he nevertheless sees its larger significance. He sees it as emblematic of "the transformation of a decentralized, nontechnical, loosely organized society, where waste and inefficiency ran rampant, into a highly organized, technical, and centrally planned and directed social organization which could meet a complex world with efficiency and purpose."[22]

The connection between the conservation movement and technological utopianism should now be evident: both recognize no necessary conflict between nature and technology, both treat the landscape scientifically, both replace parts of the natural environment with a man-made environment, and both favor efficiency. In taking conservation to such an extreme, technological utopianism demonstrates the development of the conservation movement in the real world as a reflection of tendencies there toward technological utopianism.

The crusade for scientific management manifests similar associations with technological utopianism, and in so doing provides a more encompassing, and more literal, example of the transformation of American society described by Hays. Scientific management was concerned with organizing and reorganizing human resources in a manner akin to the conservation movement's treatment of natural resources.

The leader of the crusade was Frederick W. Taylor. He sought to apply his theories primarily to the factory, but his disciples attempted to apply them to the office, the government, the school, and other institutions of American life. In his *Principles of Scientific Manage-*

ment (1911), which became the bible of the movement, Taylor enunciated the basic steps to maximum efficiency: (1) break down operations into individual tasks and in turn into specific motions; (2) time with a stopwatch the specific motions comprising each individual task; (3) eliminate superfluous and inefficient motions; (4) determine an ideal time for each remaining motion; (5) assign every worker a fixed and limited task; (6) print instruction cards for each of those tasks; (7) standardize equipment; (8) systematize purchasing and inventory controls; (9) assign foremen to oversee individual tasks; (10) reward superior performances by individual workers; and (11) create a planning unit to supervise and to improve every phase of production.[23]

The investigations that led Taylor to formulate these principles began in the 1880s, but it was not until 1910 that Taylor, a mechanical engineer until then known mostly in professional circles, won public attention. The prominent lawyer Louis Brandeis, arguing before the Interstate Commerce Commission at the time, referred favorably to Taylor's research. From then until 1915, when Taylor died, the movement for scientific management was enormously popular. It even became what Samuel Haber calls a nationwide "efficiency craze,"[24] and certain of Taylor's disciples, most notably Harrington Emerson, went as far as to equate the philosophy with the pinnacle of human progress. "If man's progress is slow," Emerson proclaimed, "it is because of wastes—wastes of everything that is precious."[25]

By 1930, however, the craze was over, and for the most appropriate, if embarrassing, reason: scientific management had proved inefficient. It had failed to reduce energy consumption and waste, to increase productivity and profits, or to spur personal morality and social control to any degree approaching the original forecasts.[26] Even Taylor's disciples were compelled to stray from the master's teachings to attain efficiency.[27] The change was most dramatic in industry, the area to which Taylor had devoted most of his efforts. There scientific management gave way in the 1920s to the subtler mechanisms of industrial psychology, which shifted the focus of "scientific" investigation from the work itself to the worker, whom Taylor had neglected—or, more accurately, had thoroughly misunderstood.[28] The burdens placed by Taylor and his disciples on ordinary workers to produce to the best of their ability—as scientific managers determined that ability to be—had naturally created resentment. Taylor and his disciples had either naively or willfully assumed that all workers were motivated solely by economic rewards and would therefore readily accept scientific management in order to maximize those rewards. Moreover, in analyzing particular tasks, they had failed to determine scientifically the average pace of operations acceptable to workers and had failed to allow for significant differences among workers regarding that pace.[29]

Scientific management pervaded government and public education as fully as it did industry. Here, too, however, its influence was ephemeral. In government as in industry, scientific management proved unworkable and above all ineffective. It could sometimes improve the mechanics of decision and policy making but, notwithstanding its proponents' claims, it could never improve the actual formulation of decisions and policies.[30] In the case of public education, scientific management proved effective but deleterious. It could lead to effective cost accounting procedures and vocational training programs but also to the undermining of the liberal arts and to an overall anti-intellectualism.[31]

Scientific management in all of its applications surely does incorporate differentiation, specialization, hierarchy, complexity, organization, and systematization among the individuals and groups on whom it is practiced; and the notion of change that it incorporates is either evolutionary or spontaneous, depending on the circumstances. Technological utopianism obviously takes scientific management to an extreme: not only in its industrial army but also in its schools (including their vocational orientation), in its government (including its apolitical outlook), in its other institutions, and in its obsession with efficiency in all realms. The citizens of the written utopias accept these restrictions willingly rather than unwillingly. And they suffer none of the physical burdens Taylor imposed on his workers. Nonetheless, the spirit and ethic of scientific management pervade technological utopianism. And as they do so they demonstrate the extent to which scientific management is a partial expression of real world trends toward technological utopianism.[32]

The various departments of the utopian industrial army represent an extreme version of the occupational or, in Hays' term, "functional" organizations that were established throughout America in the late nineteenth and early twentieth centuries.[33] Moreover, most of the occupations represented in that industrial army had their counterparts in the real world of the same period. Although far more occupational organizations exist in the real world than in the utopias—thanks to the virtual elimination from technological utopia of supposedly irrelevant (or dysfunctional) occupations such as lawyers, politicians, scholars, artists, and preachers—the predominantly technical workers and experts who do remain, engineers above all, were among the earliest professionals to organize themselves in real world America.[34]

It would be simple and tedious to list here the names and dates of just the major organizations of professionals and nonprofessionals that arose in the late nineteenth and early twentieth centuries: those,

for example, of doctors, lawyers, academics, businessmen, and workers. W. Lloyd Warner and his associates have found that of the more
than 200,000 voluntary associations—not exclusively occupational
ones—in existence in the United States by 1967, including about
12,000 national associations, approximately 75 percent had been
organized after 1900: 30 percent between 1900 and 1920, 25 percent
between 1920 and 1940, and the remainder after 1940.[35] The late
nineteenth and early twentieth centuries witnessed an unprecedented growth of occupational organizations, especially professional
organizations.

To be sure, earlier periods of American history, above all the
antebellum period, also had many organizations (that period was far
from devoid of institutions generally). As has been observed by generations of commentators, Americans have nearly always been, in the
phrase of Arthur Schlesinger, Sr., "a nation of joiners."[36] The key
point here is that the *form* of at least the occupational organizations of
the late nineteenth and early twentieth centuries, as much as their
number, size, and scope, made that period different from earlier
periods in American history. Their form was organic: a variety and a
hierarchy not merely of occupational ranks within most such organizations—that is, of levels of education, of training, of certification,
and of expertise—but also of geographical or other units within most
of them. Most earlier American occupational organizations had followed a federal pattern, with relative equality among the units in the
system and a functional rather than a substantive difference between
the head and the system's other units; but now such organizations
tended to follow the unequal organic pattern, with definite differences among the units in the system and a substantive difference
between the head and the other units.[37] The head was usually the
central office and administration, which was usually established in a
major city, often the state or national capital. The exceptions to this
generalization—such as the American Medical Association, the
Knights of Labor, and the Grange—do not refute the rule insofar as
they were either older or weaker organizations and insofar as they
later reorganized along more organic lines (as did the AMA) or disappeared (as did the Knights of Labor and the Grange).

Like the proliferation of utopian writings in general in this period,
the proliferation of organizations in general in itself reveals something about American society of the time. It reflects the development
of a society compatible in form though not fully aligned in content
with technological utopianism.

American business and industry changed primarily to enhance the
very profits abhorred by the technological utopians, who preferred to

share and spread wealth in a semisocialist scheme. Yet the means by which business and industry grew to their present shape and structure between the 1870s and 1910s are strikingly akin to those of technological utopianism: above all, by an unprecedented reliance upon administration. Before this period, American business and industry were, for various reasons, very much smaller in size and very much simpler in form. The collective technological advances in energy, in transportation, and in communications represented by the mining of anthracite coal and the invention of the railroad and the telegraph provided a new and necessary technological basis by the mid-nineteenth century for the later corporate growth. Corporate growth strategies—such as volume expansion, product diversification, geographical dispersion, technological research and development, and horizontal and vertical integration—led to the establishment of manufacturing and marketing divisions within corporations and to the creation of a general administrative unit for supervising them. These corporations then formed the first large-scale organizations of any type in American history.[38]

The purposes of these changes were greater efficiency and so greater profits. The results were greater bureaucratization and also—despite those new divisions—greater administrative centralization. Corporate administration became an "identifiable activity"[39] distinct from buying, selling, processing, transporting, and advertising; in fact, it became the principal activity of corporate officials. Bold, individualistic entrepreneurs like Andrew Carnegie and John D. Rockefeller played decisive roles in the founding of some large-scale industries, but not all. And even when such entrepreneurs were prominent in their companies, they were not always averse to adopting managerial techniques; and, in any event, they were invariably replaced by professional corporate managers.[40] In technological utopianism as in corporate management, the leaders of the industrial army are full-time administrators and superb organization men and women. If corporations, for all their power, prestige, and public relations efforts, never became quite the counterparts of the industrial army of utopia, they nevertheless became models of organization and efficiency that the American government at all levels sought to emulate.[41]

Practically since its origins in the late eighteenth century, American government at local, state, and national levels has become ever larger and more bureaucratic. But particularly since the Civil War has more and more power accrued to the government, and more and more of that power has in turn accrued to the permanent, nonelected administrators of government. As Hays has shown, the actual operation as well as organization of the government has become increasingly centralized in two directions: upward, from municipal to state to national authority, and inward, with decision-making power becoming centralized at each of these levels.[42] Furthermore, the bureaucracy at each level of government has so gained power that, especially at the national level, it has come to be regarded as virtually a fourth branch of government. Where, in 1790, there were approximately 1,000 civilian employees of the federal government, by 1800 there were roughly 3,000; by 1860 roughly 37,000; by 1880 roughly 95,000; by 1900 roughly 230,000; and by 1920 roughly 430,000.[43] Thus even if technical experts have hardly replaced professional politicians, as they usually do in technological utopianism, such experts have nevertheless won considerable actual power—as have, of course, the lawyers that the technological utopians would have banished completely. The centralization and bureaucratization of government have, moreover, led the former laissez-faire philosophy of government to be supplanted by a view of government as, in Sidney Fine's words, a "general-welfare state": the belief that the government "could benefit society by a positive exertion of its powers and that it should therefore act whenever its interposition seemed likely to promote the common well-being."[44] This view of government certainly accords with the technological utopian view of government.

The desire to end waste and corruption throughout American society—that is, to become more efficient—prompted the specific governmental changes considered here. These reforms spurred, but did not alone produce, centralization and bureaucratization of American government at all levels. Local and state reformers sought remedies for waste and corruption in an expansion of the civil service, the establishment of city commission and city manager forms of government, the substitution of at-large for ward and district representation on city councils, the formation of municipal research bureaus, the creation of state regulatory agencies, and an increase in home rule and taxing power for cities.[45]

The most successful of these several structural reforms were the establishment of city commission and city manager forms of government and the creation of municipal research bureaus. The city commission, of which Galveston's (1901) was the first but Des Moines' (1907) the most publicized, gave its five or six members supreme authority to formulate policy and also to administer it. It was patterned after a corporate board of directors (except that its members, unlike most corporation directors, actually administered the city's departments as well as formulated their policies). Its popularity subsequently waned only because the city manager form of government, which had first been tried in Staunton, Virginia, in 1908—but which gained publicity only after Dayton adopted it in 1914—proved more efficient. Here authority was concentrated in one manager rather than divided among several commissioners. The city manager government too was modeled upon a corporate structure, but with a chief executive as well as a board of directors. Many cities, beginning with Lockport, New York, in 1910, found the ideal solution in a combination of the two forms, with a manager made responsible to a commission.[46]

The municipal research bureaus collected, analyzed, and published vital statistics; they systematized budgeting, accounting, and auditing processes; they imposed inventory controls; and they examined and improved worker efficiency. In short, they centralized municipal management. The New York City Bureau, which was the oldest (1907) and the most important, became so proficient that it was soon handling county, state, and even national affairs as well as local ones. In 1921 it therefore changed its name to the National Institute of Public Administration. Modeled on the research departments of large corporations, these bureaus, like the city commission and city manager forms of government, exemplified the application of contemporary business methods and administrative arrangements to problems of urban life.[47]

Not surprisingly, most people advocating such structural reforms were allied with big businessmen economically and socially or else were big businessmen themselves. The reformers' ability to alter the structure of city and state government, much less the nature of the city or state itself, however, proved much more limited—and much more costly—than they had anticipated.[48]

The desire to end waste and corruption throughout American society which prompted the structural reforms on the local and state levels prompted similar efforts on the national level, especially after 1900.[49] National reformers, like their state and local counterparts, either were sympathetic to big business or were big businessmen.[50] Economic regulation was the foremost arena for federal government reorganization and expansion. Regulation was accomplished not only by the passage of specific legislation but, more important, by the establishment of such powerful new agencies as the Federal Reserve Bank, the Federal Trade Commission, and the Interstate Commerce Commission.[51]

American military and civilian preparations for the First World War greatly accelerated the growth of the federal government. Forced to transform a peacetime economy into a wartime one and to win public support for his policies, President Wilson established many new administrative agencies. The most important of these were the Council of National Defense (out of which grew the War Industries

Board), the War Finance Corporation, the Food and Fuel Administration, the United States Railroad Administration, the War Labor Policies Board, the Committee on Public Information, and the American Alliance for Labor and Democracy.[52] Though the agencies themselves were short-lived, many of the tasks that they undertook became periodic if not permanent responsibilities of the federal government: for example, the regulation of industrial priorities, production, and prices, and the control of public information. Furthermore, the businessmen who directed or advised the agencies sought to apply to them the same techniques of modern corporate management that their predecessors had applied to previous agencies on the local, state, and national levels.[53]

The efforts to enlarge, centralize, and bureaucratize the federal government culminated in the separate but equally unsuccessful attempts by Presidents Herbert Hoover and Franklin Roosevelt to reorganize and streamline the presidency and thereby increase its efficiency and strengthen its administrative power.[54] The commissions that they each appointed issued reports—not surprisingly, favorable to their views—that might well have been implemented, had not unanticipated events intervened. In the case of Hoover, the Depression severely diminished his popularity and power; in the case of Roosevelt, his growing popularity and power aroused his critics to vigorous opposition.

Roosevelt wanted to keep the presidency a frankly political institution, one involved as much in formulating policies as in effecting them. He unapologetically utilized the expertise of political and social scientists to further his political programs and objectives.[55] Antithetically, Hoover wished to take the presidency out of politics and to limit its role largely to the evaluation and implementation of policies and programs recommended by nonpartisan groups of private citizens. (Hoover, it may be noted, was a prominent engineer.) Such private groups would represent business, labor, agriculture, the professions, and other special—and, ideally, enlightened—interests. They would constitute a "private government," free of the bureaucratic excesses of the public one.[56] Where Roosevelt directed his advisers to design a program that he himself could implement, Hoover refused to interfere with the work of his advisers, lest he color their research or obligate himself to utilize their findings.[57] Hoover's vision of a nonpartisan, apolitical government closely matches the technological utopian ideal of government, but it is Roosevelt's concept of a highly political government, however respectful of experts, that has become—or, perhaps more accurately, remained—the reality.[58]

Despite the continuing politicization of American government at

all levels, the gap between it and government in technological utopianism has surely narrowed. True, we will see in the next chapter, it
is either naive or deceptive to call for a government of any type and
declare that it will be wholly free of politics. Yet the growth not just of
government but also of government bureaucracy and regulation since
the late nineteenth century has made substantial portions of American government purely administrative in nature, in the manner,
though hardly to the degree, of technological utopianism.

Directly related to these developments within business, industry,
and, especially, government was the rise of independent research
centers. Just as reform-minded businessmen and government leaders
had organized municipal research bureaus in the first decade of the
twentieth century, their successors in the second and third decades of
the century organized national research centers. But where the municipal bureaus were government-run, the national centers were private organizations that contracted with the government to carry out
research and recommend policies, many of which were ultimately
adopted. Furthermore, where the establishment of municipal bureaus
was an aspect of the organization and centralization of government,
the establishment of the national centers was more an aspect of the
organization and centralization of knowledge.[59] Yet the latter institutions were no less committed than the former to ending waste and
corruption throughout government, to formulating governmental
policy more efficiently and more objectively, and to removing decision-making further from politics and public opinion.[60]
    The most important of the national research centers were the Na-
tional Research Council (1916), the National Industrial Conference
Board (1916), the Twentieth Century Fund (1919), the National
Bureau of Economic Research (1920), and the Brookings Institution
(1927). The Brookings Institution incorporated the Institute for Government Research (1916), the Institute of Economics (1922), and the
Robert Brookings Graduate School of Economics and Government
(1923).[61] Like the government bureaucracy they often served, these
research centers were never as far removed from politics as they tried
to appear. Like government itself, they could never be wholly free of
politics even in their most purportedly objective activities. Yet, like
the government bureaucracy, they did narrow the gap between
technological utopianism and the real world by their emphasis on
government as administration of policy and, indeed, of society itself.
They are a further example of trends leading real world America
toward technological utopianism.

As extensive as were these various movements to reshape American
culture, they remained independent movements seeking control over

only particular spheres of life. Technological utopianism, by contrast, seeks control over all spheres of life. The reform movement of this period that comes closest to matching the boldness of technological utopianism's aim is planning—planning on the municipal, regional, and national levels. Planning's vision, however, is far less comprehensive than the utopian one. Although the idea of social planning is hardly a twentieth-century invention, it is an idea that has been put into practice, at least in the United States, largely in the twentieth century, and then not frequently. As a profession, planning, even city planning, is a still more recent phenomenon. It is a fusion of the older professions of landscaping, public health, civil and sanitary engineering, and architecture.[62]

Before 1900, planning, if it existed at all, was almost exclusively at the local level and was piecemeal or haphazard or both.[63] Streets, parks, playgrounds, sewers, water mains, transit lines, and buildings were constructed by independent and sometimes rival contractors, who cared only about the cost, not the aesthetics, of the projects they undertook. Only insofar as their separate water supply, sewerage, or park systems actually were planned could their efforts be described as examples of genuine planning. The only overall plan they might be said to have had was the uninspired grid pattern.[64] Conversely, the "City Beautiful" campaigns of the late nineteenth century, which coincided with the first stirrings of municipal reform and stressed civic pride and moral uplift, ignored virtually all commercial and utilitarian considerations in the name of nobler, aesthetic ones. Wide boulevards, open plazas, and ornate buildings were the kinds of artifices these planners sought.[65]

The handful of Americans who did balance aesthetic and technical considerations—the previously mentioned landscape architect Frederick Law Olmsted, for example—invariably found themselves blocked by political opposition, special interests, and public indifference. The carefully landscaped cemeteries, walkways, playgrounds, and, above all, parks that he and a small number of other landscape architects of the mid- and late nineteenth century designed did not significantly influence the planning of American cities as a whole, certainly not in their own day. If anything, ironically, their designs influenced the planning of suburbs intended as escapes from cities.[66]

The turning point in the development of American planning came not with the Chicago World's Fair of 1893, as has traditionally been assumed, but, as Jon Peterson has shown, with the McMillan Commission blueprint for Washington of 1900–1902, which updated while preserving L'Enfant's original plan. In design the Chicago Fair did present a model of a well-planned city, but its neoclassical patterns, suprahuman dimensions, and playful demeanor soon made its attractions obsolete rather than prophetic. It was rather the more modest and more practical design of the McMillan Commission that became the model for future city planning.[67]

If city planning before 1900 was rare, city planning after 1900 was practically common, not least because of the initiative provided by the Commission. Between 1907 and 1917, more than a hundred municipalities, including half of the nation's fifty largest cities, engaged in some form of comprehensive planning. Ninety-seven of them established permanent planning agencies, and twenty-four enacted zoning codes. Over a dozen universities introduced courses on various aspects of planning. And the National Conference on City Planning, founded in 1909, was followed in 1917 by the American City Planning Institute (now the American Institute of Planners), which, unlike the National Conference, restricted its membership to full-time professional planners.[68]

The widespread adoption of city planning naturally increased the prestige of city planners. In Roy Lubove's phrase, their role had changed "from reformer to technician"; the reforms they had originally proposed had been adopted and now required "only" implementation. Meanwhile the planners' own goal had changed from "City Beautiful" to "City Efficient."[69] Like their fellow reformers in the city managers' offices, the city commissions, and the municipal research bureaus, they were seized by a strong desire to end waste and corruption in American society and to do so as scientifically as possible. More often than other urban crusaders, they invoked the organic metaphor and model of society in the course of their efforts.[70]

Even before it came into its own, city planning involved, in addition to the improvement of existing communities, the creation of new ones that were wholly planned from the outset. The most significant of these new communities were the so-called "commuter suburbs," "company towns," and "garden cities." All three were modern industrial or semi-industrial developments set, paradoxically if not surprisingly, in semirural or suburban settings, separate from the existing cities to which comprehensive city planning rightly might have been applied. The aim, as in technological utopianism, was to avoid the extremes of pure urbanism or pure ruralism found in most actual existing communities. This goal sometimes led to the founding of new versions of the "middle landscape," that reconciliation between technology and the pastoral.[71] But in the majority of cases, the objective of the planned community was predominantly commercial and the achievement was modest.

The electrified railways of the 1880s, along with automobiles in the early 1900s, spurred the rise of commuter suburbs outside of Boston, Milwaukee, Norfolk, and other cities.[72] Although the commuter suburbs were better designed than most cities, they were usually drab and uniform, commercialism ordinarily governing their construction. Only the best planned of them—Shaker Heights, Ohio (1910s and 1920s), and the Country Club District of Kansas City, Missouri (1910s and 1920s), in particular—balanced aesthetic and environmental concerns with financial ones and thereby realized the ideal of the commuter suburb.[73]

Commercialism also governed the creation of the company towns, of which the most famous were Pullman, Illinois, built in 1880–1885, and Gary, Indiana, built in 1906–1910. Both communities were named after their founders: railroad magnate George Pullman and steel magnate Elbert Gary. Their elaborate and expensive construction soon stirred hopes of revitalizing all of industrial America. Their construction in fact was celebrated in numerous speeches and writings hailing each as a genuine utopia. The founders themselves harbored less sublime designs. They, and their counterparts elsewhere, saw the towns primarily as financial investments that, despite their enormous costs, would eventually prove profitable. The juxtaposition of worker residences with workplaces, together with corporate ownership and management of both, was intended to increase efficiency, decrease costs, reduce labor turmoil, and maximize profits. The proximity of both towns to Chicago, then the nation's industrial center, would compound the profits. When, however, excessive paternalism and insufficient benefits raised both worker discontent and costs, the Pullman Company and United States Steel simply abandoned their model cities, leaving them to find their way as ordinary industrial communities. Pullman became a working class residential suburb of Chicago, and Gary became an impoverished, blighted, small industrial city—in effect, an industrial dystopia.[74]

Only in the garden cities did planning regularly take priority over profit. The man who inspired them, the English reformer Ebenezer Howard, was a genuine visionary rather than a businessman. As he envisioned garden cities in his landmark book *Tomorrow: A Peaceful Path to Real Reform* (1898), each was to be a well-planned community of 30,000, with a balanced industrial and agricultural economy and with fields and a forest, or greenbelt, circumscribing it.[75] All of the cities were to be interlinked by rapid transit systems and superhighways and joined in turn to a regional center of 60,000. Although designed to protect rural areas from urban encroachment and simultaneously to reduce urban overcrowding and poverty, Howard's garden city system—with its meticulous planning, its regionalism, and its acceptance of urbanization and industrialization— was comparable to the megalopolis of technological utopianism.[76]

The most prominent American garden cities were Forest Hills Gardens, Long Island, New York (1911); Sunnyside Gardens, Queens,

New York (1929); Radburn, New Jersey (1929); Greenbelt, Maryland (1937); Greenhills, Ohio (1938); and Greendale, Wisconsin (1938). Like the company towns, they experienced a mixed fate. No American garden city completely fulfilled Howard's original dream, which was largely realized in the garden cities he established at Letchworth, England, in 1903 and at Welwyn, England, in 1920. The American communities that adhered longest and most faithfully to the garden city ideal were Greenbelt, Greenhills, and Greendale—the three established during the New Deal at the urging of "Brains Trust" member Rexford Tugwell, who had sought support for three thousand greenbelt cities instead of only these three. Although none of the three provided for industry, and they only partly provided for agriculture, all did include a greenbelt and well-landscaped pastoral settings. They, too, however, eventually strayed from the ideal, once they lost government sponsorship and became private enterprises.[77] Thus the actual planning of both existing cities and new ones failed to achieve the comprehensiveness, the pervasiveness, and the permanency of the technological utopian vision.

The initial success of city planning prompted efforts at planning on a larger, regional scale. The goal of regional planning was best articulated by the Regional Planning Association of America (RPAA), a small and informal group of planners, architects, economists, social philosophers, and conservationists who met regularly between 1923 and 1933.[78] The RPAA sought a balance between the wholesale urbanization and the wholesale suburbanization of regions. It took account of the social and psychological needs of citizens, not merely their physical and economic ones. Sensitive to the value of diversity among regions, the association refrained from advocating a single scheme for all regions. It even refused to endorse the garden city, which its members keenly admired. The same transportation and communications developments that allowed for centralization of society could allow for decentralization instead. In its commitment to diversity, the association strove as much to preserve existing buildings and communities of historical worth as to construct new ones.[79] Such aesthetic and humane concerns did not conflict with the association's equally strong concern for efficiency, which it pursued through detailed cost analysis of all its proposals. The RPAA simply saw no conflict between diversity and efficiency or between aesthetics and materialism or between social needs and physical ones. A properly designed social organism—a term the RPAA itself used—could integrate them all. The conflict it did see was between planning and uncontrolled individual exploitation of the land.[80] Here there could be no compromise. As in technological utopianism, planning had to reign supreme and had to be genuinely comprehensive.

The foremost example of regional planning in American history has been the Tennessee Valley Authority (TVA), which President Franklin Roosevelt established in 1933. From its inception it was seized on by proponents and opponents alike as the symbol of government intervention in economic affairs. So persistent was the opposition—especially by private utilities, which stood to lose the most from its establishment—that neither Roosevelt nor Arthur Morgan, whom he appointed chairman of TVA, was able to implement the political, social, and cultural reforms intended to accompany the economic ones. Yet, if the TVA did not produce the grassroots democracy and cooperative communitarianism that Morgan in particular had sought, it did produce numerous technical achievements: dam construction, flood control, land reclamation, crop diversification, and, above all, cheap rural electrification. These achievements surpassed those of the Progressive conservationists Hays describes and approximated those envisioned on a national scale by technological utopianism.[81] Still, regional planning efforts in real world America, like city planning efforts, never attained the comprehensiveness, the pervasiveness, and the permanency of their technological utopian counterparts.

As city planning inspired regional planning, city and regional planning inspired national planning. National planning, however, differed not only in scope but also in kind from city or regional planning. It dealt less with the physical blueprints for society and more with the political and economic preconditions for implementing those blueprints—factors that city and regional planners frequently ignored at their peril. As George Soule, the leading American advocate of national planning argued, the "partial planning" characteristic of city and regional planning could not operate effectively on the national level. Instead, national planning would, in organic fashion, have to "recognize the interdependence of agriculture and industry, of city and country. It would [have to] see that planning for an industrial plant, a city, a region, or . . . the whole nation, must be placed in some general setting which involved all these things and more as well, if conflicting results were not to arise."[82]

As Lewis Mumford noted, planning as such need not be purely technocratic or elitist—as it often was with city and regional planning, or, for that matter, with scientific management and Technocracy. Rather, national planning was—or could be made—fully compatible with democracy: "Hence planning demands for its success not an authoritarian society but a society in which free thought and voluntary action and experimental effort still play a major part in its existence." The organic interdependency of regional planning as espoused by the RPAA would, to this extent, carry over to national planning.[83]

So different was the conception of planning at the national level from the conception of it at city and regional levels that few national planners came from the ranks of city and regional ones. Where the pioneer city and regional planners were predominantly landscapers, conservationists, architects, and humanist social critics, the pioneer national planners were predominantly economists, scientific managers, businessmen, and social scientists.[84] In 1934 national planners formed their own organization—the National Economic and Social Planning Association (NESPA)—rather than work within the American City Planning Institute, the Regional Planning Association of America, or comparable bodies.[85]

National planning took root even less well in real world America than did either city or regional planning—largely because it was feared that introducing "state planning" would allow socialism into capitalist society. What arose instead, as Otis Graham points out, was a "broker state," which ultimately failed to provide solutions to national problems but which gave national planning a bad name because it was falsely portrayed as actual national planning. If the broker state hardly resembled true national planning, it resembled the extensive national planning of technological utopianism even less. Moreover, technological utopianism employs all varieties of planning—physical, economic, social, administrative, and so forth—at all levels of society and puts planning itself in the mainstream of society. Far from undermining its social order, planning helps create and sustain it.

It may be that the technological utopian ideal of a wholly planned society is not extreme, since many movements contemporary with it also advocated planning on a large scale. However, planning in technological utopianism certainly plays a more central role than has ever been approached in the real world. Again, the extremism of its conception illuminates trends in real world America toward such a genuinely comprehensive planning of society.[86]

Of all the reform crusades of the late nineteenth and early twentieth centuries, the Technocracy movement of the 1930s and 1940s most closely approximated technological utopianism. It was the one that relied most fully on technology as the panacea for the problems of American society. Yet paradoxically it was because of its preoccupation with technology that Technocracy cannot be considered an equivalent of technological utopianism. In their preoccupation with increasing efficiency and production, the overwhelming majority of Technocrats ignored the social ramifications of technological advance. Only a few of the Technocrats, whose movement attracted hundreds and even thousands of adherents, seem to have been concerned with issues like government, education, religion, culture, recreation, and social relations. I have already mentioned the most notable of their number, Harold Loeb, who qualifies as a technological utopian. Howard Scott, the leader of the Technocracy movement, is the most notable Technocrat who does not.

The origins of Technocracy are shrouded in controversy, but most of its future leaders were apparently inspired by their association between 1919 and 1921 with Thorstein Veblen, then teaching at the New School for Social Research.[87] Veblen offered a strategy, spelled out in *The Engineers and the Price System* (1921),[88] for ridding American society of the waste and extravagance he had long condemned. He advocated (1) the voluntary abdication of all absentee owners of big business; (2) their replacement by reform-minded technicians and workers; (3) the creation of a national directorate to supervise the reallocation of all goods and services; and (4) the elimination of the artificial price system based on the equally artificial monetary system. Veblen contended that these changes would increase America's industrial output by from 300 to 1,200 per cent. He insisted that if the technicians and workers were so inclined, they could, by withholding their indispensable knowledge and skills, soon paralyze the whole of American government and industry and establish a "Soviet of Technicians." But, he reluctantly conceded, they were not so inclined. Disillusioned, Veblen observed that "by settled habit the technicians, the engineers and industrial experts, are a harmless and docile sort, well fed on the whole, and somewhat placidly content with the 'full dinner-pail' which the lieutenants of the Vested Interests habitually allow them."[89]

Nevertheless, Veblen's analysis stirred the organization of the Technical Alliance in 1920. Although it lasted but a year and sought milder goals than Veblen's own—not, for example, the elimination but only the reform of the profit system—the Alliance laid the structural and ideological foundation for the Technocracy movement. Both the Technical Alliance and Technocracy were led by Scott, an enigmatic Greenwich Villager.[90]

The Technocracy movement was established in 1932, at the depth of the Depression, which, allegedly, several Scott-directed studies had predicted.[91] Its premise was that the ability to produce energy and utilize it was the true measure of human progress, for energy was necessary to run the machines and structures that produced the goods that improved life. In analyzing human progress, the Technocrats consequently used physical categories like "man-hours" rather than the conventional monetary ones Veblen had scorned.[92] They argued that until about 1700 progress had been limited by man's almost total dependency upon his body for energy production—or, in the terms I am using here in defining technology, by mankind's then limited control over nature. Only with the industrial revolution did people become capable of increasing energy production significantly. Yet just when people, especially in the United States, were at last capable of producing enough energy to satisfy the basic needs of all citizens, the efficient production of goods was being undermined by greed and waste, as Veblen had exposed.

Oddly, Technocracy proposed no solution to this dilemma. Its official introductory pamphlet frankly declared, "Technocracy proposes no solution, it merely poses the problem raised by the technological introduction of energy factors in a modern industrial social mechanism."[93] Although much of Veblen's scheme was implicitly adopted, his strategies for changing the ownership of American industry by persuasion or by force were not. Technocracy's consequently vague tactics, coupled with its vague analysis, created widespread skepticism among other experts. Moreover, the movement employed technical jargon and complex charts that invariably impressed its audiences—and perplexed them as well. As an early Technocracy pamphlet proclaimed with unintended irony, "Technocracy's scientific approach to the social problem is unique, and its method is completely new. It speaks the language of science, and recognizes no authority but the facts."[94]

Scott and other Technocrats preached ceaselessly about their "scientific" and so "foolproof" scheme for not merely ending the Depression but also effecting unprecedented and permanent efficiency and in turn abundance. In this, they were like other prophets of technological progress who were described in chapter 5. That same early pamphlet declared, "In Technocracy we see science banishing waste, unemployment, hunger, and insecurity of income forever. . . .we see science replacing an economy of scarcity with an era of abundance. . . . [And] we see functional competence displacing grotesque and wasteful incompetence, facts displacing guesswork, order displacing disorder, industrial planning displacing industrial chaos."[95]

Of the countless cures for America's Great Depression, few enjoyed as spectacular, if as spectacularly brief, a reign as Technocracy. The *Literary Digest* proclaimed in December, 1932, "Technocracy is all the rage. All over the country it is being talked about, explained, wondered at, praised, damned. It is found about as easy to explain . . . as the Einstein theory of relativity."[96] Although the movement persists even today, Technocracy's heyday lasted only from June 16, 1932, when the *New York Times* became the first influential press organ to report its activities, until January 13, 1933, when Scott, attempting to silence his critics, delivered a rambling, confusing, and uninspiring address on a well-publicized nationwide radio hookup.[97] The ill-fated speech hardly helped the cause, and shortly afterward the Continental Committee on Technocracy (CCT), a subordinate body within the movement, bolted and became a rival movement, headed by Loeb. In response, the majority of Technocrats formed Technocracy Inc., headed by Scott.[98] Initially outmaneuvered by the CCT, Technocracy Inc. persevered and eventually gained far more members, if no more influence, than the united movement had had originally. The faction-ridden CCT collapsed in October 1936.[99]

The differences between the two organizations were significant. The CCT was led by well-to-do cosmopolitans seeking not only economic reforms but also social, political, and cultural ones; Technocracy Inc. was led by lower-class technicians with exclusively economic objectives.[100] Where the CCT seemed safely removed from the specter of authoritarianism, which Veblen's "Soviet of Technicians" had always raised in some minds, Technocracy Inc. raised the specter with its militaristic demeanor, its rigid hierarchical structure, its special insignia, its special salute, its grey uniforms, and its fleet of grey automobiles. The resemblance of Technocracy Inc. to fascism (more than to communism) was not lost on commentators,[101] nor was its mounting anti-Catholicism (both in contrast to its supposed scientific rationality).[102] Indeed, having by 1940 failed to enlist what he deemed a respectable number of technical experts in his crusade, Scott tried to recruit masses of lay citizens in order to pressure the government into appointing him "Director General of Defense" for the duration of the Second World War. He disavowed violence as well as political ideology, but his tactics seemed ominous.[103] Scott's efforts failed, however, and Technocracy Inc. soon shrank to its present meager size.[104] The great expectations raised by Technocracy in its various forms were never met, and today when people think of a "technocracy," a dismal un-utopian society comes to mind.

Technocracy, then, never came to equal its technological utopian counterpart in theory or in practice. Unlike technological utopianism, Technocracy ignored the social ramifications of technological advance and actually seemed to pride itself on indifference to providing solutions to non-economic problems. If it invoked the language of technological utopianism (and other American crusades) in proclaiming an imminent "new era [of efficiency and so abundance] in the life of man" under a Technocratic regime, [105] it failed to provide a comprehensive utopian-like blueprint for that better society. I am not suggesting that technological utopianism offered superior, much less more popular, solutions to the problems that Technocracy and other reform crusades of the period wished to solve. Once again, I am pointing out that by taking their collective traits and ideals to an extreme, technological utopianism truly reveals what all such movements sought for modern America. The implicit goal was a society— more precisely, a technological society—built on an organic model.

We cannot, however, end the story here, with Technocracy as the first (if partial) organized expression of technological utopianism in American history. Nor is it enough to confirm the findings of recent historians that late nineteenth- and early twentieth-century America was a time of unprecedented zeal for organization, specialization, hierarchy, bureaucratization, efficiency, and the like. These two developments must be brought together. We must examine the assumption which arose during these years that, at least in America, technological advance could finally bring about a better if not ideal society. Or, in the context of the reconceptualization of utopianism that took place first in Europe and then in America, we must look at the transformation in the meaning of utopianism from "possible" to "probable." Both technological utopianism and the piecemeal reform crusades discussed in this chapter are tangible expressions of the belief that various technological improvements now make it possible to realize a genuinely good if not perfect American society, and that this good society actually will come about in the not-too-distant future. Unlike many earlier utopians, the technological utopians invariably gave specific dates to their prophecies, dates usually within a century of their publication. Because the piecemeal reformers saw no need for comprehensive change, they rarely offered any dates at all—but exactly because they felt that the particular alterations they sought either were already in the making or soon would be. This optimistic view, in both cases, was fueled by the appearance of recent technological improvements with broad social implications and the anticipation of further ones.

The shift from communities to writings as the principal expression of American utopianism follows the rapid advance of events. American society in the late nineteenth century, with its unprecedented complexity, no longer seemed susceptible to communitarian solutions to its problems; and the impending grand future offered an unprecedented opportunity to plan. Indeed, it made comprehensive written blueprints essential for the reconstruction process to come. Detailed guidelines, not just vague hopes, were now required for building that future.

The organization of the Technocracy movement marked a further shift toward associating utopianism with the probable. Its own aversion to utopianism notwithstanding, Technocracy did assume that a better society of some kind could be fashioned in the near future by groups of citizens working together, whether elites of technical experts or masses of laymen. This contrasted with the isolation of most of the technological utopians. Few of the utopians tried to form groups to build their utopias, and of those who did, only Bellamy—through his Nationalist crusade—had any real success. Even that crusade was not primarily concerned with technological advance, and it ultimately proved no more cohesive or influential than Technocracy. In addition to Technocracy, the New Machine, the Technical Alliance, the Utopian Society of America,[106] and other small and fleeting associations of like-minded Americans embodied similar assumptions in the 1910s, 1920s, and 1930s about the favorable prospects for organization to pursue a genuine technological society. Now that the written blueprints had been largely completed by Bellamy, Veblen, and others, all that remained, it seemed, was for younger Americans to act on them.

The excitement, the promise, the seeming inevitability of these ideas were nowhere caught more dramatically than in the world's fairs in America in the early twentieth century, and in the 1930s in particular. These popular exhibitions represent the emergence of another expression of American utopianism beyond communities and writings, another proffered solution for America's problems. Contrary to popular stereotypes, world's fairs are more than harmless diversions from everyday life, more than idyllic escapes from individual or collective problems. They are significant social and cultural artifacts that reveal much about the times and the places that produced them, if not necessarily about future times and places. Fairs must therefore be approached seriously and rooted within the mainstream of their times and places and not relegated to the periphery as exotic, hence unimportant and uninfluential, phenomena.

The first truly international fair anywhere, London's Crystal Palace Exhibition of 1851, had been followed a mere two years later by a smaller version of the same event held in New York City. There followed dozens of international extravaganzas throughout the Western world, including several in various American cities (as in Chicago in 1893).[107] World's fairs both here and abroad have continued in ever-increasing size, scope, and popularity till the present and more are being planned until at least the year 2000.

The American fairs of the 1930s were distinctive in their unprecedented dramatizations of mankind's ability to shape the future through technological advance. Previous world's fairs in both the United States and Europe had certainly had a technological bent, as they still do. This bent was present and continues in the strictly commercial and ideologically unpretentious trade fairs whose history antedates the 1851 Exhibition. But the 1933–1934 Century of Progress International Exposition in Chicago, the 1935 California-Pacific Exposition in San Diego, the 1939 Golden Gate International Exposition in San Francisco, and, above all, the 1939–1940 World of Tomorrow in New York City, even more than their successors, manifested the idealism that a veritable utopia would come about in America in the very near future—by 1960, to be exact, according to the World of Tomorrow, the only one to set a specific date.[108] Thus the fair moved the deadline for (a version of) technological utopianism from a century ahead to a few decades. No less important, the four major architects of the World of Tomorrow—Walter Dorwin Teague, Henry Dreyfuss, Raymond Loewy, and, most notably, Norman Bel Geddes— were pioneering industrial designers who readily assumed that the achievement of a technological utopia of some kind only awaited their design. Having begun their careers by designing the individual components of a new world, from appliances to vehicles to buildings, they enlarged and redirected their efforts to designing the new world itself. As in technological utopianism, this proudly man-made environment would replace much of the natural environment.[109] The chronological and psychological gap between the real and ideal realms had thus been all but bridged. A leader of the Utopian Society of America declared about the nation's future in 1942, reflecting on these and related developments, "Thus we see a continued shortening of the time between the utopian vision and its materialization. . . . The vision of the utopian thinker becomes more clear and definite as the time for materialization diminishes."[110]

The gaps between these fantasy worlds and the real one proved far wider than was anticipated by the fairs' designers, their government and corporate sponsors, and probably most of their visitors. The America of 1960 that Geddes and other designers envisioned had not come about even in 1980 and seems quite unlikely to come about by the year 2000. The physical aspects of particularly the World of Tomorrow largely have come about and, where not yet in place, readily could, given sufficient funds and desire. But the social and cultural dimensions of that utopia, which Geddes and his fellow designers assumed would inevitably result from technological progress, have been realized in only partial and even paradoxical ways, as we will see in the next chapter. The very dream of world peace through international gatherings like fairs, a feeling which permeated the World of Tomorrow in its 1939 season, was shattered by the Second World War, a technological nightmare, well before its 1940 season. As hopes for universal peace have dimmed, recent fairs have made fewer social and cultural predictions and have been less avowedly idealistic regarding "the shape of things to come." Thus the America of 1960, or 1980, resembled the World of Tomorrow's and other fairs' exhibits only in bare outline—in its sleek skyscrapers and smooth superhighways, and then primarily in its newer cities or renovated older ones. Much more, such as greater social and economic equality, remains to be filled in. The same situation holds for the collective vision of technological utopianism that I discussed in chapter 2. This is hardly surprising, as the technological utopian vision and the imagined future dramatized in the world's fairs are strikingly similar, both in their physical layout for America in the future and in their technological determinism.

What Arthur Schlesinger, Jr., wrote about the failure of the Technocracy movement relates to the fate of the fairs of the same years and in turn to the fate of technological utopianism for the entire period 1883–1933: "Caught between ribaldry and irrelevance, the technocratic dream faded as rapidly as it had arisen. Yet a residue remained: a sense of infinite technological possibility; a susceptibility to new approaches; a readiness to break with the past."[111] On the one hand, the future remained open to more successful technocratically-oriented visionaries, as for example, the industrial designers. On the other hand, the very (progressive) changeability of modern technology always threatened to undermine the most foresighted of prophets, much less those who, like the technological utopians and the industrial designers, primarily extrapolated from the present to the future. As Scott wrote with unintended irony, in criticizing Marxism in 1965, "It is well . . . to bear in mind [that] the technological progression of the next 30 minutes invalidates all the social wisdom of previous history. Technology has no ancestors in the social history of man. It creates its own."[112]

Here Scott finally proved to be an authentic, if unintentional, prophet. For if much of the collective vision of the technological utopians and the industrial designers remained unrealized by 1980, other parts of both visions—and of the implicit vision of the Technocrats as well—had long since been realized but rendered obsolete. These parts, moreover, were from among the supposedly permanent physical features of the future American society. Thus, for example, the transportation and communications devices described in chapter 2 above—from trains to telegraphs—had long since been drastically improved on or outright superseded by real world developments. Similarly, the streamlined style applied by the industrial designers to almost every item they touched—from appliances to aircraft—had become passé as early as the 1940s, as had several of the individual items themselves.[113] And no one had foreseen the importance of the revolution in electronics and information processing that began in the 1930s and has truly reshaped at least the Western world in recent years.[114] Indeed, that revolution, as I have suggested elsewhere, may have rendered all future world's fairs obsolete.[115] Why travel to distant points to see the future if satellites, computers, word processors, and other electronic innovations can now deliver the needed information to the office or factory or home? Why come in person if the latest technological advances are available on a television or other screen?[116] Thus the disjunction between the past and the present and between the present and the future, which all three groups regularly proclaimed in self-congratulatory tones, ultimately worked against them and left them along with considerable portions of their respective visions as historical relics. The same unswerving faith in technological progress that had inspired all three sets of idealists had, in this respect anyway, led all of them astray. Presumably, Scott had not anticipated in 1930 and 1940 that he would be left behind by events and, in any case, he would not acknowledge it in 1965.

It would nevertheless be simplistic and inaccurate to dismiss the Technocrats, the industrial designers, or the technological utopians— the oldest and least cohesive of the three groups—as only historical relics. Again, technological utopianism is significant not only in its own right, as a hitherto neglected strain of American culture, but also as an illumination of Technocracy (and industrial design) and the many other larger and better known developments discussed here. Here lies the relevance of the technological utopians to the Technocrats and the industrial designers and in turn to modern American history. In effect, technological utopianism functions like a microscope: by first isolating and magnifying these piecemeal reform crusades and by then bringing them into collective view, it enables us to see them in their tamer—but no less genuine—form either within or close to the mainstream of American culture of the late nineteenth and early twentieth centuries.[117] It would be misleading to contend that technological utopianism and the various traits of that organic model of society alone characterized American culture in this period, but it is no distortion of the period to argue that their values together constituted a prominent part of American culture—and still do.

Consequently, the technological utopians were, in part, prophets in their own time concerning the values, the shape, and the direction of real world America. The ultimate test, however, of their predictive powers—and of their importance to us today—is the accuracy of their visions of our own time, especially the nontechnical aspects of utopia that they foresaw as following inevitably in the wake of the technical ones. This is the subject of the next chapter.

Seven

Prophecy and Fulfillment

How accurate were the technological utopians' prophecies? To what extent were their predictions for America's future realized and, if realized, how close to the expected dates? If not yet realized, may some of the anticipated developments still come about? If some predictions apparently are not realizable, what effect will their absence have on contemporary technological society? And in all of these cases, what of the connection that the technological utopians firmly drew between technological progress and social progress?

As I indicated in the preceding chapter, many of the physical aspects of technological utopia have in fact come about and, where not now in place, readily could occur, with sufficient funds and desire. Sleek skyscrapers and smooth superhighways are now characteristic of newer or renovated American cities, but many of the other physical features that the utopians assumed would be permanent fixtures of future American society have, as also noted, long since been drastically improved on or outright superseded by real world developments. Several futuristic features of the technological utopias would now seem quaint.

What, then, of the nontechnical aspects of utopia, the predicted social by-products of technological progress? For better or for worse, contemporary technological society seems headed in—or wedded to—other directions. The innumerable technological advances that have come into being since 1933, when the last of the utopian works in this study appeared, have neither produced nor led to equivalent social advances. Certainly, many of them have made the lives of Americans less burdensome and more comfortable, but they have not

made people's lives qualitatively happier, as had been predicted by
the technological utopians and many non-utopian prophets alike.
Moreover, many now see technological progress and social progress
not as an equation but as an antithesis. Rather than leading to a utopia,
technological developments have been seen by some critics as leading
to a dystopia. George Kateb, himself a defender of utopianism, has
succinctly put it, "There is not, for the most part, skepticism about the
capacity of modern technology and natural science to execute the
most vaulting ambitions of utopianism; on the contrary, there is a
dread it will."[1] Or, as we noted at the beginning, the technological
progress once hailed by millions as the panacea for virtually all of
mankind's problems—material and social—has not only failed to
solve a number of major problems but, to thousands, has become a
principal problem in itself.

Nowhere, perhaps, has the gap between prophecy and fulfillment
proved wider—or more paradoxical—than in the related realms of
work and leisure. As countless contemporary social critics have
observed, the one realm has steadily absorbed the other and has
nearly eliminated the classical distinctions between them. The worth
of ever more activities previously seen as separate from work has
come to be judged by their usefulness to work. To this extent, real
world America has increasingly come to resemble technological
utopia.
    But with one crucial difference: many contemporary Americans
find little pleasure in their work. Yet, relentlessly, they nevertheless
work. Sebastian de Grazia has shown that the amount of time Americans actually spend working far exceeds the conventional forty hours
a week. He calculates that the time typically spent not just on the job
but preparing for it, traveling to and from it, and continuing it over-
time even exceeds the time spent working by pre-industrial laborers
like farmers and artisans, who are conventionally thought of as having
worked a much longer day than do people today.[2] As in technological
utopia so in modern America: the supposed beneficiaries of labor-
saving inventions spare themselves no labor. Likewise, the free time
Americans today do have is often, as in technological utopia, not
really free but rather is spent on "useful" activities as exhausting as
their work itself. Leisure time has become a scarce, and so precious,
commodity.[3] In effect, the machines and structures that freed men and
women from much taxing physical labor repeatedly have freed them
only for other, no more satisfying forms of labor.
    In technological utopia, work unites people with their society.
Society values their work, no matter how routine it is, and the workers
see their labor as contributing to the common good. In real world

America, however, work frequently divides people from their society. Many jobs have no dignity beyond the paycheck, and it is often difficult to see how work, especially the most routine kind of work, contributes to the common good. Where in technological utopia people are frequently portrayed as happiest doing the most routine—that is, most mechanical—tasks, in our technological society the unhappiest workers are frequently those performing just such tasks. Where work seems to have so little ultimate importance, it divides people not only from their society but also from themselves. They do not, as in technological utopia, identify with their work.[4] Work is something external to them, not something that is part of themselves. Or, if such work is a part of themselves, it is a demeaning part.

In technological utopia, work has the same status as in Calvinist doctrine: it is a "calling," a guarantor, if only psychologically, of salvation. In modern America, by contrast, work is often a chore, and a guarantor of only physical comfort and, at best, social prestige.[5] In technological utopia, people find happiness in their work. In modern America, people increasingly find happiness only in *relief* from their work, or in what their work buys. The utopian ideal of easy and constant work has actually been realized for many, but it has brought something other than utopia.

The perverse parallel between the utopian ideal and the present reality runs still deeper. Contemporary American society fulfills the utopian ideal not only in transforming leisure into further work but also in its indifference to the classical view of leisure: a time when mankind, freed from work, could devote itself to nobler pursuits, above all politics, philosophy, and the creative arts. Technological utopia has no need of these pursuits. Politics, it asserts, no longer exists because the perfect philosophy has been found. And a perfect society has little need of creative arts. Modern America obviously has need of serious political, philosophical, and artistic activities; but just as obviously it often shows small interest in them except in the most pedestrian or commercial ways, precisely because in a work-centered culture they do not "pay." Hannah Arendt, whose views complement and extend de Grazia's, notes the ironic result: "a society of laborers . . . about to be liberated from the fetters of labor," yet which "does no longer know of those other higher and more meaningful activities for the sake of which this freedom would deserve to be won."[6]

Classically, the highest form of leisure was contemplation. What current American society calls leisure is either a disguised extension of work or else an escape from work—something not found in technological utopia, where work is so pleasurable that a desire to escape from it never arises. Classically as well, leisure meant escape from work, but it did not mean sheer escape (or would-be escape), as it seems to today. It meant free time for other activities, particularly contemplation.[7] The closest contemporary approximation to contemplation is also the most dubious: university life. In substituting active, productive research for the classical ideal of unhurried reflection, the university has characteristically turned contemplation itself into one more variety of work. Indeed, colleges and universities are increasingly seen as places to learn a remunerative trade.

In developing their thesis that the work ethic has come to dominate not just American but Western society as a whole, social critics de Grazia, Arendt, and Josef Pieper each analyze the ways in which the term *work* has come to subsume more and more activities. De Grazia points out that where the term was once restricted to menial and physical occupations, it now encompasses all occupations, and in so doing reduces them all to mere jobs, to mere "work."

> Before the nineteenth century's close, if you worked, you labored or toiled, and if you did other than these, you did not *do* something; you did not work as you do today; you *were* something—a carpenter, mason, soldier, physician. One's work then was rarely called work.[8]

Arendt observes that where once the term *work* referred to private, individual activities concerned with the production of artifacts, it has gradually come to refer to mankind's public, collective activities. Previously, mankind's main public activity was action, or the relationship of human beings to one another rather than, as in work, to some product. Politics was the primary instance of action. (More about that shortly.) Today, work, precisely because it is a productive activity, is mankind's main public activity, and anything that seems nonproductive is relegated to the status of a "hobby."[9]

Like de Grazia and Arendt, Pieper pits work against leisure, which for him too means the intellectual activity of contemplation. He then notes how the contemporary term *intellectual work* reveals the subversion of the contemplative ideal. Classically, such a term would have been an oxymoron. Today, intellectual activity has lost its independence of work and has become largely absorbed by it. Pieper, following de Grazia and Arendt, deems leisure the basis of culture, and no real leisure remains.[10]

Where Pieper, de Grazia, and Arendt make leisure the basis of culture, historian Johan Huizinga, in a related critique, declares play to be its basis. By *play* Huizinga no more means relaxation than Pieper, de Grazia, and Arendt do by *leisure*. Play stands outside work, just as leisure does, and like leisure it is an escape from work. But like leisure, it is no mere escape. It is a positive activity in its own right.

When, as in its present corrupt form, it is, like leisure, sheer escapism, then it is not really play at all.

By play Huizinga means an organized game of any kind. It means organized activity, so that mere pastimes, activities without fixed rules, are not play. On the other hand, play is both an activity to be enjoyed and an activity to be performed for its own sake. An activity that is not enjoyable is work rather than play; and so is an activity done for some purpose beyond itself—for profit, for example. Insofar as play is enjoyable and serves no purpose beyond itself, it is not "serious," by which Huizinga means only that it is not practical. He does not mean that it is frivolous. Like leisure, play becomes corrupted when it becomes mere relief from work and so frivolous, or else becomes an extension of work and so serious.[11]

Play, however, is different from leisure. If play in Huizinga's special terminology is not practical and so not serious, leisure in the more conventional sense—as contemplation—is a supremely serious and practical matter. Its aim is knowledge, and the aim of knowledge is the guidance of life. To seek knowledge for its own sake, then, is to stray far from the classical ideal. In contrast to leisure, play exists for its own sake, and its aim is pleasure. It may absorb participants as fully as work or leisure, but neither in Huizinga's view nor in the common understanding can it be called a serious affair.

Dance, music, sports, law, religion, philosophy, poetry, and war are among the chief instances of play considered by Huizinga. Since these activities clearly still exist—at least in the real world, if not necessarily in technological utopia—it is not for their disappearance that Huizinga flays modern Western society. Rather, it is for their transformation, from genuine play into pseudo-play. On the one hand, Huizinga charges, some kinds of play have become largely amateurish. Dance, music, and sports have frequently degenerated into mere entertainment, mere diversion, mere escapist fancy. In this respect these activities have ceased to be arts or skills. They have become frivolous. On the other hand, law, religion, philosophy, poetry, and war have become thoroughly professionalized. They have become varieties of work in themselves. Men and women practice law not after work but *as* their work. Men and women are paid to practice law, just as they are paid to be religious, to philosophize, to write poetry, and to fight—and often to do these full-time. There also exist ever more "professional"—paid—dancers, musicians, and athletes, so that dance, music, and sports suffer the double blow of at once not being taken seriously enough and being taken all too seriously.[12]

The parallel with technological utopianism lies here, in the transformation of play into work. With play, as with leisure, the fictional inhabitants of technological utopia differ from contemporary real world citizens in needing no relief from their work, since they are happy in doing it.

Because they are so happy working, they eagerly turn their every extracurricular activity into work. No utopian activity exists for its own sake, even if the citizens of utopia deem their work "play." There exists no genuine play. The absence from technological utopia of extensive artistic and athletic activities confirms this point. The pseudo-play of much of real world America does not seem, however, much of an improvement.[13]

The fulfillment in real world America of the technological utopian ideals of work and of leisure has been no more complete—or success-ful—than the fulfillment of utopia's physical structures. As the year 2000, the date for the achievement of most of the twenty-five tech-nological utopian visions, approaches, these social and cultural dimensions of technological utopia do not seem to be approaching fulfillment. For all the ease and comfort that modern technology has provided to most Americans—and these benefits ought not be mini-mized—the individual and collective happiness that technological utopians assumed would inevitably follow has proved elusive. The proliferation of technological advances will probably not bring it any closer.

The comparisons between technological utopia and contemporary America could easily be extended to other, related realms: for exam-ple, those of knowledge, of politics, of bureaucracy, of art, and of personality. Both the utopian and contemporary societies are ob-sessed with the accumulation, analysis, and distribution of knowl-edge, and the kind of knowledge they are obsessed with is techni-cal—that is, practical—knowledge. Both societies consequently give great responsibilities to their technical experts, albeit, as noted, in varying degrees.[14] In technological utopia technical knowledge is not merely highly prized but is virtually the only kind of knowledge. In modern America, however, academic knowledge, or knowledge for its own sake, plainly persists if it does not always flourish. Thus the knowledge explosion regularly said to characterize modern Western society refers to the astronomical increase in knowledge of all kinds, from the most practical to the least.[15]

As de Grazia, Arendt, and Pieper criticize Western society not for the kind of work it values so much as for its failure to value genuine leisure above work, so other social critics fault Western society not for the particular applications it makes of technical knowledge so much as for its failure to apply completely different kinds of knowledge to improve existing society.[16] These critics would cultivate philosophi-cal, historical, and political knowledge, and not only to shore up existing society. As we will see in chapter 9, they and other social critics would even seek alternatives to existing society—just to make the lives of its members more fulfilling.[17] The issue here is not whether Western society should employ these different kinds of knowledge. We are asking only whether, any more than technological utopia, it does employ them.

In the case of politics, the renouncing of politics in technological utopia has parallels in real world America. As we saw in chapter 6, many segments of American society in the late nineteenth and early twentieth centuries became permeated with technical experts and, in so doing, sought to rise above politics (and other messy phenomena) in their various activities. They did not always succeed, and several historians have found various political and economic motivations behind the pose of objectivity. But the trend away from politics was set and has since accelerated as technological society has grown. The contemporary knowledge explosion and the consequent demand for the acquisition, analysis, and application of technical information have brought technical experts to the fore and relegated politicians— with their value-laden, unscientific, and sometimes corrupt practices—to the background.

Still, we must be wary of exaggerating current trends. We cannot minimize the persistence of politics and of politicians and the failure of anything remotely resembling the pure technocracy of technological utopianism to appear. Indeed, this failure has undermined the arguments of a number of prominent political scientists and sociologists who, especially in the late 1950s and early 1960s, maintained that American society was becoming steadily more apolitical and non-ideological and steadily more dependent on technical experts. Scholars such as Daniel Bell, Seymour Martin Lipset, Edward Shils, and Zbigniew Brzezinski, moreover, welcomed these developments. They preferred that, whenever possible, decisions should not be made on the basis of political negotiation or popular voting but strictly on technical grounds. In effect, they wished to replace politics by technology.[18]

A double critique has since been made of this stance. The more obvious critique, by Jean Meynaud and Franz Neumann, was that politics does and will continue to exist: both because no sufficient consensus on basic values and goals ever exists in American, much less Western, society and because no decision on values and goals is ever merely technical. Any decision—say, a decision to allocate resources for a particular project—inevitably involves social, economic, psychological, and other factors that make the decision more than technical, that make it political.[19]

The less obvious critique has conceded that politics has indeed

ceased to exist: not the politics of the technocratically-oriented scholars (and the technological utopians) but politics in a much grander
sense, a sense to which they are oblivious. Just as Arendt, de Grazia,
and Pieper bemoan the disappearance of real leisure from modern,
technological society, and just as Huizinga laments the disappearance
of true play, so Arendt and Sheldon Wolin, among others, bewail the
disappearance of a larger, more fundamental kind of politics. By
politics they mean an autonomous, self-sufficient activity, not one
beholden to work or other outside realms. Similarly, they do not deny
that politics of some kind remains, but they point to the disappearance of an ideal kind of politics.[20]

This ideal kind of politics they trace to the Greek *polis*. The Greeks
regarded politics as too important to be left to professional politicians.
Politics for them meant direct, open, public discussion of issues
among all citizens. Politics was quintessentially ideological, but
ideological in the sense defined in this work as the analysis of the
entire social order. For the Greeks, politics was not the reduction of
ideology to slogans and commercials produced by advertising agencies or to safely popular positions determined by pollsters. Politics
was an activity to be participated in, not a technique to be learned or
sold.

Above all, politics meant theorizing. It meant espousing a distinctive model of society and the evaluation of society in the light of that
model. It meant criticism of society as much as justification for it.
"Political scientists," observe Arendt and Wolin, were not, as today,
supposedly neutral analysts of society, concerned with only those
aspects of political life that can be analyzed objectively.[21] Rather, they
were practicing politicians who, unlike most politicians today, not
only reflected deeply on society but also attempted to act on their
reflections.

In contemporary American (though not Western) society, politics
is largely bereft of any true ideological conflicts—a judgment on
which Bell, Lipset, Shils, Brzezinski, Arendt, and Wolin all agree.
They define *ideology* either as dogmatic, extreme, blindly uncritical
thinking or, more frequently, as vapid sloganeering. Where Arendt
and Wolin regret the absence of ideology, Bell, Lipset, Shils, and
Brzezinski practically revel in it, or once did. They saw the reduction
of ideology to sloganeering, organization, and technique as a necessary step toward eliminating ideological divisions altogether and
establishing instead an ideological—for them, non-ideological—consensus. Whether or not they still hold this assumption or advocate
this development is unclear.[22] In any event, I will take up the issue of
ideology as it relates to theorizing and in turn to utopianism in
chapter 9. For now, I will only observe that the gap between politics,

or antipolitics, in technological utopia and in real world America remains wide but is gradually narrowing.

The emphasis in technological utopia on technical knowledge and on technical experts naturally leads to an acceptance of and even preference for bureaucracy. Bureaucracies exemplify the application of technical knowledge by technical experts to social problems. In technological utopia the bureaucracies are integral parts of nearly every segment of society and, like everything else, function smoothly and efficiently. The most important bureaucracy, the management hierarchy of the industrial army, affects the lives of all citizens but for that very reason is widely respected and esteemed.

Certainly there are parallels between the roles of bureaucracies in technological utopia and real world America. As we noticed frequently in chapter 6, various sectors of American society in the late nineteenth and early twentieth centuries sought bureaucracies along with the technical knowledge and the technical experts needed to operate them. If these bureaucracies were usually less centralized and less comprehensive than their technological utopian counterparts, they were nevertheless seen by their creators and promoters as crucial to the maturation of an industrialized modern America. Whatever the motives of those creators and promoters may have been, they generally assumed that their bureaucracies would function smoothly and efficiently—more so than would other forms of public and private administration.

Here, too, however, the parallels are inexact. In the real world bureaucracy has come to be viewed by many as characteristically conservative, unimaginative, and, most damning, inefficient. Far from being praised as progressive, it has increasingly been attacked as reactionary. Victor Ferkiss has rightly observed, "It is hard for many people today to recognize that bureaucracy came into the world—the Western world at least—as a liberating force, as did its concomitant, administrative centralization."[23] True, as Ferkiss notes, the recent revolution in electronics and information processing allows for decreasing the dependence on bureaucracy or, alternately, increasing the mobility of its members and patrons. This development may lessen the role of bureaucracy or improve its performance or both. Yet the common stereotype of bureaucracy, like most stereotypes, contains a core of truth that cannot lightly be dismissed. This negative perception of bureaucracy precisely illuminates the gap between the prediction of technological utopia and the reality in contemporary America.

What of the role of art in the real and ideal realms? Whether broadly or narrowly conceived, art plays a quite modest role in technological utopia. Technological utopia thus has no parallel with the actual growth in recent decades of artistic activities of all kinds, amateur and professional alike. Yet in the real world art has in many instances become a conservative force, just as in technological utopia. The state at various levels has increased its support of artistic efforts, but artists have consequently felt pressured to support the state and its values, just as the technological utopians described. Indeed, the appeal of art in technological utopia not just to an elite but to the entire population has its counterpart in the real world development of mass culture— itself made possible by technological advances in communications and transportation.

No less important, the artistic avant-garde has gradually ceased to shock, as even so antiradical an observer as Bell has conceded. Instead, with a few exceptions such as electronic music, the artistic avant-garde has been embraced by mainstream American culture and has become an integral part of it. The obsession with newness has led to uncritical acceptance of whatever is chic, regardless of its true worth. Avant-garde art has in large measure thus lost its traditional role of critic of existing society.[24]

Consequently, the issue of the role of artist in utopia, a controversy dating back to Plato, has taken a new form. Where Plato banished the poet, as representative artist, from the Republic, fearing that the poet's message and elitism would subvert the masses, the contemporary artist of whatever stripe has been brought ever closer to the center of real world technological society in hope of somehow legitimizing its progress toward utopia. Now, only dystopias like Aldous Huxley's *Brave New World* and Eugene Zamyatin's *We* fear artists; for only in their confines do artists' ideological and social stances still constitute a genuine threat to the dominant values. Yet insofar as such dystopias reflect real trends moving toward fulfillment, the conflict between the artist and utopia persists.[25]

Finally, in the case of personality, the total conformity, rigid self-control, and nearly complete suppression of emotion that characterize technological utopia also have parallels in real world technological society. Citizens of both societies emulate machines and structures in varying degrees. Again, the parallels are not exact and should not be exaggerated. Here as elsewhere, the satisfactions that the inhabitants of utopia find in a particular way of life have not necessarily been duplicated in the real world. Any number of social critics have sought ways of using America's present or potential affluence to create a society freer, less repressed, and more individualistic than that which presently exists.

The best known of these critics, Herbert Marcuse, has revised Freud's association of repression with civilization. Marcuse has associated repression with only the earlier stages of civilization rather than, as did Freud, all its stages. Specifically, he has associated repression with civilization before the modern industrial revolution, when the continuous scarcity of food, clothing, and shelter were said to require almost continuous repression of the desire for pleasure and gratification. Otherwise, civilization could not have survived. If, as Marcuse optimistically assumes, modern America has solved the problem of scarcity, its citizens need not and should not remain repressed. Some repression, he concedes, must remain in order for civilization to persist. This he calls "basic" repression. But "surplus" repression, he argues, should be abandoned.[26] Marcuse, then, like other contemporary social critics, would fault American society for failing to realize the greatest possibility of modern technology: saving people from having to live and work like machinery. Instead, technology has too often been taken as a model and people have been seen—and treated—as parts of the machinery. This criticism clearly applies to that most striking characteristic of technological utopian (and, say its critics, modern Western) society: that technology, which offers to liberate mankind from work, has in practice just as often prompted them to work all the harder.[27]

Because of these unanticipated outcomes—more precisely, para-doxes—of actual technological advances in modern times, many have sought new formulations of the complex relationship between tech-nological progress and social progress. Philosopher Nicholas Rescher has examined the findings of leading pollsters and concluded that Americans and, by extension, citizens of comparably industrialized societies generally find technological advances insufficient in them-selves to constitute perceived increases in personal happiness. By and large people do not oppose advances that enhance domestic comforts or contribute to the national welfare. But they simultane-ously yearn for the "good old days," when, despite these advances, life was supposedly happier; they also simultaneously desire ever more such advances to meet the ever rising expectations created by the most recent advances. Hence a terribly complicated, even contra-dictory, set of assumptions colors conventional Western attitudes toward technological progress and social progress.[28] What most con-cerns Rescher, however, is the growing tendency to blame technolo-gy—and science—rather than ourselves for not producing the satis-fying increase in personal happiness that they were supposed to produce. This concern is surely legitimate if, I suggest, exaggerated by him and others, such as Bell. The deeper problem, I submit, is deter-mining what to do next, once it is agreed that "Science and technology cannot deliver on the $64,000 question of human satisfaction . . . because, in the final analysis, they simply do not furnish the stuff of which real happiness is made.''[29] This leads us to that prospect of a technological plateau—and, to the last of the utopian works in this study, Harold Loeb's *Life in a Technocracy: What It Might Be Like.*

Eight

The Prospect of a Technological Plateau

Life in a Technocracy appeared in January, 1933, just before Loeb and Howard Scott parted over differences about the future of Technocracy and of Scott's leadership. The book had been written three years earlier, however, when no one would publish it. Whether no one would publish it because, as Loeb claimed, its ideas were not yet in vogue, or because, as Scott charged, the Technocracy movement would not approve its ideas, the book offered no overt disagreement with Scott's leadership—whatever Loeb's private feelings may have been.[1] Indeed, Loeb's foreword thanks Scott, and Scott alone, for inspiration through discussions held in 1919 and in 1930 and for assistance on the technical parts of the argument, which largely elaborate the Technocratic notions that we have considered in chapter 6 above and which need not be repeated here. Where, according to Loeb, "most reformers are concerned in the first place with human and humanitarian values and thereby confuse their practical measures with emotional considerations," Technocracy shrewdly treats "the production and distribution of wealth as an engineering problem."[2] As a result, Technocracy is prepared to treat the consequences of its own adoption by American society, while the misplaced priorities of most other reform crusades prevent them from ever being ready to assume actual responsibilities. The consequences for American society of the possible adoption of Technocracy are

Portions of this chapter first appeared in my article "Reconsideration: Harold Loeb's *Life in a Technocracy: What It Might Be Like* (1933)." *New Republic* 175 (October 30, 1976): 42–44, copyright 1976 The New Republic, Inc., and are reprinted by permission.

what interest Loeb most (and Scott least). *Life in a Technocracy* presumes that Technocracy is a movement primarily for economic change, and, as such, is quite practical and capable of implementation in the near future—provided that the majority of citizens, with or without the consent of their elected representatives, wanted it implemented. Loeb's book further presumes that the material abundance for all to which Scott vaguely alluded could become a reality. Like the earlier technological utopian writings, the book goes on to treat the non-economic realms affected by unprecedented universal affluence.

Because so much of *Life in a Technocracy* was described in chapter 2 as part of the composite portrait of technological utopianism, we need not consider here its vision of the physical dimensions of technological utopia, of the industrial army, of the government, and of the laws and customs. I want, however, to elaborate on the related realms of work and of leisure.

As we saw detailed in chapter 2 and again in the preceding chapter, technological utopianism emphasizes and even glorifies work. As one of the utopians, Byron Brooks, put it succinctly, "This is the age of work."[3] Other utopians even equate work with play and thus with pleasure. To be sure, hard manual work has been drastically reduced or virtually eliminated by the various new machines and structures. Indeed, the number of hours required of able adult citizens for work of any kind has been markedly reduced. The type of work performed by the vast majority of citizens is no more taxing intellectually than it is physically, because of the pervasive new automated equipment found on the assembly lines but not limited to them. Loeb himself states, "Though releasing no one from a certain minimum effort which provides for the satisfaction of physical wants, a technocracy [technological utopia] would require so small a proportion of the total of human time and energy that everyone would have ample scope to pursue any outside activity which he fancied."[4]

Although such automated equipment would presumably offer a life of ease to the inhabitants of technological utopia, they instead prefer to devote their considerable leisure time to self-improvement, whether cultural, physical, or professional. There is no relaxation, no "wasted" time. Their outside activities are invariably practical. These activities might themselves be said to constitute work. It is, however, in equating only some of these leisure activities with the practical that Loeb deviates subtly but significantly from his utopian predecessors.

Freed from "that preoccupation with economic security which has always weighted the soul of man except on a few tropical islands,"[5] the inhabitants of technological utopia would finally be able to devote their principal energies to other, higher pursuits: in Loeb's listing, education, recreation, religion, and, above all, art. Those who wished to work virtually for the sake of work could still do so; and Loeb would expect all citizens to spend at least their required working hours either operating, supervising, or conducting research on machinery. But for most citizens, their longer leisure hours would be their truly creative ones, and Loeb devotes nearly half his book to the encouragement of creative leisure pursuits. In this respect his work is unique among the technological utopian writings studied here.

Education, religion, recreation, and art could thus flourish under a technological utopian order that liberated them all from capitalist exploitation. Yet no citizen would be compelled to participate in any one or group of these activities. The only universal requirement would be an elementary education, primarily for the purposes of imparting literacy and of insuring socialization. Just because Loeb's technocracy "is not concerned with moral and other values not material," it would allow all but "definitely anti-social" religious beliefs and practices.[6] And just because "cultural schools" are "of no concern to this purely productive state," it would directly sponsor only "industrial schools" for the majority of youth needing technical training for eventual entry into one of the industrial army's ninety-two departments.[7] Citizens who wished to establish cultural schools would, however, receive ample indirect government assistance. In the industrial schools, the aim would be "knowledge for the sake of use, not knowledge for the sake of the student's satisfaction"; in the cultural schools, the objective would be the reverse: "knowledge for its own sake."[8] Similarly, because the "control of anything the appeal of which is subjective, such as the theater, or painting, should not be entrusted to the state," all artistic activities except the actual maintenance of museums, theaters, and movie houses would also be entrusted to the citizens.[9] The citizens would nevertheless again receive ample indirect assistance—including, when warranted, the production and distribution of their artistic and literary works. Only insofar as organized sports required impartial arbitration would the government supervise them; other forms of recreation would also be left to the citizens themselves.

Loeb anticipates that art, conceived broadly as painting, music, literature, theater, architecture, and handicrafts, "will probably become in a technocracy the most important field of human activity."[10] Art, once cleansed of capitalist corruption, could, ideally, "improve the lot of man on earth as emphatically in the inner psychic sphere as man's genius, directed toward conquering the outer material world, has ameliorated the conditions of his physical existence."[11] Art could restore to artists and perhaps they in turn to other citizens "that sense of imminent wonder . . . which all live things are heir to" but which—

in the process of the "rationalization" of the modern world perceived by Max Weber—has been lost in the conquest of nature and the "domestication" of man.[12] That special sense, which Loeb admits is also evident in the magic, voodooism, and ecstasies of other cultures, could gradually be coupled with voluntary and diverse forms of genetic engineering to produce "a race of man superior in quality to any now known on earth, a society more exciting, interesting, and variegated than has ever been possible, and a nation in which no individual should be unhappy or discontented for remediable causes."[13]

We need not accept Loeb's blueprint for technocracy, either in part or in whole, in order to recognize that he has surely provided a technological plateau. Having assumed that technological progress in general and Technocracy in particular could create unprecedented and permanent material abundance for all Americans, he has proposed the subsequent—or, to an extent, simultaneous—achievement of equivalent social progress. And he has defined the latter broadly enough to encompass whatever activities would prove fulfilling to the inhabitants of his utopia: "In a technocracy, the pressure, to conform to every passing mode now exerted by advertising, would be definitely lifted. As a result the citizens may find it easier to be themselves." Yet, he cautions, his ideal society could not "guarantee that individuals express themselves."[14] It could only permit and encourage them to do so.

Far from making these two modes of progress antithetical, as do many other proponents of change, Loeb sees them as complementary. The technological advances would lay the groundwork for the social advances, and the social advances that then would come about would transform the society from a pure technocracy to an authentic utopia—or so Loeb expects. As he observes about the heterogeneity of work and leisure pursuits in each community, technicians and artists alike would always be "aware of the great plant[s] in their midst by serving which they acquired such a superfluity of the good things of life."[15] The plant, of course, symbolizes the technology, which, in Loeb's vision, would be respected by citizens but not worshipped.

Whether or not we accept this scheme, we surely must concede that it is more sophisticated than its predecessors. Even its genetic and social engineering, while hardly universally appealing, are presented in less chilling a fashion than in previous works. Perhaps because he was more cosmopolitan then the other twenty-four visionaries—or perhaps because he alone of them had experienced the dashed dreams of the First World War and the Great Depression before composing his vision—Loeb thought more deeply than the rest about the social and cultural consequences of actually achieving a technological utopia in

America. Whatever the reasons for Loeb's subtle deviation from the earlier technological prophets—or, for that matter, his graphic deviation from Scott—his deviation in these respects is itself the critical issue.[16]

Loeb retreated from utopianism as early as the mid-1930s, partly out of disillusionment over the failure of his Continental Committee on Technocracy and with the (temporary) triumph of Scott's Technocracy Inc., but partly too out of a new optimism that the existing American technology and the existing American economic, social, and even political systems could produce the universal material abundance that were the necessary basis for any genuinely good society.[17] *Life in a Technocracy* remains an undeservedly neglected work which, along with the other technological utopian writings studied here, might still modestly illuminate the directions of contemporary technological society. It speaks, then, to the real present as well as to the imagined future.

The notion of a technological plateau did not, of course, originate with Loeb. Its origins are obscure, but historian Jean Gimpel has recently applied the term to medieval Western Europe, in conjunction with the aftermath of the world's first industrial revolution. In a pioneering study, Gimpel has demonstrated that the first such revolution began in Western Europe in the tenth century rather than, as commonly assumed, in England in the eighteenth century. The whole of Western Europe in the Middle Ages, he shows, was nearly as technological a society as was England alone in what should now be called the second industrial revolution. Ordinary Europeans were surrounded by machines and structures in their everyday lives: from water- or wind-powered mills to iron plows and harnesses to mechanical clocks to Gothic cathedrals. However stagnant the Middle Ages may have been in other respects, argues Gimpel, they were definitely not dark ages technologically.[18]

Gimpel attributes the rise and decline of this first industrial revolution to a variety of factors: to changes in temperature and rainfall, in agricultural output, in public health, in working class efficiency, in international relations, in the support of the Roman Catholic church, and, above all, in the psychological drive of the general populace. The specific changes need not be detailed here. Gimpel's crucial point is that medieval Western Europe had by about 1375 reached a technological plateau—a condition that Gimpel, unlike Loeb in his modern context, laments. Gimpel, much like the technological utopians aside from Loeb, believes firmly in technological advance as the principal solution to problems and sees the restriction of technological advance as a principal problem in itself.

Where, however, the technological plateau of the late thirteenth and early fourteenth centuries was followed four centuries later by the second industrial revolution in England, Gimpel fears that the plateau which he discerns in the modern West, including North America, will become a permanent condition. The cycle that he sees as Western history will cease: "We can anticipate centuries of decline and exhaustion. There will be no further industrial revolution in the cycles of our Western civilization."[19] Presumably the so-called third industrial revolution, that of computers and other electronic information processing devices, will not be sufficient to halt this predicted decline.[20]

Ironically, the example which Gimpel singles out as symbolic of "the entry of [particularly] the United States into her aging or declining era"—the refusal of the Congress in 1971 to allocate funds for the supersonic transport (the SST)—has been viewed by others as a hopeful sign.[21] It has been praised as indicating a new, if hardly pervasive, skepticism about the technological imperative: that is, the assumption, common in America, that whatever developments are technically feasible ought for just that reason be funded and produced, regardless of the economic and social costs.[22] Gimpel's insistence on the technological imperative might ultimately render that part of his study as obsolete as parts of the similarly inclined visions of the technological utopians (and the Technocrats and industrial designers of the 1930s). Indeed, any persistent disavowal of the technological imperative in the contemporary West might well disprove Gimpel's unsubstantiated assertions that Western history has been cyclical and that its cycles are now over.

Yet Gimpel's notion of a medieval technological plateau, however much he is against it, remains a valuable one. We have seen other expressions of that notion cropping up in the early stages of the second industrial revolution in reaction to its actual or projected excesses: the positive proposals for limits to technological advance offered by Owen, Fourier, and Marx and Engels, as described in chapter 4 above. None used the term explicitly, as does Gimpel, but all provided a glimpse of such a stage of technological and social development. Unlike Gimpel, all saw hope rather than despair in such a prospect. They were not the only critics of unadulterated technological advance of the late eighteenth and nineteenth centuries, but they were among the few who proposed specific means of regulating and accommodating—as opposed to merely stopping—technological advance. We need not accept all or any of their proposals to appreciate this fact.

A related example of a technological plateau—indeed, a realized plateau—coming about between the first and the second industrial revolutions has recently been unearthed by Noel Perrin. The period was between 1637 and 1855, the site was Japan, and the instrument of change was the gun. When two Europeans with guns arrived in Japan in 1543, the country was familiar only with swords, spears, and bows and arrows. Not that the Japanese were then especially placid. Quite the opposite: the nation was strong, aggressive, and warlike; this was the time of the samurai warriors. But warfare was intensely heroic in nature and was dominated by single combat. Far from resisting guns, however, the Japanese, including the samurai, readily embraced, improved, and utilized them. But the price paid for this technological progress was, at least for the ruling elite, social and cultural regress. Samurai hegemony declined, thanks to the ability of the lower orders to defeat their social superiors through expert marksmanship, and so did the culture of traditional combat. Finally, through wonderfully subtle and gradual means, the Japanese gave up their guns and, after 1637, reverted to their previous weapons. The ban on guns lasted for more than two centuries, until 1855, when U.S. Commodore Matthew Perry's "opening" of Japan returned firearms to the country, where they have remained ever since. Significantly, the Japanese continued existing technological developments and embarked upon others—all more benign than guns—while phasing out guns. There were advances in areas as diverse as silk production and water-powered crushing mills. Interestingly, gunpowder was retained for use in blasting loose ore in deep mines. Japanese restrictions on technological advances were therefore consciously selective. Here, then, was an actual technological plateau.[23]

Owen, Fourier, and even Marx and Engels wrote at just the initial and middle stages of industrialization in the West; the automated production processes envisioned by Marx and Engels were more fantasy than fact. And the samurai discussed by Perrin were dealing with the comparable stage of Japanese industrialization. Yet between the late nineteenth and mid-twentieth centuries, when technological society truly came into being in the West, there were no calls for reaching a technological plateau. If anything, there was greater polarity than ever regarding the proper role of technology in society; its abundant admirers saw no end to it, and its detractors wanted it sharply reduced.[24] Perhaps people needed more distance from the immediacy of technological advance before it could be seen as a chosen means to chosen goals.

A notable exception to the uncritical acceptance or denunciation of technology was Simone Weil (1909–1943), the French philosopher, mystic, and social critic. Weil's growing fame and reputation stem primarily from her philosophical and theological writings. Her reflections on the nature of work and of workers in modern industrialized societies extend the insights of Owen, Fourier, Marx, and Engels and anticipate those of more recent critics whom we will consider next.

Raised in a highly cultured upper middle class environment, Weil had little early contact with industrial work or workers. From December 1934 until August 1935, however, she held jobs in a series of highly automated factories in Paris to experience firsthand the lot of the working class. These jobs left her physically and emotionally exhausted, and she was forced to abandon each of them after only a few weeks. She nevertheless wrote perceptively, if briefly and fragmentarily, about the prospects for improving industrial work and workers and, in the process, implicitly advocated a technological plateau.[25]

Weil intensely disliked manual labor, especially on an assembly line. Such work, she declared, "is like a death."[26] She was, of course, not the first critic of such work to liken an ordinary factory worker to a slave. But she accepted, as many others did not, the necessity of work in general and of industrial work in particular for the twentieth-century West. And she saw in the fatigue of honest toil the "spirituality of work" and the "joys parallel to fatigue," particularly eating and resting.[27] Moreover, she perceived, as did few of her contemporaries, that management and bureaucracy were no less integral than technology to work in the twentieth century—and no less accountable than the equipment itself for the exploitation of the workers. "For the bureaucratic machine," she wrote, is "as irresponsible and as soulless as are machines made of iron and steel."[28] Her hopes for reform of industrial work of all kinds, manual and managerial alike, rested on making work less hurried, less specialized, less monotonous, less humiliating, more dignified, and more fulfilling.

As the non-Marxist Weil argued in perhaps her most famous book, *The Need for Roots* (published posthumously in 1949), "the abolition of the proletarian lot, chiefly characterized by uprootedness, depends upon the creation of forms of industrial production and culture of the mind in which workmen can be, and be made to feel themselves to be, at home."[29] By this Weil meant that, to invoke Marx and Engels more directly, the industrial workers' alienation had finally to end: alienation, that is, from their labor, from the products and the machines and structures of their labor, from their families, and from themselves. But Weil had less faith than did Marx and Engels that such problems as the excessive specialization of work could be solved through new technology. She did not oppose technology itself, including new technology. Like work, the beginning forms of technology were bound up with survival in the natural world and so were part of the realm of necessity, as Weil also called it. In its later forms, technology could lessen the physical and emotional burdens of industrial work if properly utilized—as it was not in most of the factories in which she had worked. Somewhat like Marx and Engels, Weil looked to "adjustable automatic machine[s] with a variety of uses" to help achieve the proper use of technology.[30] This equipment would perform the mindlessly repetitive functions heretofore performed by men and women. Such machinery alone, however, would not be enough to overcome the several forms of alienation.

As a starting point for additional, profounder changes, Weil proposed that industrial workers systematically learn how their particular machinery operated, how their particular tasks fitted into the overall production process of their factories, and how mechanization shaped their particular work routines. She then proposed that large factories be abolished and be replaced by small workshops owned and operated by a single worker. Such workshops would employ the modern automated equipment she endorsed. Ideally, their owners would disperse throughout the country to demonstrate to others the virtues of this new technology. She noted, "the relative ease with which energy can be transmitted in the form of electricity certainly makes a high degree of decentralization possible."[31] Eventually, individual workshop owners would either join with others in their region in one cooperative organization or base their activities in their homes. In any case, "Such workshops would not be small factories, they would be industrial organisms of a new kind, in which a new spirit could blow; though small, they would be bound together by organic ties strong enough to enable them to form as a whole a large concern."[32]

Weil hoped that the majority of industrial workers would be able to own and operate such workshops. But, reflecting on her personal inexperience before working in factories, she insisted that all prospective industrial workers receive a comprehensive vocational—and academic—education, including an apprenticeship, and pass rigorous examinations in both areas before being certified to acquire a workshop. With each workshop, she proposed, would come ownership of a home for the worker and his family and a plot of land surrounding it on which they could grow much of their food. In order for the workers to keep up with subsequent technological developments, they would periodically attend a new university for workers. There they could also acquire training to operate new equipment for the purpose of varying their work routines. This institution would likewise engage in technical research to benefit workers as well as businessmen and consumers. No social or cultural barriers would separate workers from technicians or managers in Weil's scheme.

To bring industrial workers and their families closer together, Weil proposed that the children and spouses of workers visit their parents'

and spouses' workplaces and learn more about the nature and importance of the activities there. This would be coupled with a reduction
in the traditional working hours in industry—thanks to technological
advances—that would further promote the reintegration of family and
working life. Finally, the reduction both in the hours and in the pace
of work would allow workers opportunities for contemplation—if not
in the classical sense described earlier, at least as more than an escape
from work. For they would be combining activities, contemplation
and work, that traditionally had been accorded the highest and the
lowest prestige. Here was another reflection of Weil's concept of the
"spirituality of work." Indeed, she proclaimed that "Our age has its
own particular mission, or vocation—the creation of a civilization
founded upon the spiritual nature of work."[33]

In all these respects Weil sought an implicit technological plateau.
She hoped to utilize modern technology to create, in effect, the idealized state of the artisan in pre-industrial times. That she, like Owen
and Fourier, among others, romanticized the artisans' lot—particularly their independence—is less important than that, unlike those
other theorists, she based her new communities more on industry
than on agriculture. This brought her closer to Marx and Engels, who
also did not seek a return to a mythic pre-industrial past. Yet Weil,
although she would have used government to bring about many of the
changes she sought, did not favor socialism over capitalism. Instead,
characteristically, she rejected the "conventional wisdom" and advocated a mixture of the two. She perceived, correctly, that socialism
could be as bureaucratic and as oppressive as capitalism. "Let us not
forget," she concluded, "that we want to make the individual, and not
the collectivity, the supreme value."[34]

In the past quarter century, as part of the growing criticism of unrestrained technological progress, other versions of implicit technological plateaus have appeared. Among the most prominent have been
those of Marcuse, Norman O. Brown, Paul Goodman, and other exponents of social and cultural, especially sexual, liberation.[35] Marcuse's
scheme has already been outlined in these pages. It is enough here to
observe that in recent years the problem of scarcity, even in supposedly affluent America, has surfaced again and can no longer be
assumed to have been virtually solved—as Marcuse and the others
assumed when they wrote in the 1950s and 1960s. Like the plans of
Bell and his fellow prophets of technocratic post-industrial society,
with whom Marcuse and his fellow radicals otherwise had little in
common, their schemes have been undone, for now anyway, by
unanticipated changes in the real world: by "limits to growth" of
various kinds.

Perceptions of these limits have in part prompted more overt expressions of the technological plateau: namely, the calls for smaller scale, less impersonal, more decentralized technology. The foremost prophet of this kind of technological plateau, E. F. Schumacher, was hardly antitechnological. Rather, he, his associates, and others throughout the world—not simply in the industrialized West—have sought more "appropriate" or "intermediate" forms of technology to fit different societies.[36] Much of their effort has been concentrated in less affluent, less industrialized countries, but they have not been indifferent to the concerns summarized here regarding technological advance in the industrialized West. Their often technical writings and, even more, their real world demonstration projects have complemented the philosophical reflections of Lewis Mumford, René Dubos, John Passmore, and the few other contemporary exponents of a technological plateau.[37]

I see a clear connection between these practical and philosophical proposals for a technological plateau and the Western utopian tradition. Mumford himself is a key bridge figure, and it is not accidental that the first book by this distinguished critic—and defender—of both cities and technology was *The Story of Utopias*, a critique and a defense of utopianism. Nor is it surprising that the book sees decentralization as a means of balancing stability and change. Long before Mumford attacked the modern "megamachine," he had here shown that any future utopias—or any genuinely good societies—could use technological advance to preserve and improve local communities and that technological advance need not homogenize the United States or other nations in the name of progress. The focus on local community did not entail isolation or parochialism; modern transportation and communications networks allowed mobility and exchange of persons and ideas, but not at the expense of different local (and national) cultures. As Mumford put it,

> If the inhabitants of our Eutopias [that is, authentic utopias] will conduct their daily affairs in a possibly more limited environment than that of the great metropolitan centers, their mental environment will not be localized or nationalized. For the first time perhaps in the history of the planet our advance in science and invention has made it possible for every age and every community to contribute to the spiritual heritage of the local group. . . . Our eutopians will necessarily draw from this wider environment whatever can be assimilated by the local community; and they will thus add any elements that may be lacking in the natural situation. The chief business of eutopians was summed up by Voltaire in the final injunction of Candide: Let us cultivate our garden.[38]

Later visionaries have drawn tighter connections between decentralization and utopianism. Ernest Callenbach's novel *Ecotopia*, for example, depicts the 1980 secession from the United States of the western portions of Oregon and Washington and California north of Santa Barbara to form an independent, environmentally conscious society. Most of the story, however, takes place in 1999. It is told through the dispatches and the diary of William Weston, a New York newspaper correspondent and the first official visitor ever admitted to this land of mystery. Automobiles are electric powered, plastics are biodegradable, and sewage is converted into fertilizer. Complementing these developments are virtual equality of the sexes, a female-dominated government, and ritual war games among men. The ethos of decentralization shapes institutions and activities alike. Eventually Weston, the initially skeptical reporter, decides to remain in Ecotopia. He records in his diary, "The Ecotopians do not feel 'separate' from their technology. They evidently feel a little as the Indians must have felt: that the horse and the teepee and the bow and arrow all sprang, like the human being, from the womb of nature, organically."[39] Ecotopia is a full-fledged technological plateau.

Gerard O'Neill's *The High Frontier: Human Colonies in Space* posits no technological plateau but does explicitly link decentralization—in the form of space colonies—with unlimited technological advance. A hardnosed technical work as well as a passionately romantic one, *The High Frontier* convincingly shows how space colonies could be established in the very near future and how life might be carried on within them. Callenbach, no doubt, would reject O'Neill's version of the good society, and vice versa, but the colonies seem not nightmarish, authoritarian outposts as much as pleasant, fairly free, quite large suburbs circling the earth.[40] If anything, the society portrayed here—and elaborated upon in O'Neill's *2081: A Hopeful View of the Human Future*—is so harmonious and so earnest that it seems a bit dull, in the manner of *Looking Backward* and the other technological utopias of the period 1883–1933. To his credit, O'Neill makes no claims that his societies witness the literal perfection of the species and admits that human beings are by nature flawed; his colonies offer opportunities to pursue happiness but do not guarantee it.[41] Still, I wonder if the permanent inhabitants of his colonies wouldn't find life boring and confining once the thrill of being in space had faded. The same question might apply to the inhabitants of Buckminster Fuller's proposed communities: his land-based cities covered by geodesic domes, his tetrahedronal cities floating on the sea, or his cloud-structure spheres floating in air. All three are related alternatives to existing technological society. It might be, of course, that the inhabitants of these circumscribed societies would find life in Ecotopia unappealing and dull.

Whatever their appeal, *Ecotopia*, *The High Frontier*, and *2081* stress the decentralization allowed by present and foreseeable levels of technology. So do other current visions of the future that rely on computers, word processors, satellites, and the like. Marshall McLuhan's vision of a "global village" is steadily taking shape, as various electronic marvels provide instantaneous international communications, allowing their users to work at widely dispersed offices or even in their homes. The global village is hardly utopia, indeed is hardly a village at all, but it is an interesting stage in the evolution of technological society.[42] The technical developments that make the global village possible could, rather than further McLuhan's dream of unceasing technological progress, help to bring about a genuine technological plateau.

Decentralized technology might, then, offer a practical means of achieving an actual technological plateau in the United States and elsewhere in certain situations. On a more mundane level, the growing preference in the past decade for smaller and more fuel efficient automobiles and for solar, wind, water, and even coal energy sources as opposed to oil and natural gas are examples of appropriate technology at work—if only for economic reasons.[43] So, too, are the increasing demands for improvements in mass production processes, particularly alternatives to the assembly line—and here not primarily for economic reasons.[44] On a broader scale, the proliferation of technological assessment and environmental impact studies is a still clearer example of this trend.[45] These examples show at least a more sophisticated and more balanced evaluation by increasing numbers of citizens—not only technical experts—of the complex relationship between technological progress and social progress. Such proposals avoid the shallow extremes of being for or against technology that have boxed in most discussions of technology since the second industrial revolution. So terribly complex a phenomenon can hardly be dealt with by such a stark choice.[46]

Unfortunately, those very extremes reappear in a number of recent works about America's and the world's future. The boundless optimism of a Buckminster Fuller and the stark pessimism of a Jacques Ellul are repeated not only in such writings as those, respectively, of McLuhan and Mumford (in his later book on the "megamachine") but also in the more technical treatises of, say, the upbeat Herman Kahn and the downbeat Club of Rome. The common assumption that experts will ultimately agree on basic issues once the facts are established is readily undermined by a reading of these supposedly objective works. The differences in outlook are little short of cosmic. The facts fail to speak for themselves, and those facts, let alone their application to future trends, are subject to heated, prolonged, and usually unresolved dispute. A comparison of Kahn's *The Next 200*

Years: A Scenario for America and the World and his *The Coming Boom: Economic, Political, and Social* with the Club of Rome's *The Limits to Growth*, or of Julian Simon's *The Ultimate Resource* with *The Global 2000 Report* to President Carter, amply confirms this. Future population growth, food supply, energy sources, and environmental conditions are among the principal areas of contention.[47]

To be sure, even the most technocratic of these visionaries, such as Kahn and Simon, qualify their findings by avoiding old-fashioned specific predictions. However plentiful their statistics and charts, the prophecies are invariably treated as scenarios, as one of several possible outcomes—lest, I suspect, things someday go awry. Often experts in the field are queried and requeried for their forecasts by the so-called Delphi method in order to achieve a consensus—not merely for the sake of truth but also perhaps to provide additional security in case of error. Even computer-based models are frequently treated as symbols, not predictions, of future developments. The most disturbing recent development in this area has been the wholesale reversal of the Club of Rome's pessimistic findings in *The Limits to Growth* just four years later with barely an apology—indeed, with a rather curious explanation instead.[48]

Still, a supreme faith in the anticipated future, whether utopian or dystopian, pervades these efforts. Despite their qualifications, they are not the products of humble or hesitant souls. Like the technological utopian writings examined here, they are largely extrapolations from yesterday to today to tomorrow (usually the year 2000). The experts readily concede that many pure inventions and basic breakthroughs are simply not predictable, not yet anyway, and casually let that critical issue drop.

Like the earlier technological utopians, their successors show limited interest in the past, modest concern for the nontechnical dimensions of technological advance, skepticism about politics and other messy matters, and boundless confidence in mankind's ability to remain master of both nature and technology. The distance between prophecy and fulfillment grows ever narrower in their works. Their critics, such as the Club of Rome and President Carter's advisers, are barely more sensitive to the complexities of the past, the present, and the future. Mumford is exceptional among these contemporary prophets in his knowledge of the past, his concern for the social and cultural consequences of technological change, and his reluctance to prescribe a particular utopian—or dystopian—scheme.[49] Despite his own pessimism in the 1960s and 1970s about technology's evolution, he, alone among the prophets (except for Callenbach), offers a semblance of a technological plateau.

Owen, Fourier, Marx, Engels, Weil, and Loeb are notable exceptions to this tendency to equip technology with only an off-on switch, and that is why they bear rereading in this context today. None, however, proved as "scientific" a prophet as each hoped to be—to invoke Marx and Engels' term of self-praise—in contrast to the pseudo-scientific "utopian" Owen, Fourier, and Saint-Simon.[50] Nevertheless, Marx, Engels, Weil, Loeb, and Mumford well understood, as Owen, Fourier, Saint-Simon, and many other avowed utopians—including the other technological utopians—have not, that any serious utopian vision functions properly not as a literal blueprint for the future but as a take-off point for reconsidering and possibly altering existing society. That is why we need not endorse, much less realize, every (or even any) aspect of utopian designs as multifaceted as those of Owen, Fourier, Saint-Simon, Marx, Engels, Weil, Loeb, and the other technological utopians in order to derive some value from them. Not only do all of these schemes reveal more about their own times and places than they do about ours; but the inevitable gaps in each between prophecy and fulfillment may be more illuminating than their modestly accurate predictions of the present.

What historian John F. C. Harrison, the foremost living authority on Owen, has written about that visionary applies equally to Fourier, to Saint-Simon, to Marx and Engels, to Weil, to Loeb, to the other technological utopians (insofar as they were unorthodox), and to all other utopians praised for being "ahead of their time":

> ... too many historians [and others] have had recourse to this substitute for historical analysis. The phrase begs the whole question of historical interpretation. No man is literally born out of his time; he is born in a certain period and the historian's job is to relate him to that time. ... it is [thus] necessary to reject rather facile attempts to show that Owen had a solution to our problems. The early nineteenth century was vastly different from our world and it is therefore an anachronism to suggest that Robert Owen's teachings are directly applicable to the twentieth century. ... his life and work are significant because they help us to understand the world of the early nineteenth century ... which is the direct ancestor of our present society.[51]

Hence, if a technological plateau is desirable, we need contemporary formulations of it, not uncritical reliance on previous ones, even as modern a scheme as Loeb's *Life in a Technocracy*.

The value of those earlier formulations as reflections of their own societies, however, should not be minimized. Like all other serious utopian visions, they play a vital role as vehicles of social criticism and, sometimes, of actual social change. The explication and defense of that role, particularly as it relates to contemporary technological society, remains our concern.

Nine

Appropriate Visions
The Uses of Utopianism Today

What, we ultimately must ask, is the value to contemporary technological society of the utopian visions scrutinized in this study? The
question is not new. Countless objections to utopianism in any serious form have been raised since the appearance of Plato's *Republic*
and other early visionary schemes. Utopianism has been persistently
criticized as impractical, immoral, deviant, conformist, revolutionary, reactionary, stagnant, authoritarian, and libertarian, among other
things. And utopianism has persistently found defenders against
these and other objections. At this point, an exhaustive list of these
charges and countercharges would serve no beneficial purpose.[1]
Rather, I want to deal here with the general role of utopianism today,
when, because of changes in the real world in the present century, the
grounds for criticism and defense have shifted markedly from previous times. The following arguments readily apply to technological
utopianism.

Two main objections have been raised to any serious forms of
utopianism in the twentieth century and beyond. First, insofar as
realization of utopia relies on an underlying assumption of the perfectibility of mankind, it is impractical. The technologically assisted
horrors of the present age—its world wars, its genocides, its nuclear
threats—make any hope of major improvements in human behavior
seem farfetched. The second objection is related. Insofar as technological progress has at last made some varieties of utopia realizable, they

Portions of this chapter first appeared in my article "Appropriate Visions: In
Defense of Utopianism Today," *World Future Society Bulletin* 18 (March–April 1984):
24–29, and are reprinted by permission.

are not desirable because of the human imperfections just mentioned. Given the human propensity to selfish and exploitative behavior, achievement of the sort of planning and control required by utopia would result in dystopia.

Nevertheless, I wish to defend the contemporary usefulness, indeed necessity, of serious utopianism, particularly in written form. Utopianism is a legitimate—and vital—means of offering constructive criticism of existing society in order to improve that society. But the criticism should be appropriate to the particular society being scrutinized. It should take a middle ground between the utter impracticality of utopianism as a dream of perfection and the practical endorsement of an existing society either as the best of all possible worlds or as an outright utopia as it stands. Criticism should mediate between naiveté and cynicism.

To be effective as social criticism, a utopian vision should be concrete enough to be applicable to the real world; and it should be detached enough to be truly critical. To be applicable to the real world does not mean to mirror reality, just to be relevant to it. Far from being escapist, utopianism in this sense is intended to be played back upon the real world in order to try to change it: it offers a vision of how the real world could be made more nearly perfect. Those who create appropriate visions, such as the technological utopians of this study, ought be neither dismissed as mere crackpots nor reduced to pale reflections of more conventional contemporary reformers. Rather, they should be recognized exactly as serious-minded utopians.

These distinctions do not rule out either genuinely radical or genuinely conservative utopian schemes (or utopians). They do not preclude utopias like Plato's *Republic*, which have virtually no possibility of realization because of the unbridgeable gap between the existing society and the utopian society.[2] Such utopias still offer a lesson in human nature which, despite its pessimism, can benefit the real world, if only as a warning against undue optimism. Nor do they preclude those utopias like the technological utopian works, which virtually guarantee some sort of realization because of an ever closer correspondence between existing society and utopian society. Such utopias offer a detailed route from the present to the future which, exactly because of its optimism, can provide a relief from excessive pessimism about existing society even if the route is never followed itself.

The prospect of bridging the actual and imagined worlds in this fashion does pose an intellectual problem for scholars who distinguish between earlier, "classical" utopias and recent, "modern" ones: utopias that were forlorn visions of the unattainable perfect society, such as the *Republic*, and those that are practical proposals for actually effecting perfection, such as *Looking Backward*. Such distinctions, however, are overdrawn. The *Republic* and More's *Utopia*—the two foremost classical utopias—were not absolutely self-condemned to nonrealization, and even the most optimistically realizable modern utopias—like the technological utopias—at least implicitly criticize existing society for not being more nearly utopian. To assert the moral superiority of the earlier utopias over the later ones, as these scholars do, primarily because the former had a dimmer and so more "realistic" view of human nature, is to ignore the other factors—technological advance above all—that largely account for the comparatively greater optimism of the more recent utopias.[3]

But if utopianism of whatever period is to fill the role of critic of existing society, it must do more than be both practical and detached. It must also offer an ideology: more precisely, it must offer an alternative to existing society's own ideology. In the absence of a prevailing ideology, it must be one of the rivals competing for ideological hegemony. Utopianism as alternative or rival ideology offers an understanding not just of how a social group already functions but also of how it ought to function. Utopianism criticizes in the name of a positive ideal: it at once reflects and seeks to improve the society that gave it birth. It describes a society better than existing society and attempts to stir thought and action that will close the gap between the two—despite the unlikelihood that perfection will actually be achieved.

Because full-fledged utopian visions are by their very nature more comprehensive and more explicit than any other written form of social criticism, they offer unique perspectives on the real and ideal realms. They identify problems in the real world that are not otherwise so clearly apparent, if visible at all. By taking to an extreme a proposal for how a social group ought to function, presenting it as a functioning society in an imaginary world, utopianism illuminates not only the intended improvements but also the underlying understanding of how the real world social group functions. People who respond to the utopian vision are forced to experience vicariously an alternative society and, in the process, to see existing society in a new light.[4] So technological utopianism, in its comparative conservatism, illuminates much of past and present American society. Hence the deeper significance of the extraordinary popularity of *Looking Backward*, reflecting its appearance in a cultural context ready to welcome its principal themes. A text has recently appeared that has readers write out their own utopias as a way of solving both personal and social problems.[5] This also assumes that a person's (or a culture's) ideals can illuminate the actual. Also based on this premise are current studies that match several historic utopian schemes with contemporary social science models.[6]

Utopianism and ideology are, then, both more closely connected with each other in function and more positive in purpose than commonly assumed. Even Karl Mannheim, whose classic *Ideology and Utopia* (1929) seemed to separate them absolutely, actually treated them as overlapping in function. Moreover, he treated both as illuminating as much as distorting reality. Most important, Mannheim advocated continuing the development of utopias as vehicles of social criticism. In language recalling Max Weber's gloomiest passages about "rationalization," Mannheim lamented the gradual "elimination of reality-transcending elements from our world," whose complete elimination "would lead us to a 'matter-of-factness' which [in turn] ultimately would mean the decay of the human will."[7]

Having redefined the connection between, first, utopianism and the real world, and, second, between utopianism and ideology, I want to redraw a third connection: the one between theory and practice as applied to utopianism as alternative ideology. We must see more clearly still that utopianism properly functions in existing society not as an actual blueprint for constructing an alternative society but as an intellectual device for rethinking aspects of existing society.

Marx and Engels grasped this function of utopianism particularly well. They criticized Saint-Simon, Owen, and Fourier as "utopian" socialists (they considered themselves to be "scientific" socialists) for devising literal blueprints and attempting to impose them on the future. In Engels' words, "These new social systems were foredoomed as Utopian; the more completely they were worked out in detail, the more they could not avoid drifting off into pure phantasies."[8] He and Marx were no doubt exaggerating these visionaries' vices and their own virtues, but their caution in this regard was exemplary. Despite their confidence in the scientific accuracy of their own predictions of the fall of capitalism and the rise of socialism, Marx and Engels were sufficiently wary of the unpredictable future not to specify the kind of society that would follow the expected triumph of the proletariat and the disappearance of the state.

By contrast, the technological utopians, among other modern visionaries, suffered no trepidations about predicting the future in detail. They limited their success as a result. Merely extending their present into the future, as the technological utopians did, allowed for none of the almost inevitable disjunctions between their day and ours.[9] Likewise, the attempt of other Americans to realize utopia immediately and directly by constructing small-scale communities also failed—as did the communitarian efforts of the followers of Owen and Fourier. The world had grown too complex for small communities to function as much more than escapes from larger world realities.

To say that the theoretical dimension of utopianism as alternative ideology demands that we avoid rigid guidelines for altering existing society or avoid making strict model communities, does not mean that we must avoid any application of theory, or praxis. But we must carefully redefine the relationship between theory and application. Sheldon Wolin, writing about political theory, has developed a notion of theoretical activity that can usefully be applied to utopian theory.[10]

According to Wolin, the outstanding political theorists from Plato's time to ours have successively contributed to a de facto "epic tradition" of political theory. They have each hoped to achieve "a great and memorable deed through the medium of thought." For them "the great deed is the great word." Even when, as with the *Republic*, great words for reasons such as imperfect humanity are unlikely to effect great deeds, the words themselves may endure as permanent testimony to their author's genius and goals. In this epic tradition, theorizing is not separate from practice but rather is itself a form of practice. The action that it seeks is the realization of theory: not, empirically, "to make the theory correspond to the world"; and not, philosophically, "to make the theory an elucidation of meanings extracted from the world"; but precisely "to make the world reflect a theory."[11] Theorizing in this sense represents a grand political gesture, whether or not the theory is realized.

Epic political theory, it should be added, invariably arises at a time of profound crisis in a society. Existing institutions and values, above all political ones, no longer work effectively or, in many cases, even make sense. Such a crisis "affords an opportunity for a theory to reorder the world" and does so by providing "symbolic representations of what society would be like if it could be reordered."[12] The common failure of the world to be reordered according to the theory does not preclude other, subtler and slower effects of epic political theory on the political and social imagination, as numerous examples from Plato through Marx testify. Even an unsuccessful or partially successful epic theoretical effort is nevertheless in itself an effort in the world.[13]

Utopian theorizing obviously is closely akin to political theorizing as just described. Utopian theorizing at its grandest also aims to achieve a great deed through great words, also combines rather than separates theory and practice, and also makes a grand gesture toward existing society. Moreover, utopian theorizing on this scale also arises at times of crisis in a society and likewise is a proposed solution to that crisis. It is not an accident that the ranks of great political and utopian theorists overlap in the cases of Plato, More, and Bacon. Utopian theorizing is, then, itself a form of utopian practice, whether or not the theory is realized. Certainly the *Republic, Utopia,* the *New Atlantis,* and, for that matter, *Looking Backward,* all have had considerable impact on the political and social imagination at various times. We see all the more that utopian theory is a potentially powerful alternative to the ideology of existing society.[14]

In addition to epic political theory, the "critical theory" of the so-called Frankfurt School offers another, more recent example of theoretical activity as outlined here.[15] Indeed, the leading practitioners of critical theory have spelled out more fully than Wolin the hoped-for function of theory: the negation of existing society's prevailing ideology, or "negative thinking." Max Horkheimer explains about this consciously dialectical process:

> The real social function of Philosophy [including critical theory] lies in its criticism of what is prevalent. That does not mean superficial fault-finding with individual ideas or conditions. . . . Nor does it mean that the philosopher complains about this or that isolated condition and suggests remedies. The chief aim of such criticism is to prevent mankind from losing itself in those ideas and activities which the existing organization of society instills into its members.[16]

Negative thinking, then, must be systematic and fundamental in its efforts. It must prevent the elimination of what Mannheim, in a related context, labeled those "reality-transcending elements" of our world. It must avoid what Herbert Marcuse has called the absorption of genuinely critical thinking amid the "one-dimensionalism" of existing technological society.[17] And in a culture as bland as America's is often said to be, concern for one-dimensionality is not misplaced. The "compulsive realism" that characterized the technological utopian and many other utopian writings of turn-of-the-century America is a pertinent example of this cultural complacency. So, too, more recently, is the artistic avant-garde's loss of the power to shock. It is not coincidental that, on the whole, the most imaginative utopian and dystopian writings, if not communities and world's fairs, have been European rather than American.[18]

We need not accept the Frankfurt School's critiques of particular aspects of modern society and culture, including technology, in order to apply its overall notion of theoretical activity to utopianism as alternative ideology. At its boldest, utopianism also challenges the most ingrained assumptions of existing society and, more thoroughly than either epic political theory or even critical theory, offers significant alternatives to them. Epic political theory is, understandably, concerned primarily with a new political order, and critical theory, because existing technological society is, in its view, impervious to any but the barest alternatives, finds mere outlines alone possible.[19] By contrast, serious utopianism examines all basic dimensions of existing and utopian society—values, institutions, material conditions, and inhabitants themselves—regardless of its particular attitude toward those two realms. Such alternatives, however, must be realistic enough, as we saw earlier, to be effective. They must be somehow projected back on existing society for substantial change to occur. We need these utopian dreams, for they can point the way to practice, unlike our usual sleeping dreams that we can recall only in part, if at all. If the problem with utopian blueprints is that they may be too rigid, the problem with utopian dreams is that they may not be rigid enough.[20]

In any event, those who would employ utopianism as a vehicle of social criticism must avoid falling victim to "false consciousness": that is, they must not assume that ideology has reached an end or is no longer legitimate. Those assumptions lead to an uncritical endorsement of existing society either as the Candide-like best of all possible worlds or as an outright utopia in itself.[21] Our very ability, remarked upon so frequently in recent years as to seem a truism, to shape alternative futures—due in large part to technological progress— ought prove a spur to more and more serious utopian theorizing of the sort described here.[22] Marcuse declares in an essay interestingly entitled "The End of Utopia":

> Today we have the capacity to turn the world into hell, and we are well on the way to doing so. We also have the capacity to turn it into the opposite of hell. This would mean the end of utopia, that is, the refutation of those ideas and theories that use the concept of utopia to denounce certain socio-historical possibilities.[23]

By *utopia*, of course, Marcuse means the "impossible," the conventional association of the term. When he writes of the end of utopia, therefore, he means the renewal of utopia as the "possible" and even the "probable." In true dialectical fashion, the tables have turned, or could. The source of Marcuse's modest optimism is not the old faith in technological advance but rather the recent reconsideration in many quarters of the traditional assumptions underlying that faith. Still, even Marcuse clearly would utilize technological advance to create social and cultural conditions that would transcend contemporary technological society.[24]

Existing society, then—including contemporary technological society—holds the potential for substantial change. And in the alternatives offered by utopianism to existing society's prevailing ideology may be found a starting point for effecting such change. The technological utopias that we have examined here provide no blueprints for change of that depth and magnitude. But they, and their predecessors and successors as well, do offer a unique perspective on technology, one that deserves to be taken into account as America and other technologically advanced societies ponder and chart their alternative futures.

Appendix

The American Technological Utopians
Their Backgrounds and Their Works

The following summaries give the basic information so far discovered about the twenty-five known technological utopians of the late nineteenth and early twentieth centuries. Their visions and lives were discussed in chapters 2 and 3. A comparable summary of the information available about the three known technological utopians of the early and mid-nineteenth century—whose visions and lives were sketched in chapter 5—appears immediately afterward. Wherever possible, each summary includes: (1) the place and date of birth and death; (2) the principal occcupation, or the various occupations in order of apparent importance to the utopian; (3) the relevant utopian publications, several having written more than one utopian work and many having written other, non-utopian works; and (4) the most useful sources, not necessarily the only sources, of this and further biographical information, including a handful of autobiographies. Because I used original editions of the writings listed here whenever possible, I did not cite the reprinted editions in chapters 2 and 3, except where I used them instead; I cite the latter in sections (3) and (4) here for the convenience of readers. The absence of one or more of the above sections from a number of individual summaries indicates, of course, that the appropriate data was not available. Three sources of information are cited repeatedly and are abbreviated as follows:

NCAB • *National Cyclopaedia of American Biography.* 58 vols. (New York and Clifton, N.J.: White, 1893–1979)

DAB • *Dictionary of American Biography.* 24 vols. (New York: Scribner, 1928–1974)

WWW • *Who Was Who in America*. 7 vols. (Chicago: Marquis, 1943–1976)

BACHELDER, JOHN
1. Weare, New Hampshire, 1817—Houghton, Michigan, 1906
2. Inventor, manufacturer, accountant, schoolteacher
3. *A.D. 2050. Electrical Development At Atlantis* (San Francisco: Bancroft, 1890; reprinted New York: AMS Press, 1977)
4. NCAB 12:549; DAB 1:465; *Old Folks' Day, New Boston, N.H., June 13, 1907* (New Boston, N.H.: n.p., 1907), 48–51

BELLAMY, EDWARD
1. Chicopee Falls, Massachusetts, 1850—Chicopee Falls, 1898
2. Journalist, professional writer, lawyer
3. *Looking Backward: 2000–1887* (Boston: Ticknor, 1888; reprinted [in definitive edition] Cambridge: Harvard University Press, 1967) *Equality* (New York: Appleton, 1897; reprinted New York: AMS Press, 1970)
4. NCAB 1:263–64; DAB 2:163–64; WWW, Historical:119; Bellamy, *Edward Bellamy Speaks Again: Articles, Public Addresses, Letters* (Chicago: Peerage, 1938; reprinted Westport, Conn.: Hyperion Press, 1975); Arthur E. Morgan, *Edward Bellamy* (New York: Columbia University Press, 1944; reprinted Philadelphia: Porcupine Press, 1974); Sylvia E. Bowman, *The Year 2000: A Critical Biography of Edward Bellamy* (New York: Bookman, 1958; reprinted New York: Octagon, 1979); John L. Thomas, Introduction, *Looking Backward: 2000–1887* (Cambridge.: Harvard University Press, 1967)

BRINSMADE, HERMAN HINE
1. Unknown
2. Professional writer
3. *Utopia Achieved: A Novel of the Future* (New York: Broadway, 1912; reprinted New York: Arno Press, 1971)
4. Unknown

BROOKS, BYRON ALDEN
1. Theresa, Jefferson County, New York, 1845—Brooklyn, New York, 1911
2. Inventor, manufacturer, professional writer, schoolteacher
3. *Earth Revisited* (Boston: Arena, 1893)
4. NCAB 3:319; DAB 3:74

CARYL, CHARLES WILLARD
1. Oakland, California, 1858—Los Angeles, 1926
2. Businessman, inventor, manufacturer, social worker

3. *New Era: Presenting the Plans for the New Era Union to Help Develop and Utilize the Best Resources of this Country* (Denver: n.p., 1897; reprinted New York: Arno Press, 1971)
4. Frank Hall, *History of the State of Colorado*, vol. 4 (Chicago: Blakely, 1895): 497–98; H. Roger Grant, " 'One Who Dares to Plan': Charles W. Caryl and the New Era Union," *Colorado Magazine* 51 (Winter 1974): 13–27

CHAMBLESS, EDGAR
1. 1876—New York City, 1936
2. Architect, patent investigator
3. *Roadtown* (New York: Roadtown Press, 1910)
4. Bruce Bliven, "Suicide of a Dreamer," *New Republic* 87 (July 1, 1936): 238–39; Thomas A. Reiner, *The Place of the Ideal Community in Urban Planning* (Philadelphia: University of Pennsylvania Press, 1963), 43–46; Joseph L. Arnold, *The New Deal in the Suburbs: A History of the Greenbelt Town Program, 1935–1954* (Columbus: Ohio State University Press, 1971), 86; Reynold M. Wik, *Henry Ford and Grass-roots America* (Ann Arbor: University of Michigan Press, 1972), 190; Harry Singer, "Roundtown to Roadtown: A New Physical Base for Social Relations in a Nuclear Age (Adapted from Edgar Chambless' *Roadtown*)" (manuscript in possession of author, Tamarac, Fla., 1981)

CLOUGH, FRED M.
1. Unknown
2. Unknown
3. *The Golden Age, or The Depth of Time* (Boston: Roxburgh, 1923)
4. Unknown

DEVINNE, PAUL
1. Unknown
2. Unknown
3. *The Day of Prosperity: A Vision of the Century to Come* (New York: Dillingham, 1902; reprinted New York: Arno Press, 1971)
4. Unknown

FULLER, ALVARADO MORTIMER
1. 1851–1924
2. Inventor, United States Army officer, phonetic textbook writer
3. *A.D. 2000* (Chicago: Laird and Lee, 1890; reprinted New York: Arno Press, 1971)
4. Arthur O. Lewis, Jr., Introduction, *A.D. 2000* (New York: Arno Press, 1971)

GEISSLER, LUDWIG A.
1. Unknown

2. Unknown
3. *Looking Beyond: A Sequel to* Looking Backward *by Edward Bellamy and an Answer to* Looking Further Forward *by Richard Michaelis* (New Orleans: Graham, 1891; reprinted New York: Arno Press, 1971)
4. Unknown

GILLETTE, KING CAMP
1. Fond du Lac, Wisconsin, 1855—Los Angeles, 1932
2. Inventor, businessman
3. *The Human Drift* (Boston: New Era, 1894; reprinted Delmar, N.Y.: Scholars' Facsimiles and Reprints, 1976)
   The Ballot Box (Brookline, Mass.: n.p., 1897)
   World Corporation (Boston: New England News, 1910)
   The People's Corporation (Boston: Ball, 1924)
4. DAB, Supplement 1:345–46; WWW 1:457; Melvin L. Severy, *Gillette's Social Redemption* (Boston: Turner, 1907); Severy, *Gillette's Industrial Solution* (Boston: Ball, 1908); James B. Gilbert, *Designing the Industrial State: The Intellectual Pursuit of Collectivism in America, 1880–1940* (Chicago: Quadrangle, 1972), ch. 6; Kenneth M. Roemer, Introduction, *The Human Drift* (Delmar, N.Y.: Scholars' Facsimiles and Reprints, 1976); Russell B. Adams, Jr., *King C. Gillette: The Man and His Wonderful Shaving Device* (Boston: Little, Brown, 1978)

HAYES, JEFF W.
1. Cleveland, 1858—Portland, Oregon, 1917
2. Telegraph operator and builder, telegraph company organizer and manager, newspaper editor, writer
3. "Portland, Oregon, A.D. 1999," in his *Portland, Oregon, A.D. 1999 and Other Sketches* (Portland: Baltes, 1913), 1–40
4. *Telegraphers of Today: Descriptive, Historical, Biographical*, ed. John B. Taltavall (New York: n.p., 1893), 203; *Morning Oregonian* (Portland), April 16, 1898, 10; *Portrait and Biographical Record of Portland and Vicinity* (Chicago: Chapman, 1903), 715–16; Alfred Powers, *History of Oregon Literature* (Portland: Metropolitan, 1935, 603); James D. Shand, "Blind Seer into Portland's Future," *Portland Commerce Magazine* 56 (August 3, 1973): 10–12; Howard P. Segal, "Jeff W. Hayes: Reform Boosterism and Urban Utopianism," *Oregon Historical Quarterly* 79 (Winter 1978): 345–57

HOWARD, ALBERT WALDO
1. Unknown
2. Professional writer, publisher, musician
3. (M. Auburré Hovorrè, pseud.) *The Milltillionaire* (Boston: n.p., 1895; reprinted in *American Utopias: Selected Short Fiction*, ed. Arthur O. Lewis, Jr. [New York: Arno Press, 1971])

4. Howard, *Beethoven Reincarnate and Paderewski Svengali: or, Mystery of Paderewski's Playing Revealed; and Beethoven and His Deafness* (Boston: n.p., 1901)

KIRWAN, THOMAS
1. 1829–1911
2. Electricity textbook writer
3. (William Wonder, pseud.) *Reciprocity (Social and Economic) in the Thirtieth Century, the Coming Cooperative Age; A Forecast of the World's Future* (New York: Cochrane, 1909)
4. Unknown

LOEB, HAROLD ALBERT
1. New York City, 1891—Marrakesh, Morocco, 1974
2. Professional writer, publisher, Technocrat
3. *Life in a Technocracy: What It Might Be Like* (New York: 1933)
4. Loeb, *The Way It Was* (New York: Criterion, 1959—Loeb's autobiography), esp. 12, 42–44; *New York Times* obituary, January 23, 1974, p. 40; Howard P. Segal, "Reconsideration: Harold Loeb's *Life in a Technocracy: What It Might Be Like* (1933)," *New Republic* 175 (October 30, 1976): 42–44

MACNIE, JOHN
1. Stirling, Scotland, 1836—Grand Forks, North Dakota, 1909
2. College professor (various fields, from mathematics to modern languages), mathematics textbook writer
3. (Ismar Thiusen, pseud.) *The Diothas; Or, A Far Look Ahead* (New York: Putnam, 1883; reprinted New York: Arno Press, 1971)
4. Webster Merrifield et al., "In Memoriam: Professor John Macnie," *University of North Dakota Bulletin* 1 (November 1909): 3–33; *The Quest for Utopia: An Anthology of Imaginary Societies*, ed. Glenn P. Negley and J. Max Patrick (New York: Schuman, 1952), 50–52; Louis G. Geiger, *University of the Northern Plains: A History of the University of North Dakota, 1883–1958* (Grand Forks, N.D.: University of North Dakota Press, 1958), 52–55, 99, 109

MERRILL, ALBERT ADAMS
1. Boston, 1875—Los Angeles, 1952
2. Aeronautical engineer, college professor (of aeronautical engineering)
3. *The Great Awakening: The Story of the Twenty-Second Century* (Boston: George, 1899)
4. *New York Times* obituary, June 3, 1952, p. 29; *Los Angeles Times* obituary, June 3, 1952, pt. 2, p. 1; Harold E. Morehouse, "Professor Albert A. Merrill," Flying Pioneers Biographies (manuscript in National Air and Space Museum, Smithsonian Institution, Washington, D.C., n.d.), 1–3; Richard P. Hallion, *Legacy of Flight:*

The Guggenheim Contribution to American Aviation (Seattle: University of Washington Press, 1977), 47–48, 52, 56

MORISON, GEORGE SHATTUCK
1. New Bedford, Massachusetts, 1842—New York City, 1903
2. Civil engineer, lawyer
3. Presidential Address (untitled), *Transactions, American Society of Civil Engineers* 33 (1895): 467–84
   The New Epoch as Developed by the Manufacture of Power (Boston and New York: Houghton Mifflin, 1903; reprinted New York: Arno Press, 1972); posthumously published collection of articles and addresses, several of them published previously
4. NCAB 10:129–30; DAB 13:191–92; WWW 1:866; "Memoir of George Shattuck Morison," *Transactions, American Society of Civil Engineers* 54 (1905): 513–21; George A. Morison, *George Shattuck Morison, 1842–1903: A Memoir* (Peterborough, N.H.: Peterborough Historical Society, 1940); George A. Morison and Etta M. Smith, *History of Peterborough, New Hampshire.* 2 vols. (Rindge, N.H.: Smith, 1954), 2:1180–81

OLERICH, HENRY
1. Hazel Green, Wisconsin, 1851—Omaha, Nebraska, 1927
2. Professional writer, farmer, teacher, inventor, machinist, lawyer
3. *A Cityless and Countryless World; An Outline of Practical Co-operative Individualism* (Holstein, Ia.: Gilmore and Olerich, 1893; reprinted New York: Arno Press, 1971)
   Modern Paradise; an Outline or Story of how some of the cultured will probably live, work and organize in the near future (Omaha: Equality, 1915)
   The Story of the World a Thousand Years Hence (Omaha: Olerich, 1923)
4. Olerich, "Autobiography of Professor Henry Olerich" (manuscript in possession of Professor H. Roger Grant, University of Akron [Omaha, 1921]); Olerich, autobiographical insert, *The Story of the World a Thousand Years Hence; Omaha Morning World-Herald* obituary, May 11, 1927, pp. 1–2; Olerich, "Sociologic Fundamentals" (manuscript, n.d., University of Michigan Library, Department of Rare Books and Special Collections, Labadie Collection); Cassius V. Cook, "The Life and Achievements of Henry Olerich" (manuscript, n.d., University of Michigan Library, Labadie Collection, Cassius V. Cook Papers); Olerich-Cook correspondence, 1908–19 (University of Michigan Library, Cassius V. Cook Papers); H. Roger Grant, "Interview with Mrs. Viola Storms," December 20, 1973 (manuscript, Iowa State Historical Society, Iowa

City); Grant, "Viola Olerich, 'The Famous Baby Scholar': An Experiment in Education," *The Palimpsest* 56 (May-June 1975): 88-95; Grant, "Henry Olerich and the Utopian Ideal," *Nebraska History* 56 (Summer 1975): 249-58; Grant, "Henry Olerich and Utopia: The Iowa Years, 1870–1902," *Annals of Iowa* 43 (Summer 1976): 349-61; Robert S. Fogarty, "Henry Olerich," in Fogarty, *Dictionary of American Communal and Utopian History* (Westport, Conn.: Greenwood Press, 1980), 85

SCHINDLER, SOLOMON
1. Neisse, Germany, 1842—Boston, 1915
2. Rabbi, educator, professional writer, social worker
3. *Young West: A Sequel to Edward Bellamy's Celebrated Novel Looking Backward* (Boston: Arena, 1894; reprinted New York: Arno Press, 1971)
4. NCAB 7:439–40; DAB 16:433–34; WWW 1:1087; Arthur Mann, *Yankee Reformers in the Urban Age: Social Reform in Boston, 1880–1900* (New York: Harper Torchbooks, 1966), ch. 3; Howard P. Segal, "*Young West:* The Psyche of Technological Utopianism," *Extrapolation: A Journal of Science Fiction and Fantasy* 19 (December 1977): 50–58

STUMP, D. L.
1. Unknown
2. Printing press foreman
3. *From World to World* (Asbury, Mo.: World to World, 1896)
4. Unknown

TAYLOR, WILLIAM ALEXANDER
1. Perry County, Ohio, 1837—Columbus, Ohio, 1912
2. Journalist, professional writer, lawyer
3. *Intermere* (Columbus, Ohio: Twentieth-Century, 1901; reprinted New York: Arno Press, 1971)
4. WWW 1:1221; Clement L. Martzolff, *History of Perry County, Ohio* (New Lexington, Ohio: Ward and Weilhand, 1902), 175–78

THOMAS, CHAUNCEY
1. Howland, Maine, 1822—Boston, 1898
2. Carriage maker
3. *The Crystal Button: or, Adventures of Paul Prognosis in the Forty-Ninth Century*, ed. George Houghton (Boston and New York: Houghton Mifflin, 1891; reprinted Boston: Gregg Press, 1975)
4. *The Quest for Utopia: An Anthology of Imaginary Societies*, ed. Glenn P. Negley and J. Max Patrick (New York: Schuman, 1952), 82

THURSTON, ROBERT HENRY

1. Providence, Rhode Island, 1839—Ithaca, New York, 1903
2. Mechanical engineer, college professor (of mechanical engineering) and administrator
3. Presidential Address (untitled), *Transactions, American Society of Mechanical Engineers* 1 (1880): 13–29
   "The Mission of Science," *Proceedings, American Association for the Advancement of Science* 33 (September 1884): 227–53
   "The Border-Land of Science," *North American Review* 150 (January 1890): 67–79
   "The Scientific Basis of Belief," *North American Review* 153 (August 1891): 181–92
   "The New Education and the New Civilization: Their Unity," *Scientific American Supplement* 34 (July 30, 1892): 13825–26
   "An Era of Mechanical Triumphs," *Engineering Magazine* 6 (January 1894): 456–67
   "Trend of National Progress," *North American Review* 161 (September 1895): 297–312
   "Progress and Tendency of Mechanical Engineering in the Nineteenth Century," *Popular Science Monthly* 59 (May-June 1901): 34–43, 141–51
   "Scientific Research: The Art of Revelation and of Prophecy," *Science* 16 (September 12–19, 1902): 401–4, 454–57
4. NCAB 4:479–80; DAB 18: 518–20; WWW 1:1238; "Robert Henry Thurston: In Memoriam," *Transactions, American Society of Mechanical Engineers* 25 (1904): 1113–20; William F. Durand, *Robert Henry Thurston: A Biography* (New York: American Society of Mechanical Engineers, 1929); *Exercises Commemorating the One-Hundredth Anniversary of the Birth of Robert Henry Thurston,* Cornell University, October 25, 1939 (Ithaca, N.Y.: Thurston Program Committee, 1940); Robert J. Kwik, "The Function of Applied Science and the Mechanical Laboratory during the Period of Formation of the Profession of Mechanical Engineering, as Exemplified in the Career of Robert Henry Thurston, 1839–1903" (Ph.D. diss., University of Pennsylvania, 1974)

WOOLDRIDGE, CHARLES WILLIAM

1. Hull, England, 1847—Helena, Montana, 1908
2. Physician, spiritualist philosopher
3. *Perfecting the Earth: A Piece of Possible History* (Cleveland: Utopia, 1902; reprinted New York: Arno Press, 1971)
4. WWW 1:1381

ETZLER, JOHN ADOLPHUS

1. Germany (unknown, but boyhood friend of civil engineer John Roebling, who was born in Muhlhausen, Thuringia, Germany, in 1806)—Unknown
2. Inventor, professional writer, possibly civil engineer
3. *The Paradise Within the Reach of All Men, without Labor, by Powers of Nature and Machinery; An Address to all Intelligent Men* (Pittsburgh: Etzler and Reinhold, 1833)
   The New World; or, Mechanical System to Perform the Labours of Man and Beast by Inanimate Powers, That Cost Nothing, For Producing and Preparing the Substances of Life (Philadelphia: Stollmeyer, 1841)
   Description of the Naval Automaton Invented by J. A. Etzler (Philadelphia: Gihon, Fairchild, 1841 or 1842)
   Dialogue on Etzler's Paradise: Between Messrs. Clear, Flat, Dunce and Grudge (London: O'Brien, 1843)
   Emigration to the Tropical World for the Melioration of all Classes of People of All Nations (Surrey, Eng.: Concordium, 1844)
   Two Visions of J. A. Etzler: A Revelation of Futurity (Ham Common, Surrey, Eng.: Concordium, 1844)
   All of these works have been reprinted in *The Collected Works of John Adolphus Etzler*, ed. Joel Nydahl (Delmar, N.Y.: Scholars' Facsimiles and Reprints, 1977)
4. Henry David Thoreau, "Paradise (to be) Regained," *United States Magazine and Democratic Review* 13 (November 1843): 451–63; Reverend T. DeWitt Talmage, *The Abominations of Modern Society* (New York: Adams, Victor, 1872), 283–86; D. B. Steinman, *The Builders of the Bridge: The Story of John Roebling and His Son* (New York: Harcourt, Brace, 1945), 18, 19, 20, 22, 34; Alan Trachtenberg, *Brooklyn Bridge: Fact and Symbol* (New York: Oxford University Press, 1965), 46–48; Robin Linstromberg and James Ballowe, "Thoreau and Etzler: Alternative Views of Economic Reform," *Midcontinent American Studies Journal* 11 (Spring 1970): 20–29; Patrick R. Brostowin, "John Adolphus Etzler: Scientific Utopian during the 1830s and 1840s" (Ph.D. diss., New York University, 1969); Joel Nydahl, Introduction, *The Collected Works of John Adolphus Etzler* (Delmar, N.Y.: Scholars' Facsimiles and Reprints, 1977)

EWBANK, THOMAS

1. Durham, England, 1792—New York City, 1870
2. Inventor, manufacturer, United States Commissioner of Patents, 1849–1852
3. *A Descriptive and Historical Account of Hydraulic and Other*

Machines for Raising Water, Ancient and Modern: With Observations on Various Subjects Connected with the Mechanic Arts, Including the Progressive Development of the Steam Engine (New York: D. Appleton, 1842; reprinted New York: Arno Press, 1972)

"Origin and Progress of Invention" and "The Motors: Chief Levers of Civilization," in *Report of the United States Commissioner of Patents for the Year 1849* (Washington, D.C.: United States Patent Office, 1849), pt. 1:483–96 and 497–511

The World A Workshop: or the Physical Relationship of Man to the Earth (New York: Appleton, 1855)

Thoughts on Matter and Force: or, Marvels that Encompass Us (New York: Appleton, 1858)

The Position of our Species in the Path of its Destiny—or, the Comparative Infancy of Man and of the Earth as His Home (New York: Scribner, 1860)

4. NCAB 7:559; DAB 6:227–28; WWW, Historical: 243; "Outline of the History of the United States Patent Office," *Journal of the Patent Office Society* 18 (July 1936): 149–51

GRIFFITH, MARY

1. Unknown—1877
2. Professional writer
3. "Three Hundred Years Hence," in her *Camperdown; or, News from Our Neighbourhood* (Philadelphia: Carey, Lea and Blanchard, 1836), 9–92; reprinted in *American Utopias: Selected Short Fiction*, ed. Arthur O. Lewis, Jr. (New York: Arno Press, 1971; also reprinted Boston: Gregg Press, 1975)
4. Nelson F. Adkins, "An Early American Story of Utopia," *Colophon* 1 (Summer 1935): 123–32; reprinted, in slightly revised form, as the introduction to a reprint of the book, edited by Adkins (Philadelphia: Prime, 1950), 5–20; Beverly Seaton, "Mary Griffith," in *American Women Writers: A Critical Reference Guide from Colonial Times to the Present*, ed. Linda Mainiero (New York: Ungar, 1980), 2: 183–85

Notes

Introduction

1. On the study of American technology, see the appropriate sections of Eugene S. Ferguson, *Bibliography of the History of Technology* (Cambridge: MIT Press, 1968), and of the annual supplementary bibliographies after 1968 of *Technology and Culture* (University of Chicago Press), the principal journal in the field. On the study of American utopianism, see the appropriate portions of Lyman Tower Sargent, *British and American Utopian Literature, 1516–1975: An Annotated Bibliography* (Boston: G. K. Hall, 1979), and the several bibliographic and historical surveys in *America as Utopia: Collected Essays*, ed. Kenneth M. Roemer (New York: Burt Franklin, 1981), pt. 4.

2. The extraordinary popularity and lasting influence of *Looking Backward* are well documented. See Elizabeth Sadler, "One Book's Influence: Edward Bellamy's *Looking Backward*," *New England Quarterly* 17 (December 1944): 530–55; John L. Thomas, Introduction, *Looking Backward: 2000–1887*, ed. Thomas (Cambridge: Harvard University Press, 1967), 69–88; Thomas, "Utopia for an Urban Age: Henry George, Henry Demarest Lloyd, Edward Bellamy," *Perspectives in American History* 6 (1972): 135–63; Peter H. Curtis, "Bellamy Nationalism and Later Reform Movements, 1888–1940" (Ph.D. diss., Indiana University, 1973); and Thomas, *Alternative America: Henry George, Edward Bellamy, Henry Demarest Lloyd and the Adversary Tradition* (Cambridge: Harvard University Press, 1983), chaps. 11, 15. On the question of the several similarities between *Looking Backward* and *The Diothas*, see Arthur E. Morgan, *Edward Bellamy* (New York: Columbia University Press, 1944), 239–43, and Morgan, *Plagiarism in Utopia* (Yellow Springs, Ohio: Morgan, 1944). Morgan concludes, "Bellamy probably was the inspiration of Macnie's book, notwithstanding the fact that Macnie's was first to be published" (*Edward Bellamy*, 241).

3. On the publication of technological utopian works between 1833 and 1933, see the Appendix. On the publication of utopian works generally between the founding of the American republic and the Civil War, see Joel Nydahl, "From Millennium to Utopia America," and Nydahl, "Early Fictional Futures: Utopia, 1798–1864," in *America as Utopia*, 237–53 and 254–91. On the publication of utopian works generally between the

175

Civil War and the New Deal, see Charles J. Rooney, "Post-Civil War, Pre–*Looking Backward* Utopia: 1865–1887," Roemer, "Utopia and Victorian Culture, 1888–1899," and Howard P. Segal, "Utopia Diversified: 1900–1949," in ibid., 292–304, 305–32, and 333–46. See also Sargent, *British and American Utopian Literature*; Roemer, *The Obsolete Necessity: America in Utopian Writings, 1888–1900* (Kent, Ohio: Kent State University Press, 1976), 181–213; and Roemer's "News Center" column in alternate issues of *Alternative Futures: The Journal of Utopian Studies*, beginning with the first issue of the first volume, Spring 1978, and ending with the last issue of the last volume, Spring/Summer 1981.

4. Thus few copies of their works survive, and standard references cite few of the authors. Except for *Looking Backward*, these works are absent from lists of the most popular books of the day.

5. Robert Wiebe, *The Search for Order, 1877–1920* (New York: Hill and Wang, 1967), 69. See also Thomas, Introduction, *Looking Backward*, 1–69. Arthur E. Bestor, Jr., makes a similar remark concerning the antebellum utopian communities: "Large numbers of Americans could be attracted to communitarianism because so many of its postulates were things they already believed" ("Patent-Office Models of the Good Society: Some Relationships Between Social Reform and Westward Expansion," in Bestor, *Backwoods Utopias: The Sectarian Origins and the Owenite Phase of Communitarian Socialism in America: 1663–1829*, 2d ed. [Philadelphia: University of Pennsylvania Press, 1970], 248). The article appeared originally in 1953 and is reprinted in his book, as is his 1957 article, "The Transit of Communitarian Socialism to America," which on p. 270 has a sentence almost identical to this one. On the connection between utopian communities and utopian writings, see chap. 6 below.

6. Within what may loosely be termed utopian studies, this flawed approach is still being followed, with minor modifications, by younger scholars. If *Looking Backward* alone no longer suffices to explain late nineteenth- and early twentieth-century America, now a handful of utopian works, often including Bellamy's classic, supposedly do. The results are nearly as dubious as the earlier studies that relied on *Looking Backward* alone. See, for example, George B. Thomas, "Blueprint for Tomorrow: American Novels of Future Change" (Ph.D. diss., Harvard University, 1970); A. James Stupple, "Utopian Humanism in America, 1888–1900" (Ph.D. diss., Northwestern University, 1971); and Helen J. Gardiner, "American Utopian Fiction, 1885–1910: The Influence of Science and Technology" (Ph.D. diss., University of Houston, 1978). Roemer's *Obsolete Necessity* is a welcome exception to this trend. See his intelligent discussion of the problem, pp. xii–xiii and chap. 1.

7. On the historical and intellectual origins of *ideology* as described here, see George Lichtheim, "The Concept of Ideology," *History and Theory* 4 (1965): 164–95, and Emmet Kennedy, *A Philosophe in the Age of Revolution: Destutt de Tracy and the Origins of "Ideology"* (Philadelphia: American Philosophical Society, 1978). On contemporary confirmations of ideology as a normative concept in every culture, see Clifford Geertz, "Ideology as a Cultural System," in *Ideology and Discontent*, ed. David E. Apter (New York: Free Press, 1964), 47–76, and Laurie B. Brown, *Ideology* (Baltimore: Penguin Books, 1973). By contrast, Lewis S. Feuer, *Ideology and the Ideologists* (New York: Harper and Row, 1975), is a recent, sophisticated, but ultimately unconvincing critique of ideology as an invariable distortion of reality.

8. Ironically, as Lichtheim and Kennedy separately show, the savants of the Institut de France themselves easily meshed normative statements with empirical ones. Because they believed so firmly in the rationality not only of mankind but also of history, whose logic and meaning they consequently expected to discern, they attempted to establish ideology as a value-free science of mind, a "science of ideas." In the process they, and later "ideologues" as well, often equated their particular ideology with ultimate and universal truth. Ideology thereby acquired the basis for its present pejora-

tive connotation within the Institut itself. On the evolution of *ideology*, see also the complete entry on the term in Raymond Williams, *Keywords: A Vocabulary of Culture and Society* (New York: Oxford University Press, 1976), 126–30.

9. On the use, or misuse, by American scholars of these terms, neither of them precisely defined even in their original French, see Geertz, "Stir Crazy," *New York Review of Books* 24 (January 26, 1978): 3; David Hall, "Intellectual History as the History of Mentalities," and William R. Taylor, "Toward a History of Perception," in *Intellectual History Group Newsletter* 1 (Spring 1979): 14–16 and 16–18; James A. Henretta, "Social History as Lived and Written," *American Historical Review* 84 (December 1979): 1299; and Robert Darnton, "Intellectual and Cultural History," in *The Past Before Us: Contemporary Historical Writing in the United States*, ed. Michael Kammen (Ithaca, N.Y.: Cornell University Press, 1980), 346.

10. Although the origins of the ideas of progress and of progress as technological progress will be examined in chaps. 4 and 5, I want to note one point here. Several historians have argued that these ideas were, in the American context, profoundly conservative because America was for various reasons proclaimed an actual, not merely potential, utopia by its earliest European settlers. Most of their descendants continued to see it in that way. Technological advances and other later developments, these historians contend, enhanced America's utopian character but were not themselves primarily responsible for it. This "conservative" interpretation of progress goes beyond that of the technological utopians, who, as will become clear in chap. 2, saw America as a potential, even probable, utopia but not yet as an existing one. The conservatism of the technological utopians' vision stems from their perception of American society as able to achieve utopia through evolutionary rather than revolutionary changes. On this conservative interpretation of American history in general, see Arthur A. Ekirch, Jr., *The Idea of Progress in America, 1815–1860* (New York: Columbia University Press, 1944), 13, 37, 267; Daniel J. Boorstin, *The Genius of American Politics* (Chicago: University of Chicago Press, 1953), esp. chap. 1; Rush Welter, "The Idea of Progress in America: An Essay in Ideas and Methods," *Journal of the History of Ideas* 16 (June 1955): 401–15, esp. 405–406; and Henry Steele Commager, "America and the Enlightenment," in *The Development of a Revolutionary Mentality* (Washington, D.C.: Library of Congress, 1972), 23. On the argument of these historians that America was a veritable utopia from the seventeenth century on, see chap. 5, n. 5.

11. Thus even if the technological utopian works that followed *Looking Backward* had been better written and less derivative, they would not necessarily have avoided obscurity. Their message had already been broadcast and warmly received.

12. On this persistence, even flourishing, of utopian communities long after the Civil War, see Robert S. Fogarty, "American Communes, 1865–1914," *Journal of American Studies* 9 (August 1975): 145–62. Before Fogarty's article appeared, it was generally assumed that such communities were, before their renewal in the 1960s, primarily an antebellum American phenomenon. Chaps. 5 and 6 discuss aspects of both the antebellum and the postbellum communities.

13. See the summary and provocative critique of such antitechnological views in Samuel C. Florman, *The Existential Pleasures of Engineering* (New York: St. Martin, 1976), esp. chaps. 4–6, and Florman, *Blaming Technology: The Irrational Search for Scapegoats* (New York: St. Martin, 1981).

14. On the presence of utopianism of varying kinds in Asia, see Jean Chesneaux, "Egalitarian and Utopian Traditions in the East," *Diogenes* 62 (Summer 1968): 76–102, and the essays by Seiji Nuita, George B. Bikle, Jr., and Harold Gould in *Aware of Utopia*, ed. David W. Plath (Urbana: University of Illinois Press, 1971). On its virtual absence in Latin America, see the essay by Joseph L. Love in *Aware of Utopia*. Frank E. and Fritzie P. Manuel observe that "the profusion of Western utopias has not been equaled in any other culture" (*Utopian Thought in the Western World* [Cambridge: Harvard University

Press, 1979], 1) but that, within Europe, utopianism "for four centuries . . . has remained predominantly English and French" (Introduction, *French Utopias: An Anthology of Ideal Societies*, ed. and tr. Manuel and Manuel [New York: Schocken, 1971], 1). See also *Utopian Thought in the Western World*, 14–15.

15. Elisabeth Hansot argues that all modern utopias are conservative insofar as, unlike earlier ones, they differ relatively little from existing society, and because, like the technological utopias examined in this study, they are expected to be realized. See Hansot, *Perfection and Progress: Two Modes of Utopian Thought* (Cambridge: MIT Press, 1974), 16. I disagree with this and other aspects of Hansot's argument, as I make clear in chap. 9 and in n.3 there.

16. As Gorman Beauchamp has shown, even some seemingly "escapist" utopias—to use Lewis Mumford's distinctions—can, like avowedly serious "reconstructionist" utopias, serve as vehicles of social criticism. Their popularity as escape mechanisms from the real world is a de facto critique of the societies which produce them. Despite their absence of rigid blueprints, they can still contribute to the improvement of those actual societies. See Beauchamp, "The Dream of Cockaigne: Some Motives for the Utopias of Escape," *Centennial Review* 24 (Fall 1981): 345–62, esp. 361–62.

Chapter 1

1. On the changing meanings of terms like these, see the many examples in Raymond Williams, *Culture and Society, 1780–1950* (New York: Harper Torchbooks, 1966), ix–xviii, and in Williams, *Keywords: A Vocabulary of Culture and Society* (New York: Oxford University Press, 1976). Concerning the history of technology, see G. Hollister-Short, "The Vocabulary of Technology," in *History of Technology*, ed. A. Rupert Hall and Norman Smith (London: Mansell, 1977), 2:125–55. On the simplified and imprecise use of *order* and *community* by several recent historians of late nineteenth- and early twentieth-century America (and other periods as well), see Robert F. Berkhofer, "The New or the Old Social History?" *Reviews in American History* 1 (March 1973): 21–28, and Thomas Bender, *Community and Social Change in America* (New Brunswick, N.J.: Rutgers University Press, 1978). A pertinent exception here is Jean B. Quandt, *From the Small Town to the Great Community: The Social Thought of Progressive Intellectuals* (New Brunswick, N.J.: Rutgers University Press, 1970). In fact, technological utopianism represents an extreme version of "the promise of technology" (67) that—in various forms of communications, from analytical surveys to newspaper stories to motion pictures—the nine prominent "Progressive" intellectuals examined by Quandt used to ease their transition from the small towns in which they grew up to the "great society" in which they either were then supposedly living or hoped to live.

2. Chronologically, these works are: Edwin M. J. Kretzmann, "German Technological Utopias of the Pre-War Period," *Annals of Science* 3 (October 15, 1938): 417–30; Hugo A. Meier, "The Technological Concept in American Social History, 1750–1850" (Ph.D. diss., University of Wisconsin, 1950), 487; Richard Gerber, *Utopian Fantasy: A Study of English Utopian Fiction since the End of the Nineteenth Century* (London: Routledge and Kegan Paul, 1955), 51; Walter H. G. Armytage, *The Rise of the Technocrats: A Social History* (London: Routledge and Kegan Paul; Toronto: University of Toronto Press, 1965), 170; Northrop Frye, "Varieties of Literary Utopias," in *Utopias and Utopian Thought: A Timely Appraisal*, ed. Frank E. Manuel (Boston: Beacon, 1967), 30; François Bloch-Lainé, "The Utility of Utopias for Reformers," ibid., 214; Hall, "Cultural, Intellectual, and Social Foundations, 1600–1750," in *Technology in Western Civilization*, ed. Melvin Kranzberg and Carroll W. Pursell, Jr. (New York:

Oxford University Press, 1967), 1:108; Mark R. Hillegas, *The Future as Nightmare: H. G. Wells and the Anti-Utopians* (New York: Oxford University Press, 1967), 10, 153; John L. Thomas, Introduction, Edward Bellamy, *Looking Backward: 2000–1887* (Cambridge: Harvard University Press, 1967), 44; M. I. Finley, "Utopianism Ancient and Modern," in *The Critical Spirit: Essays in Honor of Herbert Marcuse*, ed. Kurt H. Wolff and Barrington Moore, Jr. (Boston: Beacon, 1968), 13; Jack Douglas, Introduction, *Freedom and Tyranny: Social Problems in a Technological Society*, ed. Douglas (New York: Knopf, 1970), 13–14; William Kuhns, *The Post-Industrial Prophets: Interpretations of Technology* (New York: Weybright and Talley, 1971), 2; Mulford Q. Sibley, *Technology and Utopian Thought* (Minneapolis: Burgess, 1971); Christopher Lasch, *The World of Nations: Reflections on American History, Politics, and Culture* (New York: Knopf, 1973), xii; Kenneth M. Roemer, "The Yankee(s) in Noahville," *American Literature* 45 (November 1973): 435; Carolyn Symonds, "Technology and Utopia," in *The Future of Work*, ed. Fred Best (Englewood Cliffs, N.J.: Prentice-Hall, 1973), 179; George Basalla, "Museums and Technological Utopianism," in *Technological Innovation and the Decorative Arts*, ed. Ian M. G. Quimby and Polly Anne Earl (Charlottesville: University Press of Virginia, 1974), 355–73; Russel B. Nye, *Society and Culture in America, 1830–1860* (New York: Harper and Row, 1974), 277; Evelyn Torton Beck, "Sexism, Racism and Class Bias in German Utopias of the Twentieth Century," *Soundings* 58 (Spring 1975): 119, 120; John F. Kasson, *Civilizing the Machine: Technology and Republican Values in America, 1776–1900* (New York: Grossman/Viking Press, 1976), 230; Roemer, *The Obsolete Necessity: America in Utopian Writings, 1888–1900* (Kent, Ohio: Kent State University Press, 1976), 114; William E. Akin, *Technocracy and the American Dream: The Technocrat Movement, 1900–1941* (Berkeley and Los Angeles: University of California Press, 1977), xiii, chap. 8; R. Jackson Wilson, "Experience and Utopia: The Making of Edward Bellamy's *Looking Backward*," *American Studies* 11 (April 1977): 45; Langdon Winner, *Autonomous Technology: Technics-Out-of-Control as a Theme in Political Thought* (Cambridge: MIT Press, 1977), 33; Williams, "Utopia and Science Fiction," *Science-Fiction Studies* 5 (November 1978), 207; I. F. Clarke, *The Pattern of Expectation, 1644–2001* (New York: Basic Books, 1979), 170; Manuel and Fritzie P. Manuel, *Utopian Thought in the Western World* (Cambridge: Harvard University Press, 1979), 777; Judith A. Merkle, *Management and Ideology: The Legacy of the International Scientific Management Movement* (Berkeley and Los Angeles: University of California Press, 1980), 143; Jean Pfaelzer, "The Impact of Political Theory on Narrative Structures," in *America as Utopia: Collected Essays*, ed. Roemer (New York: Burt Franklin, 1981), 119–21; Daniel Walden, "The Two Faces of Technological Utopianism: Edward Bellamy and Horatio Alger, Jr.," *JGE: The Journal of General Education* 33 (Spring 1981): 24–30; and Anthony F. C. Wallace, *The Social Context of Innovation: Bureaucrats, Families, and Heroes in the Early Industrial Revolution, as Foreseen in Bacon's New Atlantis* (Princeton, N.J.: Princeton University Press, 1982), 3.

3. The *Oxford English Dictionary* ([Oxford: Clarendon Press, 1933], 11:485–86) and all other studies on the subject ascribe coinage of the term *utopianism* to Thomas More's *Utopia*, which first appeared, in Latin, in 1516. As I show in chap. 4, the concept has undergone several significant changes since. A useful introduction to the history of the term *utopianism* and its applications is Roger L. Emerson, "Utopia," in *Dictionary of the History of Ideas*, ed. Philip P. Wiener (New York: Scribner, 1973), 4:458–65. On the critical study of utopianism in the West, see the summary in Manuel and Manuel, 10–12.

4. For an elaboration of these points in the context of nineteenth-century British and French utopianism, see Barbara Goodwin, *Social Science and Utopia: Nineteenth-Century Models of Social Harmony* (Sussex, Eng.: Harvester Press, 1978), chap. 1 and pp. 63–69. On the relativity and incompleteness of perfection in many utopian schemes, see Lyman Tower Sargent, "The 'Utopian Tradition' " (paper in possession of

author), 14–16. For related discussions of the several prices inhabitants of utopia often pay for these improvements, see Gorman Beauchamp, "Utopia and Its Discontents," *Midwest Quarterly* 16 (Winter 1975): 161–74, and Sargent, "A Note on the Other Side of Human Nature in the Utopian Novel," *Political Theory* 3 (February 1975): 88–97.

5. Frank Manuel, "Toward a Psychological History of Utopias," in *Utopias and Utopian Thought*, 70.

6. George Kateb, *Utopia and Its Enemies* (New York: Free Press, 1963), 6 n.6. On this point, see also Manuel and Manuel, 7. On the definition of utopia outlined in this and the preceding two paragraphs, see also Sargent, "Utopia—The Problem of Definition," *Extrapolation: A Journal of Science Fiction and Fantasy* 16 (May 1975): 137–48, and J. C. Davis, *Utopia and the Ideal Society: A Study of English Utopian Writing, 1516–1700* (New York: Cambridge University Press, 1981), chap. 1. On the differences, in light of this comprehensive definition, between utopian and science fiction writings, see Sargent, "Eutopias and Dystopias in Science Fiction: 1950–75," in *America as Utopia*, 347–48, 356.

7. According to Victor Ferkiss (*Technological Man: The Myth and the Reality* [New York: Mentor Books, 1970], 48), the term *technology* was coined by a German, Johann Beckmann, in 1772. But according to the *Oxford English Dictionary* (11: 137), the term, as "the terminology of a particular art or subject," may be traced back to 1658, and as "a discourse or treatise on an art or arts" or as "the scientific study of the practical or industrial arts," the term may be traced back to 1615. In any case the coinage of the term should not be ascribed, as it has sometimes been, to Jacob Bigelow's *Elements of Technology* (Boston: Hillard, Gray, Little, and Wilkins, 1829). The evolution of the term in America nevertheless began with Bigelow's work; see chap.5 below. For accounts of the European origins, see Robert P. Multhauf, "Some Observations on the State of the History of Technology," *Technology and Culture* 15 (January 1974): 1–2, and Reinhard Rürup, "Reflections on the Development and Current Problems of the History of Technology," ibid. 15 (April 1974): 167–68. On the need for broader but also more precise conceptions of technology along the lines sketched in the following paragraphs, see Kranzberg and Pursell, "The Importance of Technology in Human Affairs," in *Technology in Western Civilization* 1: 7–11, and George H. Daniels, "The Big Questions in the History of American Technology," *Technology and Culture* 11 (January 1970): 1–21. A useful introduction to the history of both the term *technology* and its applications is D. S. L. Cardwell, "Technology," in *Dictionary of the History of Ideas* 4:357–65.

8. David P. Billington, "Structures and Machines: The Two Sides of Technology," *Soundings* 57 (Fall 1974): 275.

9. Ibid., 276, 275. The idea of structures as housing machines is part of a critique by another civil engineer to a revised and expanded version of Billington's 1974 publication. See Billington, "Technology and the Structuring of Cities," and Mario G. Salvadori, "The Aesthetics of Technology: In Response to David P. Billington," in *Small Comforts for Hard Times: Humanists on Public Policy*, ed. Michael Mooney and Florian Stuber (New York: Columbia University Press, 1977), 182–98 and 199–203. Although not himself an engineer, Winner usefully expands some of the points raised in this paragraph so far by including "tools, instruments, machines, appliances, weapons, [and] gadgets" as "apparatus," or "the physical devices of technical performance" (11). The use of *apparatus* would not, of course, preclude the use of *structures* and *machines*.

10. Billington touches briefly upon this further dimension of technology as hardware—or, perhaps more precisely, this addition to technology as hardware—early in his 1974 article: "Structures and machines characterize a basic similarity underlying all technology. Often referred to as technique, this similarity represents a way of organizing thought for action that emphasizes the efficiency of means in producing structures and machines, or more generally the means of mastering nature" ("Struc-

tures and Machines," 275). On technology as not just technical skills but outright knowledge, see Edwin T. Layton, Jr., "Technology as Knowledge," *Technology and Culture* 15 (January 1974): 31–41. In broadening the conception of this aspect of technology, Winner's work is again helpful. He places "the whole body of technical activities—skills, methods, procedures, routines—that people engage in to accomplish tasks . . . under the rubric *technique*" (12). Once again, Winner's conception enlarges rather than replaces the others discussed here.

11. These differences will be discussed in chap. 6. For an overview of the impact of electronics on the technological imagination, see James W. Carey and John J. Quirk, "The Mythos of the Electronic Revolution," *American Scholar* 39 (Spring 1970): 219–41, and 39 (Summer 1970): 395–424.

12. On the historical and contemporary relationships between technology and science, see Kranzberg and Pursell, "The Importance of Technology in Human Affairs," 5–6; Winner, 21–25; Layton, 31–41; James K. Feibleman, "Pure Science, Applied Science, Technology, Engineering: An Attempt at Definitions," *Technology and Culture* 2 (Fall 1961): 305–17; Peter F. Drucker, "The Technological Revolution: Notes on the Relationship of Technology, Science, and Culture," ibid., 342–51; Derek J. de Solla Price, "Is Technology Historically Independent of Science? A Study in Statistical Historiography," ibid. 6 (Fall 1965): 553–68; Price, "On the Historiographic Revolution in the History of Technology," ibid. 15 (January 1974): 44–47; and Rürup, 186–87. As the relationship between technology and science grew closer in the nineteenth century, however, the relationship between pure and applied science grew closer as well. The latter relationship in American (and European) history nevertheless remains unclear and the subject of considerable scholarly dispute. See Nathan Reingold, "Alexander Dallas Bache: Science and Technology in the American Idiom," *Technology and Culture* 11 (April 1970): 163–77; Reingold, "American Indifference to Basic Research: A Reappraisal," in *Nineteenth-Century American Science: A Reappraisal*, ed. Daniels (Evanston, Ill.: Northwestern University Press, 1972), 38–62; and Arthur P. Molella and Reingold, "Theories and Ingenious Mechanics: Joseph Henry Defines Science," *Science Studies* 3 (October 1973): 323–51.

13. See Jacques Ellul, *The Technological Society*, tr. John Wilkinson (New York: Vintage Books, 1964). Using *civilization* instead of *society*, John W. Oliver, a pioneering historian of American technology, employs the term much more positively when he states: "American Civilization is fundamentally a technological civilization" (*History of American Technology* [New York: Ronald, 1956], 627). As noted next in the discussion of *culture*, the association of *civilization* with material developments is not uncommon.

14. On the general relationship between technological change and social change, see Daniels, "The Big Questions in the History of American Technology," 1–21; Price, "On the Historiographic Revolution in the History of Technology," 42–48; Rürup, esp. 187–93; and Eugene S. Ferguson, "Toward a Discipline of the History of Technology," *Technology and Culture* 15 (January 1974): 20–29.

15. Ellul equates virtually everything in the contemporary world with technology—or *technique*, as he terms it—and in turn equates technology with omnipotence and the enslavement of mankind. No other phenomena seem to matter any more, and mankind's prospects for restoring a measure of control over technology are seen as dim. Without examining Ellul's work in depth, it should be noted that Ellul, like the early historians of American technology whom Daniels criticizes, fails to consider the extent to which society and so mankind have an impact on technology. See Daniels, "The Big Questions in the History of American Technology," 1–21. Billington ("Structures and Machines," 275, 286 n.1), criticizes Ellul for equating technology with machines alone and ignoring its structures. On the general question of the autonomy of technology in the twentieth century, see Charles L. Sanford, "Technology and Culture at the End of

the Nineteenth Century: The Will to Power," in *Technology in Western Civilization* 1: 726–39, esp. 728; Siegfried Giedion, *Mechanization Takes Command: A Contribution to Anonymous History* (New York: Norton, 1969), esp. Conclusion; Ferguson, 20–26; and, above all, Winner. In his comprehensive definition of technology, Winner correctly includes the social uses of hardware, of technical skills, and of technical knowledge—again what he calls *apparatus* and *techniques*. He uses the terms *organization*, or "all varieties of technical (rational-productive) social arrangements," and *network*, or "those large-scale systems that combine people and apparatus linked across great distances" (12). Kranzberg and Pursell ("The Importance of Technology in Human Affairs," 6) also include "organization" of activity in defining technology and so, like Winner, correctly include its functions as well as its forms.

16. Perhaps the best example of this variety of definitions of *culture* is A. L. Kroeber and Clyde Kluckhohn, *Culture: A Critical Review of Concepts and Definitions* (1952; reprinted New York: Vintage Books, n.d.). The book leaves unresolved the meaning of the term but does declare that, as of 1952 at least, "as yet we have no full theory of culture. We have a fairly well-delineated concept, and it is possible to enumerate conceptual elements embraced within that master concept. But a concept, even an important one, does not constitute a theory" (357).

17. Williams, *Culture and Society*, xiv. See also the entry "Culture" in his *Keywords*, 76–82.

18. John Demos, Introduction, *Remarkable Providences: 1600–1760*, ed. Demos (New York: Braziller, 1972), 1, and Warren Susman, Introduction, *Culture and Commitment: 1929–1945*, ed. Susman (New York: Braziller, 1973), 2.

19. As Williams puts it, "The history of the idea of culture is a record of our reactions, in thought and feeling, to the changed conditions of our common life" (*Culture and Society*, 295). On the changing notions of culture in America in the 1920s and 1930s, see Susman, Introduction, *Culture and Commitment*; Susman, "The Thirties," in *The Development of an American Culture*, ed. Stanley Coben and Lorman Ratner (Englewood Cliffs, N.J.: Prentice-Hall, 1970), 179–218; *Critics of Culture: Literature and Society in the Early Twentieth Century*, ed. Alan Trachtenberg (New York: Wiley, 1976); and, more peripherally, Thomas Reed West, *Flesh of Steel: Literature and the Machine in American Culture* (Nashville: Vanderbilt University Press, 1967).

20. In 1869 Matthew Arnold, the English exponent and simultaneous critic of *culture*, associated the term with perfection: "culture being a pursuit of our total perfection by means of getting to know, on all the matters which most concern us, the best which has been thought and said in the world" (*Culture and Anarchy*, ed. J. Dover Wilson [Cambridge: Cambridge University Press, 1960], 6). Ironically, Arnold associated technological advance with cultural decline. On Arnold's conception of culture, see Williams, *Culture and Society*, chap. 6.

21. Neil Harris, *The Land of Contrasts: 1880–1901*, ed. Harris (New York: Braziller, 1970), 2. For an illuminating review of the series of which the volumes edited by Harris, Demos, and Susman are a part, see Laurence Veysey, "A New Record of American Civilization," *Reviews in American History* 1 (September 1973): 318–30. Veysey notes the emphasis on technology in most of the eight volumes in the series but takes the conventional approach and considers technology's impact on American society, not the way that American society has shaped technology.

22. *Culture* may, as indicated, encompass material objects, but not necessarily all material objects—only those that either reflect or generate what Harris labels "social generalizations" or do both. On contemporary approaches to material culture, see the essays in *Material Culture and the Study of American Life*, ed. Quimby (New York: Norton, 1978).

23. See n. 13 above for an example of the use of *technological civilization*. The historian Charles A. Beard used the term positively in his "Is Western Civilization in Peril?" *Harper's* 157 (August 1928): 265–73. On the common use of the term in the

1930s, see Susman, "The Thirties," 188. On the evolution of this term alone, see the entry "Civilization," in Williams, *Keywords*, 48–50. *Civilization*, it must be emphasized, cannot necessarily be equated with *culture*. As Kroeber and Kluckhohn wrote in 1952, "there was first a phase in which the two [*culture* and *civilization*] were contrasted, with culture referring to material products and technology; then a phase in which the contrast was maintained but the meanings reversed, technology and science being now called civilization; and, beginning more or less concurrently with this second phase, there was also a swing to the now prevalent non-differentiation of the two terms, as in most anthropological writing" (25)—but not necessarily elsewhere. For a notable example of the association of *civilization* with the material realm and of *culture* with the artistic and intellectual, see John A. Kouwenhoven, *The Arts in Modern American Civilization* (New York: Norton, 1967), chap. 1. By contrast, see George E. Mowry's casual and undefined use of *technological culture* to encompass various developments in the 1920s in his collection of primary sources, *The Twenties: Fords, Flappers, and Fanatics*, ed. Mowry (Englewood Cliffs, N.J.: Spectrum Books, 1963), 43.

24. For extensive elaboration, see *Evolutionary Thought in America*, ed. Stow Persons (New Haven, Conn.: Yale University Press, 1950), esp. chap. 1 by Robert Scoon, "The Rise and Impact of Evolutionary Ideas," and the entries "Evolution" and "Revolution" in Williams, *Keywords*, 103–105 and 226–30.

25. The process of change will not usually be the cause-and-effect process of the mechanical social order that is detailed, with the alternate organic social order, in chap. 5. For extensive elaboration, see Cynthia Eagle Russett, *The Concept of Equilibrium in American Social Thought* (New Haven, Conn.: Yale University Press, 1966).

26. For extensive elaboration, see Sumner H. Slichter, "Efficiency," and H. S. Person, "Waste," in *The Encyclopaedia of the Social Sciences*, ed. Edwin R. A. Seligman (New York: Macmillan, 1931), 5: 437–39, and (1935), 15: 367–69. See also chap. 6 below, nn. 23 and 26, on the meanings of *efficiency* for scientific management. Significantly, neither entry appears in the 1968 *International Encyclopedia of the Social Sciences*. For modest elaboration, see James Weinstein, *The Corporate Ideal in the Liberal State, 1900–1918* (Boston: Beacon, 1968), xiii–xiv; Otis L. Graham, Jr., *The Great Campaigns: Reform and War in America, 1900–1928* (Englewood Cliffs, N.J.: Prentice-Hall, 1971), 155; and Melvin G. Holli, "Urban Reform in the Progressive Era," in *The Progressive Era*, ed. Lewis L. Gould (Syracuse, N.Y.: Syracuse University Press, 1974), 143–44, 146.

27. See Donald Hoke, "Conference Report: The Rise of the American System of Manufactures, 1800–1870: Smithsonian Institution, March 1978," *Technology and Culture* 21 (January 1980): 67–70, and the book growing out of the conference, *Yankee Enterprise: The Rise of the American System of Manufactures* (Washington, D.C.: Smithsonian Institution Press, 1981). See also the varied uses of *system* in the volume that popularized the term, reprinted as *The American System of Manufactures: The Report of the Committee on the Machinery of the United States, 1855, and the Special Reports of George Wallis and Joseph Whitworth, 1854*, ed. Nathan Rosenberg (Edinburgh: Edinburgh University Press, 1969). A careful reading of the report supports the historians who see *system* as applying even at this early date to more than machines and structures and economics alone. On the application of *system* in these various respects to twentieth-century technology, see the fine summary in Drucker, "Technological Trends in the Twentieth-Century," in *Technology in Western Civilization* 2: 19–22. Drucker stresses the need for conceptualization of any *system* before its implementation. And on the application of *system* in these same respects to the whole of the history of technology, see the excellent summary in Thomas P. Hughes, "The Order of the Technological World," in *History of Technology*, ed. Hall and Smith (London: Mansell, 1980), 5: 1–16.

28. For excellent case studies in this application of *system* to whole societies,

starting with cities, see Graham R. Taylor, *Satellite Cities: A Study of Industrial Suburbs* (New York: Appleton, 1915); Zane L. Miller, "Daniel Drake, the City, and the American System," in *Physician to the West: Selected Writings of Daniel Drake on Science and Society*, ed. Henry D. Shapiro and Miller (Lexington: University Press of Kentucky, 1970), xxiii–xxxiv; Miller, "Cincinnati: A Bicentennial Assessment," *Cincinnati Historical Society Bulletin* 34 (Winter 1976): 231–49; and, most revealingly, Geoffrey Giglierano, "The City and the System: Developing a Municipal Service, 1800–1915," *Cincinnati Historical Society Bulletin* 35 (Winter 1977): 223–47. See also the broader but more technical studies in Allan R. Pred, *The Spatial Dynamics of U.S. Urban-Industrial Growth, 1800–1914: Interpretive and Theoretical Essays* (Cambridge: MIT Press, 1966). On the application of *system* to the entire United States, see the brief but illuminating Introduction of Samuel P. Hays to *Building the Organizational Society: Essays on Associational Activities in Modern America*, ed. Jerry Israel (New York: Free Press, 1972). These publications and, more important, the primary sources on which they are based, would certainly qualify Russett's contention, "The idea that society, or a particular aspect of society such as political life, can be analyzed as a system of interdependent parts is by now a commonplace of social research, but it has come to be so only rather recently" (118). Similarly, the absence of an entry on *system* in any form from *The Encyclopaedia of the Social Sciences* of the 1930s and the presence of a lengthy entry on "Systems Analysis" in *The International Encyclopedia of the Social Sciences* of the 1960s may not accurately reflect the intellectual currents of the pre-1930s period.

29. For a clear and concise explanation of Weber's concept of rationalization, see Julien Freund, *The Sociology of Max Weber*, tr. May Ilford (New York: Vintage Books, 1969), 17–24. See also the narrower but still wide-ranging conception of the term in Moritz Julius Bonn, "Rationalization," in *The Encyclopaedia of the Social Sciences* (1934), 13: 117–20. On Weber's views of the United States, see Henry Walter Brann, "Max Weber and the United States," *Southwestern Social Science Quarterly* 25 (June 1944): 18–30.

Chapter 2

1. Kenneth M. Roemer, *The Obsolete Necessity: America in Utopian Writings, 1888–1900* (Kent, Ohio: Kent State University Press, 1976), 3. See also Virgil L. Lokke, "The American Utopian Anti-Novel," in *Frontiers of American Culture*, ed. Ray B. Browne et al. (West Lafayette, Ind.: Purdue University Studies, 1968), 123–24, 141–42. Edward Bellamy indicated clearly his intention to make *Looking Backward* as logical and as realistic as nonfiction and as appealing as a reputable novel in his essays "Why I Wrote *Looking Backward*" (1890) and "How I Wrote *Looking Backward*" (1894), both reprinted in Bellamy, *Edward Bellamy Speaks Again! Articles, Public Addresses, Letters* (Chicago: Peerage Press, 1937), 199–203 and 217–28, esp. 223–24.

2. On contemporary reformers and utopians other than the technological utopians, see Daniel Aaron, *Men of Good Hope: A Story of American Progressives* (New York: Oxford University Press, 1951), chaps. 3, 5–7; Frederic Cople Jaher, *Doubters and Dissenters: Cataclysmic Thought in America, 1885–1918* (New York: Free Press, 1964), chaps. 1–2; Robert Wiebe, *The Search for Order, 1877–1920* (New York: Hill and Wang, 1967), chaps. 3, 6; R. Jackson Wilson, *In Quest of Community: Social Philosophy in the United States, 1860–1920* (New York: Wiley, 1968); John L. Thomas, "Utopia for an Urban Age: Henry George, Henry Demarest Lloyd, Edward Bellamy," *Perspectives in American History* 6 (1972): 135–63; John F. Kasson, *Civilizing the Machine: Technology and Republican Values in America, 1776–1900* (New York: Grossman/Viking, 1976), chap. 5; Roemer, esp. chaps. 1–3; and Thomas, *Alternative America: Henry*

George, *Edward Bellamy, Henry Demarest Lloyd and the Adversary Tradition* (Cambridge: Harvard University Press, 1983).

3. Charles W. Wooldridge, *Perfecting the Earth: A Piece of Possible History* (Cleveland: Utopia, 1902), 11. As indicated in chap. 5, this use of "cause and effect" reflects the persistence of a mechanical metaphor and model of society that had generally disappeared by this time.

4. Robert Thurston, "Trend of National Progress," *North American Review* 161 (September 1895): 312. As Bellamy wrote in his Postscript to the second edition of *Looking Backward*, the work, "although in form a fanciful romance, is intended, in all seriousness, as a forecast, in accordance with the principles of evolution, of the next stage in the industrial and social development of humanity. . . ." (*Looking Backward: 2000–1887*, ed. Thomas [Cambridge: Harvard University Press, 1967], 312. And as King Camp Gillette wrote about one of his works, "It is not a dream. It is reality" (*World Corporation* [Boston: New England News, 1910], 98). See also Thurston, "The Mission of Science," *Proceedings, American Association for the Advancement of Science* 33 (September 1884): 233–34; Gillette, *The Human Drift* (Boston: New Era, 1894), 131; Charles W. Caryl, *New Era: Presenting the Plans for the New Era Union to Help Develop and Utilize the Best Resources of this Country* (Denver: n.p., 1897), 16; Edgar Chambless, *Roadtown* (New York: Roadtown, 1910), 172; and Gillette, *The People's Corporation* (Boston: Ball, 1924), 126–29. Here, as elsewhere in this chapter, I present the most illuminating quotations in the text and cite other, equally representative quotations in the notes. Throughout the chapter the sequence of citations reflects the chronological order of publication. The absence of references to the works of all twenty-five technological utopians for every point made in this chapter reflects an absence, in those works not mentioned, of apt quotations or descriptions; it does not reflect a contrary position on those particular points.

5. Of the twenty-five advocates of technological utopia, all but five specify a date before the year 2000, and all but two specify the United States (in part or whole or as part of North America). Roemer, chaps. 2 and 3, esp. pp. 32–33 and 54–55, reaches similar conclusions about the nature of time and place in his numerically much larger sample. Because *The Obsolete Necessity* is easily the foremost study of late nineteenth- and early twentieth-century American utopianism in general, in this chapter I repeatedly compare Roemer's findings with mine.

6. Thomas Kirwan (pseud. William Wonder), *Reciprocity (Social and Economic) in the Thirtieth Century, the Coming Cooperative Age; A Forecast of the World's Future* (New York: Cochrane, 1909), 115, and Bellamy, *Equality* (New York: Appleton, 1897), 305–306. "It is not," says Byron Brooks, "a revolution, but an evolution" (*Earth Revisited* [Boston: Arena, 1893], 102–103). "The process of ascent [toward utopia]," says William A. Taylor, "is slow and purely evolutionary" (*Intermere* [Columbus, Ohio: Twentieth-Century, 1901], 41). See also Bellamy, *Looking Backward*, 127–28. Many, though not all, of the works Roemer treats also stress evolutionary rather than revolutionary change; see Roemer, chap. 4. Here, as elsewhere in this chapter, I specify whether the person quoted is the author (one of the technological utopians) or one of his fictional creations (one of the inhabitants of technological utopia).

7. John Bachelder, *A.D. 2050. Electrical Development at Atlantis* (San Francisco: Bancroft, 1890), 4, and Alvarado M. Fuller, *A.D. 2000* (Chicago: Laird and Lee, 1890), 24. See also John Macnie (pseud. Ismar Thiusen), *The Diothas; Or, A Far Look Ahead* (New York: Putnam, 1883), 1–3; Bellamy, *Looking Backward*, 103–108; Ludwig A. Geissler, *Looking Beyond: A Sequel to* Looking Backward *by Edward Bellamy and an Answer to* Looking Further Forward *by Richard Michaelis* (New Orleans: Graham, 1891), 18; Chauncey Thomas, *The Crystal Button: or, Adventures of Paul Prognosis in the Forty-Ninth Century*, ed. George Houghton (Boston and New York: Houghton Mifflin, 1891), 1–9; Brooks, 1–5; Henry Olerich, *A Cityless and Countryless World; An*

Outline of Practical Co-operative Individualism (Holstein, Ia.: Gilmore and Olerich, 1893), 16–25; D. L. Stump, *From World to World* (Asbury, Mo.: World to World, 1896), 6; Albert A. Merrill, *The Great Awakening: The Story of the Twenty-Second Century* (Boston: George, 1899), 9–10; Paul Devinne, *The Day of Prosperity: A Vision of the Century to Come* (New York: Dillingham, 1902), 34–37; George S. Morison, *The New Epoch as Developed by the Manufacture of Power* (Boston and New York: Houghton Mifflin, 1903), 128–31; Kirwan, 10–11; Herman H. Brinsmade, *Utopia Achieved: A Novel of the Future* (New York: Broadway, 1912), 1–3; and Fred M. Clough, *The Golden Age, Or the Depth of Time* (Boston: Roxburgh, 1923), 9–17.

8. Roemer, chap. 2, esp. pp. 25–32, discerns a similar difficulty among his utopians of connecting history with the end of history.

9. Wooldridge, 325, and Gillette, *World Corporation*, 240. Chambless envisions Roadtown as "'A New Heaven and a New Earth' here on this God plowed and human harrowed planet in this the early years of the Twentieth Century" (172).

10. Morison, 28. See also Olerich, 354, and Bellamy, *Equality*, 4, 382. Lokke, 146, surely exaggerates these attitudes toward the past by characterizing them as hatred as opposed to indifference. On the implications for literary form, style, and character development (or lack thereof) of this antihistorical outlook within American utopian writings overall, see Jean Pfaelzer, "The Impact of Political Theory on Narrative Structures," in *America as Utopia: Collected Essays*, ed. Roemer (New York: Burt Franklin, 1981), 119–120.

11. Thurston, "The Mission of Science," 236; Gillette, *People's Corporation*, 32; and Chambless, 27–28. See also Macnie, 78; Bachelder, 25; Morison, 31; and Olerich, *Modern Paradise; an Outline or Story of how some of the cultured will probably live, work and engage in the near future* (Omaha: Equality, 1915), 57.

12. Solomon Schindler, *Young West: A Sequel to Edward Bellamy's Celebrated Novel* Looking Backward (Boston: Arena, 1894), 45. As Chambless puts it, "Roadtown eliminates all possible waste" (26–27). See also Macnie, 78, 88; Bellamy, *Looking Backward*, 117; Bachelder, 23–24; Fuller, 245, 247; Olerich, *Cityless and Countryless World*, 81–82, 342–44; Gillette, *Human Drift*, 91, 108–109; Albert W. Howard (pseud. M. Auburré Hovorrè), *The Milltillionaire* (Boston: n.p., 1895), 10; Stump, 42, 44–46, 97; Bellamy, *Equality*, 40; Merrill, 94–95; Taylor, 140–41; Devinne, 76–78, 107–109, 129; Wooldridge, 138, 143; Kirwan, 126–27; and Chambless, 64–68, 165–67.

13. Brooks, 43. Devinne and Howard provide almost equally vivid pictures of domesticated technology: "You must have noticed the universal cleanliness which prevails everywhere—in the streets, in our houses, on railway, and in street cars. . . . Our whole city, you might say, is a big garden without dust or dirt" (Devinne, 191–92). And: "Under such a bardic regime all cities become gradually transformed into immense palaces nicely intermingled with fragrant gardens and luxuriant parks—there being no dirty streets or unsightly habitations of any description, the cities being interlined with Triple Highways, canopied and as secure from all storms and dust and heat as the marbled hall of any palace" (Howard, 5). See also Fuller, 143–44; Thomas, 61, 83; Caryl, 16; Merrill, 12, 39; Kirwan, 129; and Chambless, 53–55, 135. Here, as in the remainder of this chapter, all references to Thomas are to utopian Chauncey, not historian John.

14. Thomas, 96, and Stump, 2. Morison sees man's unprecedented capacity "to manufacture power" as the beginning of what he calls "The New Epoch": "The manufacture of power means that *wherever needed* we can now produce practically unlimited power; whatever the measure of a single machine, that machine can be used to make a greater one; we are no longer limited by animal units, confined by locations of waterfalls, nor angered by the uncertain power of wind" (4–5). See also Thomas, 114, 125; Howard, 9; and Thurston, "Progress and Tendency of Mechanical Engineering in the Nineteenth Century," *Popular Science Monthly* 59 (May 1901): 35.

15. See Leo Marx, *The Machine in the Garden: Technology and the Pastoral Ideal in America* (New York: Oxford University Press, 1964). See also Howard P. Segal, "Leo Marx's 'Middle Landscape': A Critique, A Revision, and An Appreciation," *Reviews in American History* 5 (March 1977): 137–50.

16. Howard, 9. As Gillette puts it, ". . . there will be only a few cities on the North American continent, probably only three or four of great size, where the larger part of the population will live" (*People's Corporation*, 165–66). See also Fuller, 245, 304–305; Thomas, 73; Gillette, *Human Drift*, 87; Stump, 31; Caryl, 102–105; Merrill, 37; Devinne, 67–68; Kirwan, 182–83; Chambless, 20; Gillette, *World Corporation*, 220–22; and Jeff W. Hayes, "Portland, Oregon, A.D. 1999," in his *Portland, Oregon, A.D. 1999 and Other Sketches* (Portland: Baltes, 1913), 6–7. The utopians themselves do not use the word *megalopolis*, which aptly describes their vision and was first used in 1961 to characterize the Northeastern United States. See Jean Gottmann, *Megalopolis: The Urbanized Northeastern Seaboard of the United States* (Cambridge: MIT Press, 1961). Roemer, pp. 44–45 and 54–55 and chap. 8, finds a similar orientation toward the city and the suburb and away from "virgin land" among his utopians. On the relationship in nineteenth-century American utopianism between machine and garden, see also Donald C. Burt, "Utopia and the Agrarian Tradition in America, 1865–1900" (Ph.D. diss., University of New Mexico, 1973), chaps. 3–6, and Burt, "The Well-manicured Landscape: Nature in Utopia" in *America as Utopia*, 175–85.

17. On planning in American history, see chaps. 5 and 6.

18. Devinne, 68, and Gillette, *Human Drift*, 89.

19. Devinne, 56–57, and Gillette, *Human Drift*, 97. See also Macnie, 6; Bellamy, *Looking Backward*, 115; Fuller, 143; Thomas, 15–16, 51–52, 7; Brooks, 31; Olerich, *Cityless and Countryless World*, 53–59; Howard, 7; Stump, 12; Caryl, 19, 103–105; Devinne, 55, 62, 127–28; Brinsmade, 24–25, 30–31; Hayes, 6; and Gillette, *People's Corporation*, 167–69.

20. Howard, 7. See also Macnie, 33, 127–28; Fuller, 245; Thomas, 49–50, 57; Brooks, 30–31; Olerich, *Cityless and Countryless World*, 59, 116; Gillette, *Human Drift*, 92–93, 128; Caryl, 101, 104; Devinne, 56; Chambless, 53–55; and Hayes, 6.

21. Taylor, 77; Macnie, 31; and Thomas, 121. Chambless goes as far as to foresee skyscrapers—laid on their sides rather than in the air—spanning the entire American countryside: "I would extend the blotch of human habitations called cities out in radiating lines. I would surround the city worker with the trees and grass and woods and meadows and the farmer with all the advantages of city life" (20). See also Caryl, 103–105; Brinsmade, 16–17; Hayes, 5; Olerich, *Modern Paradise*, 42, 154–55; and Clough, 33, 39–41.

22. Gillette, *People's Corporation*, 125, 173; Macnie, 85; and Thomas, 114. A citizen in Devinne's novel boasts, "All our farms are operated in accordance with the rules of science. Agriculture is a science highly developed. . . . Modern agriculture and forestry are reduced to the exactitude of factory work" (201–202). And Bellamy's West recounts, "Still we swept on mile after mile, league after league, toward the interior, and still the surface below presented the same parklike aspect that had marked the immediate environs of the city. Every natural feature appeared to have been idealized and all its latent meaning brought out by the loving skill of some consummate landscape artist, the works of man blending with the face of Nature in perfect harmony. . . . 'How far does this park extend?' I demanded at last. 'There seems no end to it.' 'It extends to the Pacific Ocean,' said the doctor. 'Do you mean that the whole United States is laid out in this way?' 'Not precisely in this way by any means, but in a hundred different ways according to the natural suggestions of the face of the country and the most effective way of co-operating with them" (*Equality*, 296). See also Macnie, 320; Geissler, 38, 43–45, 51; Thomas, 114; Olerich, *Cityless and Countryless World*, 61–63, 120–122; Gillette, *Human Drift*, 113; Schindler, 28; Howard, 9; Stump, 31, 38,43; Caryl,

117; Taylor, 79, 90–91; Kirwan, 129; Chambless, 38–39, 90–101, 121; Gillette, *World Corporation*, 146–48; Hayes, 17; Olerich, *Modern Paradise*, 45, 73, 101; Clough, 47–48; and Gillette, *People's Corporation*, 189–90.

23. On the use of electricity in technological utopia, see Macnie, 20–21, 30, 32–35; Bellamy, *Looking Backward*, 168; Bachelder, 23–24; Fuller, 151, 309–14; Thurston, "The Border-Land of Science," *North American Review* 150 (January 1890): 75; Geissler, 37–38; Thomas, 50, 63; Brooks, 46–47; Olerich, *Cityless and Countryless World*, 53, 61; Gillette, *Human Drift*, 110; Howard, 8–9; Stump, 38, 45; Caryl, 16; Taylor, 76–77, 128; Kirwan, 9–10; Chambless, 102; Brinsmade, 17–18; Hayes, 4; and Clough, 21.

24. Clough, 40. "Science," says Gillette, "is annihilating space" (*People's Corporation*, 19). See also Chambless, 33–35, 161–62, and Gillette, *World Corporation*, 221–22.

25. On the means of transportation and communication, see Macnie, 20–21, 30, 32–35, 41; Bellamy, *Looking Backward*, 164–65; Bachelder, 23–24; Fuller, 136–37, 150–51, 171, 239–40, 347; Thurston, "Border-Land of Science," 73–79; Geissler, 24–25, 28, 38, 85–102; Thomas, 50, 55, 65, 128, 181–88, 193–206, 240–42, 263–65; Brooks, 30, 33, 94–96, 154; Olerich, *Cityless and Countryless World*, 59, 116, 153, 158–59; Gillette, *Human Drift*, 91–92; Howard, 7–8; Stump, 10, 12, 38, 44, 46–49, 63, 106–107; Caryl, 16, 104; Merrill, 15–16, 39, 96–97, 121–22, 188–89, 211–12, 269; Taylor, 55–61, 76–77; Devinne, 56, 69–70, 195–96; Wooldridge, 139; Kirwan, 12, 15–17, 33–34, 189–95, 208–10; Chambless, 40–49, 56–57, 70–71; Brinsmade, 5, 8, 21, 33; Hayes, 4–5, 37; Clough, 40, 176–77; and Gillette, *People's Corporation*, 60.

26. Olerich, *Cityless and Countryless World*, 131, and Bellamy, *Looking Backward*, 211. On these technological advances, see Macnie, 41–44, 101–104, 132–34, 168–69; Bellamy, *Looking Backward*, 160–61, 168; Bachelder, 23–24; Thurston, "Border-Land of Science," 75–76; Geissler, 17; Thomas, 65, 84–85, 128–29; Brooks, 53; Olerich, *Cityless and Countryless World*, 53, 159, 328; Gillette, *Human Drift*, 91, 124; Howard, 7–8; Stump, 45, 82; Caryl, 100, 117; Merrill, 16; Taylor, 140–41; Devinne, 66, 106–109, 133–34; Wooldridge, 140; Kirwan, 22–23, 65, 86, 89–90; Chambless, 68, 72, 82–84, 105, 153; Brinsmade, 12, 20–22; Hayes, 36–37; and Gillette, *People's Corporation*, 60.

27. Gillette, *Human Drift*, 89, and *World Corporation*, 232. See also Macnie, 6, 168; Bellamy, *Looking Backward*, 168–69; Thomas, 231; Brooks, 53; Gillette, *Human Drift*, 24, 26, 124; Howard, 22; Stump, 33–35, 44, 82; Caryl, 100; Merrill, 10, 15–16, 94, 97; Devinne, 66, 95, 108; Chambless, 81–83, 165–67; and Gillette, *People's Corporation*, 167–68.

28. Bellamy, *Equality*, 54, and Clough, 34. See also Thurston, "Mission of Science," 236; Olerich, *Cityless and Countryless World*, 131, 142; Gillette, *Human Drift*, 23, 110; Stump, 37–38, 98; Caryl, 100; Taylor, 88; Kirwan, 116–24; and Chambless, 105–107.

29. Howard, 17. See also Macnie, 83–85, 320; Fuller, 248–51; Thomas, 103–104; Geissler, 38–39, 51; Olerich, *Cityless and Countryless World*, 61; Gillette, *Human Drift*, 87–89; Schindler, 267; Howard, 9–10, 15–16; Bellamy, *Equality*, 68–69, 298–99; Taylor, 32; Devinne, 200–203; Wooldridge, 3–8, 244–47; Chambless, 166–67; Hayes, 7, 18, 29, 35; Clough, 52–54, 107–108, 124; and Gillette, *People's Corporation*, 171–72.

30. Brinsmade, 39, and Gillette, *People's Corporation*, 152, 177. See also Macnie, 12–13, 120; Geissler, 17; Thomas, 61; Stump, 37–38; Bellamy, *Equality*, 55, 249; and Devinne, 72. Roemer observes that one of the favorite utopian schemes for "transforming the average American into an ideal individual implied that it was possible to plan, create, and control a new identity for mankind" (75).

31. The technological utopian vision does not, however, entail overwork. As Harold Loeb says, "Though releasing no one from a certain minimum effort which provides for the satisfaction of physical wants, a technocracy [technological utopia] would require so small a proportion of the total of human time and energy that everyone would have ample scope to pursue any outside activity which he fancied" (*Life in a Technocracy: What It Might Be Like* [New York: Viking, 1933], 60). On daily work in

utopia, see also Macnie, 12; Bachelder, 37; Geissler, 17; Olerich, *Cityless and Country-less World*, 173; Schindler, 35; Howard, 5; Caryl, 116; Taylor, 87; Devinne, 75; Kirwan, 61; Chambless, 114–15; Brinsmade, 35–36, 39; Clough, 140; and Loeb, 55. On leisure activities in utopia, see Macnie, 12; Thurston, "Mission of Science," 236, 249–50; Bellamy, *Looking Backward*, 163–66, 196, 221–23, 240; Olerich, *Cityless and Country-less World*, 54, 116, 324, 348–50; Schindler, 35; Caryl, 100, 136; Devinne, 110–11; Wooldridge, 176; Kirwan, 38; Chambless, 123–26; Brinsmade, 40; Clough, 45–46; and Loeb, 104–78. Roemer, 93–95, 150–51, discerns a similar emphasis among his utopians on making effective use of leisure time.

32. Olerich, *Cityless and Countryless World*, 74; Thomas, 61, 126; Olerich, *Cityless and Countryless World*, 350; Bellamy, *Equality*, 34; and Brooks, 45. See also Geissler, 38–39, and Devinne, 178.

33. Olerich, *Cityless and Countryless World*, 6, and Gillette, *People's Corporation*, 121. See also Thurston, "Mission of Science," 228; Bellamy, *Looking Backward*, 178; Geissler, 4; Thomas, 122; Brooks, 55, 134; Stump, 37; Caryl, 16; Devinne, 227; Kirwan, 5, 201–202; Chambless, 116–22; Hayes, 39–40; Olerich, *Modern Paradise*, 15; Clough, 172–73; and Gillette, *People's Corporation*, 19–20, 93–95. Roemer, 76–77, also stresses the pragmatic emphasis on cooperation among his utopians.

34. Gillette, *People's Corporation*, 161.

35. Chambless, 132. See also Macnie, 12, 115; Thurston, "Mission of Science," 237, 244–46; Bellamy, *Looking Backward*, 129–36; Bachelder, 34; Thurston, "The New Education and the New Civilization—Their Unity," *Scientific American Supplement* 34 (July 30, 1892): 13825–26; Brooks, 92; Olerich, *Cityless and Countryless World*, 324; Schindler, 69–70, 72–73; Howard, 10; Devinne, 90, 172–78; Wooldridge, 127–28; Kirwan, 42–43, 46–48; and Hayes, 10. Roemer, 121–23, 149, finds a similar emphasis on technical subjects among his utopians.

36. Macnie, 114, and Gillette, *World Corporation*, 134. Loeb, for instance, justifies including "general knowledge" in the curriculum on the grounds that "it is nearly as important for real efficiency as the essential knowledge of the particular industrial processes." He immediately qualifies even this mild concession: "But such general knowledge would be subordinate to the practical end. It would be knowledge for the sake of use, not knowledge for the sake of the student's satisfaction" (114).

37. Thomas, 126, and Gillette, *Human Drift*, 121. See also Stump, 50; Merrill, 343; and Gillette, *World Corporation*, 239.

38. Thurston, "Progress and Tendency," 38, and Stump, 36. See also Macnie, 156–62; Thurston, "Mission of Science," 246–47; Bellamy, *Looking Backward*, 131–32, 171–82; Geissler, 50; Thomas, 243–44; Olerich, *Cityless and Countryless World*, 94; Gillette, *Human Drift*, 21–29, 117–21; Schindler, 136; Caryl, 101, 134–35; Wooldridge, 27; Chambless, 131; Gillette, *World Corporation*, 142; and Gillette, *People's Corpora-tion*, 120–22, 141–49.

39. Kirwan, 53; Bellamy, *Looking Backward*, 133; and Gillette, *World Corporation*, 87. "Every worker," says Gillette, "will begin at the broad base of the industrial pyramid, . . . choosing agriculture, manufacturing, or any other pursuit, according to his own taste. He will gradually rise to a higher class of employment, as children rise to a higher class in our school system" (*People's Corporation*, 148). See also Macnie, 119–20; Bellamy, *Looking Backward*, 133–34; Geissler, 66; Thomas, 178–79; Olerich, *Cityless and Countryless World*, 217–18; Schindler, 47, 52; Stump, 21, 101; Caryl, 111; Devinne, 219–21; Kirwan, 53, 197; Chambless, 131, 163–67; Brinsmade, 36–38; and Gillette, *People's Corporation*, 144–47, 181–84.

40. Gillette declares, "The greater the knowledge the higher the individual will rise in the corporate system" (*People's Corporation*, 181). See also Brooks, 55.

41. Many authors adopt Henry George's single-tax scheme as a means of preventing the accumulation of excessive wealth, unlike other late nineteenth-century utopians,

who saw the single tax as a panacea in itself. On the accumulation of money and property, see Macnie, 115; Bellamy, *Looking Backward*, 166-68; Thomas, 268–70; Brooks, 98–101; Olerich, *Cityless and Countryless World*, 156, 230; Howard, 5; Stump, 29; Merrill, 117; Chambless, 163–65, 168; Brinsmade, 73–80; and Gillette, *People's Corporation*, 149.

42. The utopians reject socialism not only in name but also in fact. They seek a balance between government and private ownership. Chambless predicts, "In Roadtown the lamb of socialism shall lie down with the leopard of individuality and a child of the common good shall lead them" (116–17); and Brooks notes, "We find the happy medium between individual ownership of all property and common ownership of all property, in the communal ownership of certain properties and personal proprietorship of others" (101). See also Macnie, 12, 115; Fuller, 168, 269–70; Geissler, 20; Thomas, 266, 268; Olerich, *Cityless and Countryless World*, 156–57, 222, 267; Stump, 30; Bellamy, *Equality*, 29–30, 91, 117; Merrill, 206–207, 264–67; Taylor, 44–45; Kirwan, 54; Chambless, 116–19, 139–41; Hayes, 12; Clough, 41; and Gillette, *People's Corporation*, 140–41, 180–82. Roemer, 77–78, 91–92, finds a similar balance among his utopians. Even Lyman Tower Sargent, in his survey, "Capitalist Eutopias in America," finds few advocates of completely unregulated capitalism. Rather, most of the visionaries he cites—including three of my technological utopians (Bachelder, Caryl, and Fuller)—sought regulated capitalism and limitations on private ownership of property. His inclusion of those three thus complements rather than contradicts my inclusion of them. See Sargent's essay in *America as Utopia*, 192–205. On the darker side of these technocratic schemes, see Arthur Lipow's provocative *Authoritarian Socialism in America: Edward Bellamy and the Nationalist Movement* (Berkeley and Los Angeles: University of California Press, 1982).

43. Loeb, 75. See also Geissler, 87–89; Howard, 17; Bellamy, *Equality*, 273–74; and Taylor, 91–92.

44. Olerich, *Cityless and Countryless World*, 245, 257. Declares Thomas: "Our representatives are what the name implies: they simply represent the best talent that is available for the office, talent that has been specially chosen, cultivated, and trained for the purpose of adapting it to the duties of that particular office" (257). Gillette adds: "Politics, which we recognize as a necessary governmental part of our competitive, industrial system, will have no place under 'World Corporation'. . . . There will be no voting, [and] no political campaigns. . . ."(*World Corporation*, 54). See also Bellamy, *Looking Backward*, 129–30; Thomas, 252–53; Stump, 93; Chambless, 142–43; Hayes, 137–38; and Gillette, *People's Corporation*, 56.

45. Loeb, 102–103. Loeb's description of political government continues: "Its offices would be elective, thereby titillating the egos of those who like to think they are running things. Prominent clowns will, doubtless, be frequently elected. Incumbents will not be released from their productive duties. In fact they will surely make a great to-do about them. How much more dignified would our chief executive have been if he had spent four hours a day steering the scoop of a steam shovel! And how much more useful!" (103). Roemer, 135–38, finds his utopians as antipolitical as those described here.

46. The laws themselves have been reduced to little more than a set of moral precepts, whose meaning and application are self-evident. Thus technological utopia has no lawyers and few court cases. "Honesty, simplicity, and efficiency," goes a typical "Code of Common Sense" (Thomas, 249). "Mechanics," claims Chambless, "is the foundation of all that is good and bad in civilization, law the paint on the finished structure" (168). See Fuller, 263–64; Olerich, *Cityless and Countryless World*, 250–53; Stump, 87, 94; Merrill, 327; Devinne, 161–62, 240; and Kirwan, 80, 135–36. Roemer, 137, 146, discerns a similar attitude toward law among his utopians.

47. Howard explains: "We have no need of the service of the ancient 'lawyers,' a profession only concomitant with warring factors and elements of ignorance, and the existence of which is virtually impossible in a Millennial State—since, did they exist, it would not be a Millennial State" (19). See also Bellamy, *Looking Backward*, 226–31; Fuller, 262; Bachelder, 36; Stump, 95; Taylor, 92; Kirwan, 46–48; Hayes, 15–16; and Gillette, *People's Corporation*, 169, 185. Roemer, 145–46, also finds dislike of lawyers among his utopians.

48. On the technical nature of utopian government, see Thurston, "Mission of Science," 246–47, 252; Bellamy, *Looking Backward*, 216–20; Fuller, 230–33; Thomas, 261; Brooks, 134–38; Olerich, *Cityless and Countryless World*, 244–45; Schindler, 196, 202–203, 267; Gillette, *Human Drift*, 120–21; Caryl, 147; Devinne, 49–51, 236–38; Morison, 47; Kirwan, 80; Hayes, 7–8; Clough, 146–48, 162, 174–76; and Gillette, *People's Corporation*, 141–43, 148–49, 163–64.

49. Thurston, "Progress and Tendency," 37; Morison, 39–40; and Gillette, *People's Corporation*, 130. See also Thurston, "An Era of Mechanical Triumphs," *Engineering Magazine* 6 (January 1894): 467, and Gillette, *People's Corporation*, 59, 129.

50. Thurston, "Mission of Science," 233. Another citizen of utopia avers: "Literature, music, the fine arts, and religion are but the superficial embellishments of the age in which they exist, and are moulded by it, but in themselves . . . exert no influence on progress," which is indebted instead "to science" (Merrill, 342). See also Macnie, 82–83; Geissler, 28–29; Brooks, 72–75; Howard, 30; Bellamy, *Equality*, 247; Devinne, 63; Hayes, 10; and Olerich, *Modern Paradise*, 129.

51. Merrill, 139, and Thurston, "Trend of National Progress," 310. See also Bachelder, 76, and Brooks, 151.

52. Roemer, 138, 150, finds a similar emphasis upon utilitarianism and mass appeal in the cultural activities of his utopians. See also Lokke, 143–46.

53. See also Brooks, 120–22; Olerich, *Cityless and Countryless World*, 439–40; Howard, 12, 14; and Bellamy, *Equality*, 255.

54. Bellamy, *Equality*, 344–45; Howard, 12; and Kirwan, 49–50. See also Macnie, 177, 179, 185–86; Bellamy, *Looking Backward*, 272–74; Bachelder, 37–38; Thurston, "The Scientific Basis of Belief," *North American Review* 153 (August 1891): 181–92; Brooks, 114–15, 121; Olerich, *Cityless and Countryless World*, 439, 447; Stump, 21; Caryl, 231; Merrill, 314, 343; Taylor, 112; Devinne, 160; Kirwan, 50, 75, 206; Chambless, 169; Clough, 156; and Loeb, 106.

55. Brooks, 122; Thurston, "Scientific Research: The Art of Revelation and of Prophecy," *Science* 16 (September 12, 1902): 402; and Morison, 75–76. Roemer, 96–99, describes his utopians' attitude toward religion, which is closely akin to that sketched in this paragraph.

56. Macnie, 45.

57. See Macnie, 147–48; Bachelder, 35–36; Thomas, 89–90; Brooks, 93; Olerich, *Cityless and Countryless World*, 99, 321, 332–35, 342–44, 351; Schindler, 38, 44; Howard, 6, 13, 21–22; Stump, 29, 67–69; Bellamy, *Equality*, 49, 51, 285; Devinne, 88–89, 92–93, 104, 161–62; Kirwan, 66–67; and Hayes, 16.

58. Nevertheless, women in technological utopia can vote, hold office, hold jobs, and attend college—and so possess a greater degree of equality than did women at the time most of the utopians wrote. Chambless observes, "The Roadtown woman will be free to do anything and everything she chooses except home drudgery" (84). On the role of women, see Macnie, 121–22, 172–73; Bellamy, *Looking Backward*, 169, 262–71; Geissler, 17; Olerich, *Cityless and Countryless World*, 90, 263–69; Bellamy, *Equality*, 43–44, 130–31, 153; Merrill, 246–47; Taylor, 47, 93–94; Devinne, 127, 131–33, 221–22, 233; Kirwan, 67–68, 124; Chambless, 74–88, 112–13; Gillette, *World Corporation*, 238–39; Hayes, 9, 12; and Olerich, *Modern Paradise*, 50, 91. Roemer, 124–33, finds a

similar improvement in the role of women in the writings of his utopians. Lokke, 141–42, suggests that the very inferiority of women in real world turn-of-the-century America—their being thought more emotional than men—ironically made them superior to men, in the eyes of utopian novelists (nearly all of them men) of all stripes, as potential converts to those utopian ideologies. Women, he contends, were widely viewed as the means by which more rational men, the true powers in American society, would see the light and embrace the cause.

59. Says Thomas: "No diseased or deformed person who is liable to communicate serious imperfections of any kind to offspring is ever allowed to marry" (68). Bachelder is blunter: "the deaf, dumb, blind and idiotic are not permitted to marry" (35). See also Stump, 110; Kirwan, 98–99, 153–54; Hayes, 13–14; and Loeb, 174–78. Roemer, 70–75, 82, observes a like emphasis upon physical, mental, and moral purification in the writings of his utopians. The general health of the great majority of inhabitants of technological utopia is extraordinarily good. "Sickness," says Stump, "is an unnatural condition" (79). Boasts Clough: "We have the promise, at no distant time, of reaching a deathless life" (155). See also Macnie, 10; Thurston, "Mission of Science," 252; Bellamy, Looking Backward, 240–41; Thomas, 38, 66–68; Olerich, Cityless and Countryless World, 82, 226; Taylor, 53, 118; Devinne, 111, 190; Kirwan, 37, 88; Chambless, 133; Hayes, 12–13; and Clough, 46.

60. The emotions portrayed, especially love, are, in the fashion of the age, so romanticized as to seem wholly artificial. Roemer, 124–25, finds a parallel situation in the writings of his utopians.

61. On cremation, see Macnie, 285–87; Bachelder, 23–24; Olerich, Cityless and Countryless World, 432–34; Schindler, 90–95; Howard, 28; Devinne, 190; Wooldridge, 202–205; Kirwan, 88–89; and Hayes, 31.

62. Schindler, 111. See also Brooks, 121, 316–17; Olerich, Cityless and Countryless World, 434, 436, 443; Merrill, 310; and Taylor, 116–17.

63. By coincidence, there have recently appeared several works on eighteenth- and nineteenth-century European utopianism, each of which argues that their particular utopian writers were pioneering social scientists—and usually, consciously so. These works have drawn connections between their subjects, some of them hitherto dismissed as "mere" utopians, and contemporary social scientists. See Keith Michael Baker, Condorcet: From Natural Philosophy to Social Mathematics (Chicago: University of Chicago Press, 1975); Bernard Cazes, "Condorcet's True Paradox, or, The Liberal Transformed into Social Engineer," Daedalus 105 (Winter 1976): 47–58; Ghita Ionescu, Introduction, The Political Thought of Saint-Simon, ed. Ionescu and tr. Valence Ionescu (London: Oxford University Press, 1976), 1–57; and, above all, Barbara Goodwin, Social Science and Utopia: Nineteenth-Century Models of Social Harmony (Sussex, Eng.: Harvester Press, 1978). The titles of some of Goodwin's chapters—"Social Maladies and Utopian Remedies," "Control in Utopia," "Cohesion in Utopia," "Utopian Values," and "Utopian Social Science"—reflect an approach which could well be applied to other expressions of utopianism, not least technological utopianism. Indeed, the issue raised by the first title was treated in the Introduction. The issues raised by the remaining titles have been treated more implicitly in the present chapter. My point is that technological utopianism is as social scientific in its objectives as the expressions of utopianism examined by these four scholars. It was unnecessary to make such a statement in the present or preceding chapters, but in chap. 9 I will raise this question of the nature of utopianism again and return to the four works mentioned here.

Chapter 3

1. The Appendix lists the twenty-five utopians and the basic information that I have been able to obtain about both their backgrounds and their works.

2. All three settled permanently in the United States as either children or young men and wrote and published their utopian works here.

3. As noted below, at least ten of the utopians showed a youthful interest in technology. Their interests are ascribable in part to their fathers' occupations and other local influences but in part also to their own temperaments.

4. It is nevertheless worth noting that at least nineteen of the twenty-five utopians published books and articles on other subjects in addition to those on utopias used in this study. The literary quality of their non-utopian writings is not very high either.

5. See Kenneth M. Roemer, *The Obsolete Necessity: America in Utopian Writings, 1888–1900* (Kent, Ohio: Kent State University Press, 1976), 9–12.

6. For twelve utopians the father's occupation has been determined, and eleven of them could certainly be said to have achieved a higher status than their fathers did. The twelfth maintained his father's status.

7. The sources of these biographical comments are listed in the Appendix. I have cited a mere handful of those sources in the remainder of this chapter, and then only when essential to provide the sources of particular points.

8. On Thurston's compartmentalized personality, see Robert J. Kwik, "The Function of Applied Science and the Mechanical Laboratory during the Period of Formation of the Profession of Mechanical Engineering, As Exemplified in the Career of Robert Henry Thurston, 1839–1903" (Ph.D. diss., University of Pennsylvania, 1974), 213, 228, 240, 293; on Thomas', see Ormond Seavey, Introduction, Chauncey Thomas, *The Crystal Button: or, Adventures of Paul Prognosis in the Forty-Ninth Century*, ed. George Houghton (Boston: Gregg Press, 1975), vi; and on Bellamy's, see John L. Thomas, Introduction, Edward Bellamy, *Looking Backward: 2000–1887*, ed. Thomas (Cambridge: Harvard University Press, 1967), 5–6, 69–70, 86, 88.

9. On Gillette's compartmentalized personality, see Russell B. Adams, Jr., *King C. Gillette: The Man and His Wonderful Shaving Device* (Boston: Little, Brown, 1978), 28–29, 81–88, 131–37, and, more briefly, James B. Gilbert, *Designing the Industrial State: The Intellectual Pursuit of Collectivism in America, 1880–1940* (Chicago: Quadrangle, 1972), 162. Gilbert, 178–79, and Roemer, Introduction, King C. Gillette, *The Human Drift* (Delmar, N.Y.: Scholars' Facsimiles and Reprints, 1976), xii–xx, do see some connections between Gillette's two worlds, as does, to a much lesser extent, Adams, 85–88, 131–33, 136–37.

10. Why the technological utopians chose utopian writings rather than communities to effect such change will be taken up in chap. 6.

11. In the doctoral dissertation from which this study grows, I included several of these professionals who provided fragmentary evidence of technological utopian beliefs. I have reluctantly dropped them from this study just because, on reflection, these fragments fell far short of the comprehensive blueprints provided by all of the twenty-five visionaries who remain. And yet I feel certain that those excluded were genuine technological utopians as well. For their names, backgrounds, and writings, see Howard P. Segal, "Technological Utopianism and American Culture, 1830-1940" (Ph.D. diss., Princeton University, 1975), Appendix. A related example of a de facto technological utopian, a figure I did not include in my dissertation but a contemporary of those I did, is Elmer Sperry (1860–1930). See Thomas Parke Hughes, *Elmer Sperry: Inventor and Engineer* (Baltimore: Johns Hopkins University Press, 1971), esp. 311–12. On the attitudes toward the future of engineers and other professionals involved in advancing technology, see chaps. 5 and 6.

12. For a similarly balanced treatment of another generation of American reformers, some of them avowed utopians, see Ronald G. Walters, *American Reformers, 1815–1860* (New York: Hill and Wang, 1978), Preface, Introduction, and Afterword. Walters says cogently, "Like most American reformers, they often envisioned themselves as part of some grand procession stretching across the centuries. . . . At other moments they (again like later reformers) regarded themselves as outsiders whose

critical distance from their society enabled them to see its flaws. . . . Yet there are flaws to the idea that reformers and radicals are greatly different from their non-reformist peers. Whatever illusions crusaders may have about their uniqueness, they are products of a milieu" (213).

13. Roemer, *The Obsolete Necessity*, 6. On a similar mixture of hopes and fears for the future among "the fifty most frequent predictors" (318) of technological advances—technological utopians Bellamy and Thurston among them—writing in this same period, see George Wise, "Technological Prediction, 1890–1940" (Ph.D. diss.,Boston University, 1976). Few other members of this diversified group were utopians of any stripe.

14. On this point, see the discussion in chaps. 7 and 8 of how far the technological utopians' prophecies were realized and the significance of this for real world happiness.

15. For an excellent example of such psychological reduction, although sophisticated in analysis, humane in purpose, and applicable to *some* utopians throughout history, see Paul Watzlawick, John Weakland, and Richard Fisch, "The Utopia Syndrome," in their *Change: Principles of Problem Formation and Problem Resolution* (New York: Norton, 1974), chap. 5. For an equally excellent rebuttal of such psychological reductionism, one readily applicable to disciplines beyond literature, the author's specialty, see Frederick Crews, *Out of My System: Psychoanalysis, Ideology, and Critical Method* (New York: Oxford University Press, 1975), chaps. 1, 9.

16. Some of these real world developments will be explored in chaps. 5 and 6, as will some of the other, more conventional solutions offered to counteract them. On these anxieties, see Neil Harris, "Utopian Fiction and Its Discontents," in *Uprooted Americans: Essays to Honor Oscar Handlin*, ed. Richard L. Bushman et al. (Boston: Little, Brown, 1979), 211–44, and Roemer, *The Obsolete Necessity*. Like Roemer's book, Harris' essay is a notable contribution to utopian studies, but it is unclear on the extent to which (science and) technology, in the eyes of these writers, would simultaneously solve and produce problems.

17. See Segal, "*Young West*: The Psyche of Technological Utopianism," *Extrapolation: A Journal of Science Fiction and Fantasy* 19 (December 1977): 50–58.

18. On the scientific evidence for anal and oral characters, in fact as well as theory, see Paul Kline, *Fact and Fantasy in Freudian Theory* (London: Methuen, 1972), chaps. 2–5, 13, and Seymour Fisher and Roger P. Greenberg, *The Scientific Credibility of Freud's Theories and Therapy* (New York: Basic Books, 1977), chap. 3 and pp. 399–404. Both books make clear, however, that the existence of such characters does not necessarily confirm the Freudian theories of their origins. David Shapiro has broadened Freud's notion of the anal character by expanding its existing psychological dimensions into one of several "neurotic styles"—here, the "obsessive-compulsive style." See his *Neurotic Styles*, Austin Riggs Center Monograph Series no. 5 (New York: Harper Torchbooks, 1973), chap. 2. For case studies of two prominent Americans with personalities similar to those of the fictional inhabitants of technological utopia, if not necessarily of their creators, see Anne Jardim, *The First Henry Ford: A Study in Personality and Business Leadership* (Cambridge: MIT Press, 1970), and especially Sudhir Kakar, *Frederick Taylor: A Study in Personality and Innovation* (Cambridge: MIT Press, 1970).

19. On these dimensions of Bellamy's personality, see Thomas, Introduction, *Looking Backward*, 5–6, 86; on Gillette's, see Adams, 15, Gilbert, 161, and William A. Titus, "King C. Gillette" (manuscript in possession of author, Sheboygan, Wis., 1948), 2; on Morison's, see "Memoir of George Shattuck Morison," *Transactions, American Society of Civil Engineers* 54 (1905): 514, 518, 519, 521, and George A. Morison, *George Shattuck Morison, 1842–1903: A Memoir* (Peterborough, N.H.: Peterborough Historical Society, 1940), 4–5, 8, 18–21; on Olerich's, see his "Autobiography" (manuscript in

possession of Professor H. Roger Grant, University of Akron, Omaha, 1921), Cassius V. Cook, "The Life and Achievements of Henry Olerich" (manuscript, University of Michigan Library, Department of Rare Books and Special Collections, Labadie Collection, Cassius V. Cook Papers, n.d.), and Grant, "Interview with Mrs. Viola Storms," December 20, 1973 (manuscript, Iowa State Historical Society, Iowa City); on Thomas', see Seavey, Introduction, *The Crystal Button*, vi, and Glenn P. Negley and J. Max Patrick, biographical note, excerpt from *The Crystal Button*, in *The Quest for Utopia: An Anthology of Imaginary Societies*, ed. Negley and Patrick (New York: Schuman, 1952), 82; and on Thurston's, see William F. Durand, *Robert Henry Thurston: A Biography* (New York: American Society of Mechanical Engineers, 1929), 18, 25–26, 160, and Durand, "Robert Henry Thurston," *Dictionary of American Biography* 18 (New York: Scribner, 1936): 519. In any case, as Frank E. and Fritzie P. Manuel observe, "There is a sense in which the mental act of creating a utopian world, or the principles for one, is psychologically a regressive phenomenon for an individual. In this respect the utopian is kindred to the religious, scientific, and artistic creators who flee to the desert, suffer psychic crises, become disoriented by the contradictions between accepted reality and the new insights of which they have a glimmer" (*Utopian Thought in the Western World* [Cambridge: Harvard University Press, 1979], 27).

20. On the hopes and especially the fears of late nineteenth- and early twentieth-century American utopian writers in general, see Harris, 211–44. Harris is not himself guilty of reductionism. He is reacting against earlier analyses of these writings, which overemphasized their optimism and rarely peered beneath the surface for traces of any other moods. Harris discusses several of the writers and works in this study and indeed stresses the importance of technology for a majority of the visionaries he treats. In contrast, R. Jackson Wilson, "Experience and Utopia: The Making of Edward Bellamy's *Looking Backward*," *American Studies* 11 (April 1977): 45–60, is guilty of reductionism, at least concerning that one book. Wilson overemphasizes Bellamy's fears of industrialization, of the laboring masses, of factories, and of large cities. More important, he misreads both *Looking Backward* and *Equality* as seeking a pastoral, placid pre-industrial past rather than, as I have discussed in chap. 2 and in n. 15 there, a new middle landscape. Wilson's essay is reprinted, with a few additions and a handful of other changes, as the Introduction to a new edition of *Looking Backward* (New York: Modern Library/Random House, 1982), vii–xxxiv.

21. On American utopianism functioning in this manner, see John L. Thomas, "Utopia for an Urban Age: Henry George, Henry Demarest Lloyd, Edward Bellamy," *Perspectives in American History* 6 (1972): 135–39, 145, 148; John F. Kasson, *Civilizing the Machine: Technology and Republican Values in America, 1776–1900* (New York: Grossman/Viking Press, 1976), chap. 5; Francine C. Cary, "The World a Department Store: Bradford Peck and the Utopian Endeavor," *American Quarterly* 29 (Fall 1977): 370–84; Robert D. Thomas, *The Man Who Would Be Perfect: John Humphrey Noyes and the Utopian Impulse* (Philadelphia: University of Pennsylvania Press, 1977); Wilson, 45–60; Harris, 211–44; Frederick E. Pratter, "The Mysterious Traveler in the Speculative Fiction of Howells and Twain," in *America as Utopia: Collected Essays*, ed. Roemer (New York: Burt Franklin, 1981), 78–90; and John Thomas, *Alternative America: Henry George, Edward Bellamy, Henry Demarest Lloyd and the Adversary Tradition* (Cambridge: Harvard University Press, 1983), 14–15, 50, 100–101, 118–20, 123, 170, 187, 286, 305–306, 332–33, 347, 357–58, 366. For a confirming example from European utopianism, see Mark Poster, *The Utopian Thought of Restif de la Bretonne* (New York: New York University Press, 1971), chaps. 7–8. See also the more general discussion of utopianism as play and, as such, as a suspension of disbelief about mankind and society in Elisabeth Hansot, *Perfection and Progress: Two Modes of Utopian Thought* (Cambridge: MIT Press, 1974), 201–203.

22. As Frank Manuel observed in an influential essay in 1965, utopianism is more

than wish fulfillment or anxiety "fulfillment." It is also a form of dreaming, with all that that implies regarding pleasant and unpleasant thoughts and feelings in integrated or fragmentary states of semiconsciousness or unconsciousness. See Manuel, "Toward a Psychological History of Utopias," in *Utopias and Utopian Thought: A Timely Appraisal*, ed. Frank Manuel (Boston: Beacon, 1967), 69–98. Curiously, Manuel resisted applying psychology to the utopians he discussed in this very essay. And where he did apply it to utopianism—in the *magnum opus* written with his wife, *Utopian Thought in the Western World*—he unfortunately provided a number of glaring examples of the reductionism he had warned about in 1965. See the perceptive review of the book by Gorman Beauchamp in *Michigan Quarterly Review* 19 (Spring 1980): 263–64.

23. Manuel, "Toward a Psychological History of Utopias," 70. On the psychology of utopianism, see, in addition to Manuel's essay, his *The Prophets of Paris* (New York: Harper Torchbooks, 1965), 202–205; Manuel and Manuel, Introduction, *French Utopias: An Anthology of Ideal Societies*, ed. and tr. Manuel and Manuel (New York: Schocken, 1971), 3–4, 11–12; Thomas, Introduction, *Looking Backward*, 1–88; Poster, chaps. 7–8; Thomas, *The Man Who Would Be Perfect*; Richard Bienvenu, "Utopia?" in *France and North America: Utopias and Utopians*, ed. Mathé Allain (Lafayette: Center for Louisiana Studies, University of Southwestern Louisiana, 1978), 165–72; Bienvenu, "Reflections: European Utopias and Society," and Michael Fellman, comment on Bienvenu's paper (paper delivered at annual meeting, American Historical Association, San Francisco, December 1978); and Thomas, *Alternative America*.

Chapter 4

1. On technological advances in Europe, see T. S. Ashton, *The Industrial Revolution: 1760–1830* (New York: Oxford University Press, 1964); Phyllis Deane, *The First Industrial Revolution* (Cambridge: Cambridge University Press, 1969); David S. Landes, *The Unbound Prometheus: Technological Change and Industrial Development in Western Europe from 1750 to the Present* (Cambridge: Cambridge University Press, 1969); and Malcolm I. Thomis, *The Luddites: Machine-Breaking in Regency England* (New York: Schocken, 1972).

2. *Utopia*'s dependence on Plato's *Republic* has been well established, although scholars disagree over its nature and extent. See, for example, Edward Surtz, Introduction, Thomas More, *Utopia*, ed. Surtz (New Haven, Conn.: Yale University Press, 1964), xii–xviii; J. H. Hexter, *More's Utopia: The Biography of an Idea* (New York: Harper Torchbooks, 1965), 27, 50, 62, 82–85; the Introduction by William Nelson and the essays by R. W. Chambers, Ernest Barker, and Ernst Cassirer in *Twentieth Century Interpretations of Utopia: A Collection of Critical Essays*, ed. Nelson (Englewood Cliffs, N.J.: Spectrum Books, 1968); James Steintrager, "Plato and More's *Utopia*," *Social Research* 36 (Autumn 1969): 357–72; and Hexter, *The Vision of Politics on the Eve of the Reformation: More, Machiavelli, and Seyssel* (New York: Basic Books, 1973), 31, 40, 43–44, 53–54, 68, 88–89, 102–103, 120–23. Comparisons between the two works often conclude that the *Republic* is less realistic and more abstract than *Utopia*. See, for example, Surtz, xiii; Hexter, *More's Utopia*, 50; and Steintrager, 367. The basis for this conclusion is that Socrates professes to be describing an imaginary utopia, where More describes his utopia as if it actually existed. The form chosen by Plato hardly means that he consigns utopia to the imagination and not to a possible reality. For my purposes, the difference between the *Republic* and *Utopia* is not that Plato's vision is less real than More's. Rather, it is that Plato's vision historically inspired few others—until, that is, More; and More, in contrast, inspired a whole utopian tradition, eventually made retroactive to Plato. More made utopianism a full-fledged mode of thought. On that

tradition, see Frank E. Manuel, Introduction, *Utopias and Utopian Thought: A Timely Appraisal*, ed. Manuel (Boston: Beacon, 1967), vii–viii; Hexter, *The Vision of Politics*, 117–18; Robert M. Adams, Preface, *Utopia: Norton Critical Edition*, ed. Adams (New York: Norton, 1975), viii–ix; Manuel and Fritzie P. Manuel, *Utopian Thought in the Western World* (Cambridge: Harvard University Press, 1979), 12–15; J. C. Davis, *Utopia and the Ideal Society: A Study of English Utopian Writing, 1516–1700* (New York: Cambridge University Press, 1981), Introduction and chap. 2; and Lyman Tower Sargent, "The 'Utopian Tradition' " (paper in possession of author), 1–23. On Plato's relatively limited influence before *Utopia* appeared, see John Ferguson, *Utopias of the Classical World* (Ithaca, N.Y.: Cornell University Press, 1975), chaps. 9–10, 12–13, 18–20.

3. To quote Hexter: "More did not believe that men were so hopelessly corrupt as to be forever incapable of a decent social order. Or, to state the same thing positively, More did believe human nature to be sufficiently malleable so that under the proper external—Marx might have said objective—conditions it was capable of attaining to the Good Society" (*More's Utopia*, 58). See also Hexter, "Thomas More: On the Margins of Modernity," *Journal of British Studies* 1 (November 1961): 20–37; Surtz, xxvii–xxx; and Hexter, *The Vision of Politics*, 126–27, 132, 136. Regardless of the evidence of optimism presented in the next paragraph, Judith Shklar ("The Political Theory of Utopia: From Melancholy to Nostalgia," in *Utopias and Utopian Thought*, 101–15) and Elisabeth Hansot, *Perfection and Progress: Two Modes of Utopian Thought* (Cambridge: MIT Press, 1974), chap. 4, proclaim More a relentless pessimist. More's failure to explain exactly how man's present nature is to be ameliorated implies no despair over the possibility of its being ameliorated. See Hexter, *More's Utopia*, 58–59, 122–24.

4. On the nature of the agrarian paradise, see P. H. Epps, "The Golden Age," *Classical Journal* 29 (January 1934): 292–96; J. O. Hertzler, "On Golden Ages: Then and Now," *South Atlantic Quarterly* 39 (July 1940): 318–29; Frank Manuel, "The Golden Age: A Mythic Prehistory for Western Utopia," in his *Freedom from History and Other Untimely Essays* (New York: New York University Press, 1971), 69–88, reprinted, with considerable additions, as chap. 2 of Manuel and Manuel; Renato Poggioli, "The Oaten Flute," in his *The Oaten Flute: Essays on Pastoral Poetry and the Pastoral Ideal* (Cambridge: Harvard University Press, 1975), 1–41; Manuel and Manuel, chap. 1; Gorman Beauchamp, "The Dream of Cockaigne: Some Motives for the Utopias of Escape," *Centennial Review* 24 (Fall 1981): 345–62; Davis, 20–26; and Sargent, "The 'Utopian Tradition,' " 1–23. The Manuels observe, "Utopia is a hybrid plant, born of the crossing of a paradisaical, otherworldly belief of Judeo-Christian religion with the Hellenic myth of an ideal city on earth. The naming took place in an enclave of sixteenth-century scholars excited about the prospect of a Hellenized Christianity" (15). On More's freedom from nostalgia, see Hexter, *The Vision of Politics*, 131–32.

5. Within a hundred years of its founding, Christianity as a whole had ceased to expect an imminent millennium. Nevertheless, sects proclaiming its imminence arose periodically, and as an intellectual tradition millenarianism became brashest in the fifteenth and sixteenth centuries, which encompassed More's lifetime.

On the history of millenarian movements and millenarian thought, see Loraine Boettner, *The Millennium* (Philadelphia: Presbyterian and Reformed Church, 1958); *Millennial Dreams in Action: Essays in Comparative History*, ed. Sylvia Thrupp (The Hague: Mouton, 1962), especially the essays by Thrupp and George Shepperson; Ernest L. Tuveson, *Millennium and Utopia: A Study in the Background of the Idea of Progress* (New York: Harper Torchbooks, 1964); Norman Cohn, *The Pursuit of the Millennium: Revolutionary Millenarians and Mystical Anarchists of the Middle Ages*, 3d ed. (New York: Oxford Galaxy Books, 1970); Manuel and Manuel, "Sketch for a Natural History of Paradise," *Daedalus* 101 (Winter 1972): 83–128; Tuveson, "Millenarianism," in

Dictionary of the History of Ideas, ed. Philip P. Wiener (New York: Scribner, 1973), 3:223–25; John G. Gager, *Kingdom and Community: The Social World of Early Christianity* (Englewood Cliffs, N.J.: Prentice-Hall, 1975), chap. 2; Davis, 31–36; and Theodore Olson, *Millennialism, Utopianism, and Progress* (Toronto: University of Toronto Press, 1982). The distinction between utopianism and millenarianism is not absolute, particularly when the language of utopianism is religious or quasireligious, as it in fact is in American technological utopianism. Nevertheless, the principal agent of change in a given scheme—man or God—is usually clear. On this point, see also chap. 5, n. 75.

6. Hexter, *More's* Utopia, 63. On the historical background of this conceptual change, see Leslie C. Tihany, "Utopia in Modern Western Thought: The Metamorphosis of an Idea," in *Ideas in History: Essays Presented to Louis Gottschalk by His Former Students,* ed. Richard Herr and Harold T. Parker (Durham, N.C.: Duke University Press, 1965), 21–22, and Manuel, *Utopias and Utopian Thought,* vii–viii.

7. More, *Utopia,* 128.

8. Ibid., 137.

9. See Manuel, "Pansophia, A Seventeenth-Century Dream of Science," in *Freedom from History,* 89–113; the essay is reprinted, with considerable changes, as chap. 7 of Manuel and Manuel, *Utopian Thought in the Western World.*

10. On Tommaso Campanella, Johann Andreae, and Francis Bacon collectively, see, besides ibid., Robert P. Adams, "The Social Responsibilities of Science in *Utopia, New Atlantis,* and After," *Journal of the History of Ideas* 10 (June 1949): 374–98; René Dubos, *The Dreams of Reason: Science and Utopias* (New York: Columbia University Press, 1961), chap. 2; Richard F. Jones, *Ancients and Moderns: A Study of the Rise of the Scientific Movement in Seventeenth-Century England,* 2d ed. (St. Louis: Washington University Press, 1961); Judah Bierman, "Science and Society in *The New Atlantis* and Other Renaissance Utopias," *Publications of the Modern Language Association of America* 78 (December 1963): 492–500; Frederick B. Artz, *The Development of Technical Education in France, 1500–1800* (Cambridge: MIT Press, 1966), 2–7; Nell Eurich, *Science in Utopia: A Mighty Design* (Cambridge: Harvard University Press, 1967), chap. 4; Marie L. Berneri, *Journey through Utopia* (New York: Schocken, 1971), chap. 2; Manuel and Manuel, *Utopian Thought in the Western World,* chaps. 9–11; and Davis, chaps. 3, 5. On Campanella, Andreae, and Bacon separately, see Bernardino M. Bonansea, *Tommaso Campanella: Renaissance Pioneer of Modern Thought* (Washington, D.C.: Catholic University of America Press, 1969); Felix E. Held, *Johann Valentin Andreae's* Christianopolis: An Ideal State of the Seventeenth Century (Ph.D. diss., University of Illinois, 1914; reprinted Urbana: Graduate School, University of Illinois, 1914); Benjamin Farrington, *Francis Bacon: Pioneer of Planned Science* (New York: Praeger, 1963), chap. 13; J. Weinberger, "Science and Rule in Bacon's Utopia: An Introduction to the Reading of the *New Atlantis,*" *American Political Science Review* 70 (September 1976): 865–85; Weinberger, Introduction, Bacon, *The Great Instauration and New Atlantis,* ed. Weinberger (Arlington Heights, Ill.: AHM, 1980), vii–xxix; and Anthony F.C. Wallace, *The Social Context of Innovation: Bureaucrats, Families, and Heroes in the Early Industrial Revolution, as Foreseen in Bacon's* New Atlantis (Princeton, N.J.: Princeton University Press, 1982).

11. For descriptions of these and other technological advances, see Campanella, *The City of the Sun,* tr. Thomas W. Halliday, in *Peaceable Kingdoms: An Anthology of Utopian Writings,* ed. Robert L. Chianese (New York: Harcourt Brace Jovanovich, 1971), 8–41; Andreae, *Christianopolis,* tr. Held, in Held, 131–280; and Bacon, *The New Atlantis,* in *Ideal Commonwealths,* ed. Henry Morley (London: Routledge, 1896), 171–213. Both Dubos and Sanford A. Lakoff ("The Third Culture: Science in Social Thought," in *Knowledge and Power: Essays on Science and Government,* ed. Lakoff [New York: Free Press, 1966], 5–14) note that, at the time Andreae and Bacon wrote, the

mere idea, let alone the practice, of applied group research of this kind barely existed. Yet Berneri remarks of Bacon, the most imaginative of the Pansophists, "Many have been dazzled and have talked of his extraordinary prophetic vision, but most of the inventions he describes as having been achieved had occupied the attention of the philosophers and scientists of the Renaissance and of many others long before their time" (133). See also Wallace, 11–21.

12. Andreae, 173; Bacon, 181.

13. The role that these societies give to Christianity varies inversely with the role they accord technology—The City of the Sun being the most overtly religious society and The New Atlantis the least. But this should pose no problem, for religion and technology complement rather than conflict with each other. Even Bacon, moreover, put limits on scientific and technological development. Frank Manuel ("Toward a Psychological History of Utopias," in Utopias and Utopian Thought, 77) explains: "He still thought of science as a body of knowledge that could be acquired in a finite period of time through assiduous cultivation of his [inductive] method. . . . there is no suggestion of a wildly dynamic republic of science, innovating endlessly, such as [the Marquis de] Condorcet depicted in his Fragment sur l'Atlantide, a commentary on Bacon, some two hundred years later." For an interpretation of Bacon's work that treats his scientific and technological concerns virtually as ends in themselves, see Langdon Winner, Autonomous Technology: Technics-Out-of-Control as a Theme in Political Thought (Cambridge: MIT Press, 1977), 21–25, 135–39.

14. Manuel, "Toward a Psychological History of Utopias," 72–79. As Manuel cautions, his periodization here and elsewhere in the essay is "somewhat arbitrary—it is certainly meant to be illustrative rather than definitive" (71). Hence the appropriateness of my extending another of Manuel's categories—"open-ended utopianism"—later in this chapter.

15. On the nature of pleasure in these utopias of calm felicity, see ibid., 75–79, and Surtz, xiii–xiv.

16. See Condorcet, Sketch for a Historical Picture of the Progress of the Human Mind, tr. June Barraclough (1795; reprinted New York: Noonday, 1955), and Fragment sur l'Atlantide, ou efforts combinés de l'espèce humaine pour le progrès des sciences, in his Oeuvres, ed. A. Condorcet O'Connor and M. F. Arago (Paris: Firmin Didot Frères, 1847), 6:597–606. Published originally in 1804, the latter work has been translated in part as Fragment on the New Atlantis; or, Combined Efforts of the Human Species for the Advancement of Science, tr. Keith Michael Baker, in Condorcet: Selected Writings, ed. Baker (Indianapolis: Bobbs-Merrill, 1976), 283–300. See also Frank Manuel, The Prophets of Paris (New York: Harper Torchbooks, 1965), chap. 2, reprinted largely intact as chap. 20 of Manuel and Manuel, Utopian Thought in the Western World; Baker, Condorcet: From Natural Philosophy to Social Mathematics (Chicago: University of Chicago Press, 1975), 340–41 and chap. 6; Baker, Introduction, Condorcet: Selected Writings, xxxiv–xxxvii; and Bernard Cazes, "Condorcet's True Paradox, or, The Liberal Transformed into Social Engineer," Daedalus 105 (Winter 1976): 47–58.

17. On the consequently ahistorical nature of Condorcet's visions, especially compared with those of fellow contemporary utopians Henri de Saint-Simon and Auguste Comte, both of them discussed below, see Baker, Condorcet: From Natural Philosophy to Social Mathematics, 344, 371.

18. Manuel, "Toward a Psychological History of Utopias," 79–85. See n. 14 above about the flexibility of Manuel's time period boundaries. A relatively minor but well-known and oft-cited utopian work that preceded Condorcet's and that is more transitional than his in its attitudes towards scientific and technological progress is Louis Sébastian Mercier's Memoirs of the Year Two Thousand Five Hundred, tr. W. Hooper (London: Robinson, 1772). Originally published two years earlier as L'An 2440, the

book envisions several advances, such as malleable glass and inextinguishable lamps, among other improvements. But it envisions nontechnological changes as well, such as the reduction of crime and of international conflict which, unlike technological utopia, are not dependent on technological advance. The book's overall significance seems to me to have been exaggerated. Its importance is as a transition between utopianism of calm felicity and of open-endedness.

19. On Saint-Simon's influence and disciples, see J. F. Normano, "Saint-Simon and America," *Social Forces* 11 (October 1932): 8–14; Felix Markham, Introduction, Henri de Saint-Simon, *Social Organization, The Science of Man and Other Writings,* ed. and tr. Markham (New York: Harper Torchbooks, 1964), xxxiii–xliv; Manuel, *Prophets of Paris,* chaps. 3–4, reprinted largely intact as chaps. 25–26 of Manuel and Manuel, *Utopian Thought in the Western World;* Artz, 186, 242–43; Robert B. Carlisle, "Saint-Simonian Radicalism: A Definition and a Direction," *French Historical Studies* 5 (Fall 1968): 430–35; Georg G. Iggers, *The Cult of Authority: The Political Philosophy of the Saint-Simonians,* 2d ed. (The Hague: Nijhoff, 1970); Iggers, New Preface and Introduction, *The Doctrine of Saint-Simon: An Exposition, First Year, 1828–1829,* ed. and tr. Iggers (New York: Schocken, 1972), v–xlvii; Carlisle, "The Birth of Technocracy: Science, Society, and Saint-Simonians," *Journal of the History of Ideas* 35 (July–September 1974): 445–64; Keith Taylor, Introduction, *Henri Saint-Simon (1760–1825): Selected Writings on Science, Industry, and Social Organisation,* ed. and tr. Taylor (London: Croom Helm, 1975), 13–61; and Ghita Ionescu, Introduction, *The Political Thought of Saint-Simon,* ed. Ionescu, tr. Valence Ionescu (London: Oxford University Press, 1976), 1–57. As several of these works make clear, Saint-Simon's disciples frequently deviated from the teachings of their master. On the impact of the French Revolution on Saint-Simon and his disciples, see David Higgs, "Nostalgia, Utopia, and the French Revolution," in *France and North America: Utopias and Utopians,* ed. Mathé Allain (Lafayette: Center for Louisiana Studies, University of Southwestern Louisiana, 1978), 25–32.

20. As I explained in chap. 1 above, the relationship between science and technology in Saint-Simon's day was tenuous. A manifestation of that relationship, as I will show in the next chapter, was the lack of familiarity in the West with the very term *technology.* For different reasons, as I pointed out in the Introduction, the term *ideology* was also relatively unfamiliar in Saint-Simon's time. Yet ideology, like technology, can be applied to his thought without in any way distorting it.

21. See Saint-Simon, *Social Organization, The Science of Man and Other Writings,* 1–27, 69–80. Ghita Ionescu argues that Saint-Simon's utopia was nevertheless an avowedly political society, in marked contrast to the avowedly apolitical ethos of most technocratic societies, including technological utopia. According to Ionescu, Saint-Simon envisioned the diffusion and virtual disappearance of power but the retention of politics, which ordinarily presupposes the existence of power; he "believed that politics, which in an Aristotelian way he saw as the dialectics of decision-making, were perennial." But at this transitional stage in history—from feudalism to industrialization—"he saw the need for a fundamental change in politics: from the . . . 'politics of power,' to the . . . 'politics of abilities' " (Introduction, *The Political Thought of Saint-Simon,* 11–12). Baker, however, offers a more conventional interpretation of the visions of both Saint-Simon and Comte in a penetrating comparison with Condorcet's vision. There is, he claims, "a profound difference between the social mathematics envisaged by Condorcet and the technocratic social science later developed by Saint-Simon and Comte. Condorcet clearly shared with these later 'prophets of Paris' the desire to replace the political conflicts of the revolutionary period with a scientific politics. But while for Saint-Simon and Comte this involved exiling politics in favor of the rational administration of things, their predecessor resolutely refused to take this step. . . . He wished to render political choice as rational as scientific argument, but he nevertheless

insisted on the fundamental differences between the two communities to which they were appropriate" (Condorcet: From Natural Philosophy to Social Mathematics, 340). Yet Condorcet, confesses Baker, did expect that, even if politics in some form remained, "With the growth of knowledge and the spread of enlightenment, power would wither away and social and political conduct would become rational and scientific" (ibid., 386)—the very position Ionescu attributes to Saint-Simon. As I will argue in chap. 7, the expected disappearance of either politics or power, much less both, from even the purest technocratic scheme, such as technological utopianism, is a naive and unrealistic hope reflecting a misunderstanding of the nature and function of politics and power alike.

22. A partial exception is Saint-Simon's plan of 1814 for reorganizing the existing European governments into several national parliaments and one grand international parliament. These bodies would have supreme authority over all national and international affairs. To insure the highest quality of representation, Saint-Simon insisted that their members come only from "men of business, scientists, magistrates, and administrators" (Saint-Simon, Social Organization, 47).

23. See ibid., 81–116, and Manuel, Prophets of Paris, 127.

24. On Comte's views, see Manuel, Prophets of Paris, chap. 6, reprinted largely intact as chap. 30 of Manuel and Manuel, Utopian Thought in the Western World; Gertrud Lenzer, Foreword and Introduction, Auguste Comte and Positivism: The Essential Writings, ed. Lenzer (New York: Harper Torchbooks, 1975), xiii–lxviii; and Kenneth Thompson, Introduction, Auguste Comte: The Foundation of Sociology, ed. Thompson (New York: Wiley, 1975), 3–35.

25. Although Comte wished to persuade his readers otherwise, there are both stylistic and substantive differences between these two major works. The second is less cautious and more venturesome than the first in both tone and content, especially in its application of sociological principles to social problems. The fundamental themes sketched here are, however, the same in both works. Comte's works may therefore be treated as one.

26. Comte dismissed liberty and equality as "antisocial" and as "incompatible with any real organization." Individuals, he argued, "should be regarded, not as so many distinct beings, but as organs of one Supreme Being" (System of Positive Polity [Système de politique positive] tr. J. H. Bridges et al., 1 [New York: Burt Franklin, 1968]: 305, 304, 291).

27. On Comte's influence and disciples, see W. M. Simon, European Positivism in the Nineteenth Century: An Essay in Intellectual History (Ithaca, N.Y.: Cornell University Press, 1963); Richmond L. Hawkins, Auguste Comte and the United States (1816–1853) (Cambridge: Harvard University Press, 1936); and Hawkins, Positivism in the United States (1853–1861) (Cambridge: Harvard University Press, 1938).

28. Comte, System of Positive Polity, 1–63, and Manuel, Prophets of Paris, 292. In addition, Comte, like Saint-Simon, grew periodically disillusioned with scientists and other technically trained persons. He accused them of overspecialization in their work and of indifference toward society outside of it. Where, however, Saint-Simon began turning away from such persons at the end of his life, Comte believed that they could be reeducated to overcome these limitations.

29. See Raymond Williams, Culture and Society, 1780–1950 (New York: Harper Torchbooks, 1966), chaps. 4, 7, and Herbert L. Sussman, Victorians and the Machine: The Literary Response to Technology (Cambridge: Harvard University Press, 1968), chaps. 1, 3, 4. On Thomas Carlyle and the nature of his desired organic society, see chap. 5 below.

30. See David K. Cohen, "Lemontey: An Early Critic of Industrialism," French Historical Studies 4 (Spring 1966): 290–303, and Cohen, "The Vicomte de Bonald's Critique of Industrialism," Journal of Modern History 41 (December 1969): 475–84.

Iggers finds a similarly intense concern for restoring social and cultural order among the (alleged) disciples of Saint-Simon, even at the expense of the technological advances that, following their master, they otherwise put foremost. See Iggers, New Preface and Introduction, *The Doctrine of Saint-Simon*, v–vii, ix–xiii, xxvi, xlii. On the impact of both the industrial and the democratic revolutions on English and French social critics, see Robert A. Nisbet, *The Sociological Tradition* (New York: Basic Books, 1966), chaps. 1–2. And on the impact of especially the French Revolution on the utopian imagination of the same period, see Bronislaw Baczko, *Lumières de l'utopie (Critique de la Politique)* (Paris: Payot, 1978).

31. Samuel P. Huntington, "Conservatism as an Ideology," *American Political Science Review* 51 (June 1957): 470.

32. On the overlapping of conservative and radical critiques of contemporary technological society, see the illuminating review essay by Edward Shorter, "Industrial Society in Trouble: Some Recent Views," *American Scholar* 40 (Spring 1971): 330–48.

33. The best single study of Robert Owen and the Owenite movement is John F. C. Harrison, *Quest for the New Moral World: Robert Owen and the Owenites in Britain and America* (New York: Scribner, 1969). Harrison's list of other secondary and of primary sources is enormous. But see also in particular Arthur E. Bestor, Jr., *Backwoods Utopias: The Sectarian Origins and the Owenite Phase of Communitarian Socialism in America: 1663–1829*, 2d ed. (Philadelphia: University of Pennsylvania Press, 1970); V. A. C. Gatrell, Introduction, Robert Owen, *A New View of Society and Report to the County of Lanark*, ed. Gatrell (Baltimore: Penguin Books, 1970), 7–81; *Robert Owen: Prince of Cotton Spinners: A Symposium*, ed. John Butt (Newton Abbot, Eng.: David and Charles, 1971); *Robert Owen's American Legacy: Proceedings of the Robert Owen Bicentennial Conference*, ed. Donald E. Pitzer (Indianapolis: Indiana Historical Society, 1972); and Manuel and Manuel, *Utopian Thought in the Western World*, chap. 28. My summary of Owen's life, views, and achievements derives primarily from these six works. The New Lanark Mills continued in operation until 1968. New Harmony became a conventional Midwestern small town but has been extensively renovated in recent years. The other, less famous Owenite communities suffered varying degrees of transformation into small towns and outright neglect.

34. On Pullman and Gary, see chap. 6, n. 74.

35. Gatrell, 43.

36. Owen, *A New View of Society*, 124, 129. As John R. Hume observes, however, Dale's own considerable contributions to New Lanark's economic and social success were, as here, often conveniently overlooked by Owen. See Hume, "The Industrial Archaeology of New Lanark," in *Robert Owen: Prince of Cotton Spinners*, 247.

37. On the need to distinguish between imagined and actual versions of such key components of that pre-industrial community as work, leisure, and play, see chap. 7, nn. 7 and 13. And as John Butt's Introduction to *Robert Owen: Prince of Cotton Spinners*, 15–16, reminds us, New Lanark was no idyllic community but rather a well-ordered business organization.

38. Owen, *Report to the County of Lanark*, 226, and *A New View of Society*, 96. In the *Report*, 253, Owen does use *machine* in an implicit social context. But his intention there, as Harrison makes clear, was to preserve—or restore—the values associated with pre-industrial Britain. In effect, he felt he had to advocate a return to agriculture in order to create model communities that only then would be capable of absorbing the shocks of industrialization. Owen's retreat from industrialization was thus not wholesale. See Harrison, 56–58, 190–192. The Manuels therefore go too far, in *Utopian Thought in the Western World*, 681–82, in characterizing Owen's later visions as overwhelming agrarian and even anti-industrial.

39. The best single study of Charles Fourier and the Fourierist movement is Manuel, *Prophets of Paris*, chap. 5, reprinted, with some changes, as chap. 27 of Manuel

and Manuel, *Utopian Thought in the Western World*. But see also such excellent supplementary works as Nicholas V. Riasanovsky, *The Teaching of Charles Fourier* (Berkeley and Los Angeles: University of California Press, 1969), and Jonathan Beecher and Richard Bienvenu, Introduction, *The Utopian Vision of Charles Fourier: Selected Texts on Work, Love, and Passionate Attraction*, ed. and tr. Beecher and Bienvenu (Boston: Beacon, 1971), 1–75. My summary of Fourier's life, views, and achievements derives primarily from these three works. On Fourier's brief and uneventful correspondence with an Owenite disciple (each tried, unsuccessfully, to convert the other), see Beecher and Bienvenu, 18. As they and additional scholars note, Fourier, in the process of seeking disciples, often lashed out at those of both Owen and Saint-Simon, his foremost ideological rivals, for allegedly stealing—and thereafter distorting—his ideas. On the relations between Owenites and Fourierists, see Harrison, 244–45.

40. See, by contrast, Gatrell, 41–42, concerning Owen's views on passion and diversity.

41. On Fourier's subtle attitudes toward industry, manufacturing, technology, and work, see the illuminating discussions in Riasanovsky, 190, 196–202, and Beecher and Bienvenu, 27–28, 32–34, 43–46, 51.

42. On Karl Marx and Friedrich Engels' reservations about providing detailed blueprints of the future, see chaps. 8 and 9 below.

43. Marx, *The Poverty of Philosophy*, ed. Engels (New York: International, 1963), 109. Here, as elsewhere, I have treated the views of Marx and Engels as fundamentally alike despite minor differences between them emphasized by some scholars and downplayed by others. Recent and reputable interpretations of Marx and Engels as technological determinists can be found in John McMurtry, *The Structure of Marx's World-View* (Princeton, N.J.: Princeton University Press, 1978), chap. 8; William H. Shaw, *Marx's Theory of History* (Stanford, Cal.: Stanford University Press, 1978), chap. 2; and Shaw, " 'The Handmill Gives You the Feudal Lord': Marx's Technological Determinism," *History and Theory* 18 (1979): 155–76. Similarly solid examples of the antithetical interpretation are in Robert C. Tucker, *The Marxian Revolutionary Idea* (New York: Norton, 1969), 14–17, and Nathan Rosenberg, "Marx as a Student of Technology," *Monthly Review* 28 (July–August 1976): 56–77. See also Marx's letter of December 28, 1846, to P. V. Annenkov, a nonsocialist Russian intellectual, in Marx and Engels, *Correspondence, 1846–1895: A Selection with Commentary and Notes*, ed. and tr. Dona Torr (New York: International, 1934), 5–18, and Kostas Axelos, *Alienation, Praxis, and Techne in the Thought of Karl Marx*, tr. Ronald Bruzina (Austin: University of Texas Press, 1976).

44. They were not, however, unaware of some of the problems that that simple distinction might create. See Marx's letter to Engels of January 28, 1863, in their *Correspondence*, 141–44.

45. Marx, *The Grundrisse*, ed. and tr. David McLellan (New York: Harper and Row, 1971), 132, his italics. He continues: ". . . labor . . . is itself only a limb of the system, whose unity exists not in the living workers but in the living (active) machinery, which seems to be a powerful organism when compared to their individual, insignificant activities" (133). See also Marx, *Capital*, ed. Engels (New York: International, 1967), 1:381–82, 384–85. Owen, as noted, used the term "living machinery" in describing New Lanark.

46. Marx, *Grundrisse*, 117. In *Capital*, 1:381, he uses "mechanical monster."

47. They expressed this position perhaps most clearly in *The German Ideology*, in *The Marx-Engels Reader*, ed. Tucker, 2d ed. (New York: Norton, 1978), 155–57, 162–63.

48. Marx, *Grundrisse*, 124. See also ibid., 148.

49. See Marx, "On the Realm of Necessity and the Realm of Freedom," in *The Marx-Engels Reader*, 439–41. The passage comes from *Capital*, vol. 3.

50. Marx and Engels, *The German Ideology*, in ibid., 160. As Marx wrote in *Grund-*

risse, "It is no longer a question of reducing the necessary labour time in order to create surplus labour, but of reducing the necessary labour of society to a minimum. The counterpart of this reduction is that all members of society can develop their education in the arts, sciences, etc., thanks to the free time and means available to all" (142). On the issue of work and leisure in a highly automated socialist society, see Michael Harrington, "Leisure as the Means of Production," and Wlodzimierz Brus, "The Implications of Modern Technology for Socialism: Comments on Michael Harrington's Paper," in *The Socialist Idea: A Reappraisal*, ed. Leszek Kolakowski and Stuart Hampshire (New York: Basic Books, 1974), 153–63 and 164–69.

51. They discuss technicians and technical abilities briefly in *Capital* 1:420.

52. Marx, "Speech at the Anniversary of the *People's Paper*," in *The Marx-Engels Reader*, 577–78. On Marx's (and Engels') vision of the future, see Shlomo Avineri, "Marx's Vision of Future Society and the Problem of Utopianism," *Dissent* 20 (Summer 1973): 323–31.

53. On Etienne Cabet, see Berneri, 219–35, and Christopher H. Johnson, *Utopian Communism in France: Cabet and the Icarians, 1839–1851* (Ithaca, N.Y.: Cornell University Press, 1974). Published originally with a longer title, the novel appeared in a popular revised edition simply as *Voyage en Icarie* (Paris: Mallet, 1842).

54. Many of the other European utopians whose visions treated but did not concentrate on technology are discussed briefly by the Manuels in their Introduction to *French Utopias: An Anthology of Ideal Societies*, ed. and tr. Manuel and Manuel (New York: Schocken, 1971), 1–16.

55. None of these European utopians, it should be added, believed in human perfectibility. Like most serious utopians, as noted in the Introduction, they looked to various social improvements to improve mankind as much as possible within the inherent limitations of the species. By contrast, as described in chap. 2, the technological utopians believed in virtual perfection through genetic as well as social engineering. In addition to the earlier citations for these visionaries, see, for Condorcet, Saint-Simon, and Comte, Manuel, *Prophets of Paris*, 46–48, and for Condorcet alone, Baker, *Condorcet: From Natural Philosophy to Social Mathematics*, 93–95. For these European utopians as a whole, see Hansot, "Reflections on War, Utopias, and Temporary Systems," in *Small Comforts for Hard Times: Humanists on Public Policy*, ed. Michael Mooney and Florian Stuber (New York: Columbia University Press, 1977), 249.

Chapter 5

1. On the European origins of the idea of progress, see J. B. Bury, *The Idea of Progress: An Inquiry into Its Origin and Growth* (New York: Macmillan, 1932); Sidney B. Fay, "The Idea of Progress," *American Historical Review* 52 (January 1947): 231–47; Bruce Mazlish, "The Idea of Progress," *Daedalus* 92 (Summer 1963): 447–61; Georg G. Iggers, "The Idea of Progress: A Critical Reassessment," *American Historical Review* 71 (October 1965): 1–17; W. Warren Wagar, "Modern Views of the Origins of the Idea of Progress," *Journal of the History of Ideas* 28 (March 1967): 55–70; Nathan Rotenstreich, "The Idea of Historical Progress and Its Assumption," *History and Theory* 10 (1971): 197–221; Wagar, *Good Tidings: The Belief in Progress from Darwin to Marcuse* (Bloomington: Indiana University Press, 1972), chaps. 1–2; E. R. Dodds, "Progress in Classical Antiquity," and Morris Ginsberg, "Progress in the Modern Era," in *Dictionary of the History of Ideas*, ed. Philip P. Wiener (New York: Scribner, 1973), 3:623–33 and 633–50; Robert Nisbet, *History of the Idea of Progress* (New York: Basic Books, 1980); and Theodore Olson, *Millennialism, Utopianism, and Progress* (Toronto: University of Toronto Press, 1982). On the application of the idea of progress to America and on America as a potential utopia, see Michael Kraus, "America and the Utopian Ideal in

the Eighteenth Century," *Mississippi Valley Historical Review* 22 (March 1936): 487–504; Arthur A. Ekirch, Jr., *The Idea of Progress in America, 1815–1860* (New York: Columbia University Press, 1944); Rutherford E. Delmage, "The American Idea of Progress, 1750–1800," *Proceedings of the American Philosophical Society* 91 (October 1947): 307–14; Henry Nash Smith, *Virgin Land: The American West as Symbol and Myth* (Cambridge: Harvard University Press, 1950), chap. 11; Rush Welter, "The Idea of Progress in America: An Essay in Ideas and Method," *Journal of the History of Ideas* 16 (June 1955): 401–15; Durand Echeverria, *Mirage in the West: A History of the French Image of American Society to 1815* (Princeton, N.J.: Princeton University Press, 1957); Hans Huth, *Nature and the American: Three Centuries of Changing Attitudes* (Berkeley and Los Angeles: University of California Press, 1957); Clarke A. Chambers, "The Belief in Progress in Twentieth-Century America," *Journal of the History of Ideas* 19 (April 1958): 197–224; Loren Baritz, "The Idea of the West," *American Historical Review* 66 (April 1961): 618–40; Edmundo O'Gorman, *The Invention of America: An Inquiry into the Historical Nature of the New World and the Meaning of Its History* (Bloomington: Indiana University Press, 1961); George H. Williams, *Wilderness and Paradise in Christian Thought* (New York: Harper, 1962); Louis B. Wright, *The Dream of Prosperity in Colonial America* (New York: New York University Press, 1965); Howard Mumford Jones, *O Strange New World: American Culture: The Formative Years* (New York: Viking Compass Books, 1967), chaps. 1–2; Ernest L. Tuveson, *Redeemer Nation: The Idea of America's Millennial Role* (Chicago: University of Chicago Press, 1968); Roderick Nash, *Wilderness and the American Mind*, 2d ed. (New Haven, Conn.: Yale University Press, 1973), chaps. 1–4; David W. Marcell, *Progress and Pragmatism: James, Dewey, Beard, and the American Idea of Progress* (Westport, Conn.: Greenwood Press, 1974); Welter, *The Mind of America, 1820–1860* (New York: Columbia University Press, 1975), 3–9, 17–21, 24–25, 34–35, 113–15; Ronald G. Walters, *American Reformers, 1815–1860* (New York: Hill and Wang, 1978); *Antebellum American Culture: An Interpretive Anthology*, ed. David B. Davis (Lexington, Mass.: Heath, 1979), Introduction and pt. 4; Joel Nydahl, "From Millennium to Utopia Americana," in *America as Utopia: Collected Essays*, ed. Kenneth M. Roemer (New York: Burt Franklin, 1981), 237–53; Lyman Tower Sargent, "Utopianism in Colonial America," *History of Political Thought* 4 (Winter 1983): 483–522.

2. For details of these revisions in the ideas just listed, see the works in n. 1 above on America in the eighteenth, nineteenth, and twentieth centuries.

3. Given these revisions by the technological utopians of those originally European ideas, it is hardly surprising that none of their works cite the European utopians discussed in the preceding chapter. Similar revisions by other Americans may account for the apparently few citations of those Europeans elsewhere in America as well. In any event, the question of influence of ideas across generations and across cultures is an enormously complex and thus far unresolved issue in history and related disciplines. The mere presence or absence of citations is among the least interesting aspects of the debate over influence. Yet neither the paucity of citations nor the complexity of this issue should lead us to infer a general American innocence of European influence. Bacon, it seems, was well known and widely respected in nineteenth-century America. See Merle Curti, *The Growth of American Thought*, 2d ed. (New York: Harper, 1951), 333–35, and Ekirch, 106 (though Ekirch adds that Bacon's works "were probably more reverenced than read" [ibid.]). Andreae, however, was also influential, at least in seventeenth-century America. See Henry A. Pochmann, *German Culture in America: Philosophical and Literary Influences, 1600–1900* (Madison: University of Wisconsin Press, 1957), 20–21. On the difficulties in using printed materials and citations of them as indices of the circulation of ideas, see Robert Darnton, "Reading, Writing, and Publishing in Eighteenth-Century France: A Case Study in the Sociology of Literature," in *Historical Studies Today*, ed. Felix Gilbert and Stephen R. Graubard (New York: Norton, 1972), 238–80, and David D. Hall, "The World of Print and Collective Mentality

in Seventeenth-Century New England," in *New Directions in American Intellectual History*, ed. John Higham and Paul K. Conkin (Baltimore: Johns Hopkins University Press, 1979), 166–80. Just as the technological utopian writings, except for *Looking Backward*, as I explained in the Introduction, were neglected in part because their fundamental ideas were already popular, so these European writings met obscurity in part for the same reason.

4. On America as an existing paradise with abundant natural resources, see Hugh Honour, *The New Golden Land: European Images of America from the Discoveries to the Present Time* (New York: Pantheon, 1975). The European images that Honour brilliantly recreates through pictures and words do not preclude other European—and, more important, American—images of America as a potential advanced society, as the following pages will make clear.

5. Daniel J. Boorstin, for example, suggests that newly arrived Europeans and, in due time, native-born Americans simply took stock of their natural resources and proclaimed them utopia. Frederick Jackson Turner and his disciples make the same connection but restrict those natural resources to free land alone. See Boorstin, *The Genius of American Politics* (Chicago: University of Chicago Press, 1953), esp. chap. 1, and David M. Potter, *People of Plenty: Economic Abundance and the American Character* (Chicago: University of Chicago Press, 1954), chap. 7, esp. p. 149, quotation from Turner. On the supposed (geographical) "givenness" of American values, see also Ekirch, 13, 37, 267; Welter, "The Idea of Progress," 405–406; and Henry Steele Commager, "America and the Enlightenment," in *The Development of a Revolutionary Mentality* (Washington, D.C.: Library of Congress, 1972), 23. On the profound conservatism of this approach to American history, see the Introduction above, n. 10. The gap between appreciation of abundant natural resources and proclamation of them as utopia cannot be bridged by the naive Baconian views of American values held by Boorstin, Turner, and others: that is, that values are culled from the environment rather than read into it. Ironically, in arguing for contemporary "conceptual art" in America as a conscious response to the "limits to growth" that I discuss in chap. 8, John Wilmerding makes the same conservative assumption about the geographical givenness of American values for earlier periods of American history as these historians do. See his "The End of Growth As an American Value? A View through the Arts," in *Growth in America*, ed. Chester L. Cooper (Westport, Conn.: Greenwood Press, 1976), 154–72.

6. Historian Francis Jennings puts it well, "American society is the product not only of interaction between colonists and natives but of contributions from both." Yet "Civilization was not brought from Europe to triumph over the Indians; rather the Indians paid a staggering price in lives, labor, goods, and lands as their part in the creation of modern American society and culture. . . ." For, finally, "The American land was more like a widow than a virgin. Europeans did not find a wilderness here; rather, however involuntarily, they made one. . . . The so-called settlement of America was a resettlement, a reoccupation of a land made waste by the diseases and demoralization introduced by the newcomers" (*The Invasion of America: Indians, Colonialism, and the Cant of Conquest* [New York: Norton, 1976], vi–vii, 41, 30.

7. This fact has generally been ignored by those historians who read nineteenth- and twentieth-century paeans to abundance back into earlier periods of our history. Ironically, contemporary concerns about possible future scarcity in a society only recently proclaimed "affluent" account for whatever reconsideration of past affluence has occurred.

8. Jennings observes, "European explorers and invaders discovered an inhabited land. Had it been pristine wilderness then, it would possibly be so still today, for neither the technology nor the social organization of Europe in the sixteenth and seventeenth centuries had the capacity to maintain, of its own resources, outpost colonies thousands of miles from home" (15). See also Jennings, 33–34. On the re-

visionist interpretations of American abundance summarized in this and succeeding paragraphs, see Daniel M. Fox, *The Discovery of Abundance: Simon N. Patten and the Transformation of Social Theory* (Ithaca, N.Y.: Cornell University Press, 1967); Simon N. Patten, *The New Basis of Civilization*, ed. Fox (Cambridge: Harvard University Press, 1968); Michael Kammen, "From Scarcity to Abundance—to Scarcity? Some Implications for the American Tradition from the Perspective of a Cultural Historian," in Kenneth E. Boulding, Kammen, and Seymour Martin Lipset, *From Abundance to Scarcity: Implications for the American Tradition* (Columbus: Ohio State University Press, 1978), 37–63; and Zane L. Miller, "Scarcity, Abundance, and American Urban History," *Journal of Urban History* 4 (February 1978): 131–55. These works do not themselves draw the connection between technological advance and utopian fulfillment that I have drawn, but they do enable me to put that connection in sharper historical perspective. Potter's *People of Plenty* is a landmark pioneering study of American abundance but suffers from a false assumption of persistent abundance throughout American history.

9. Joseph Whitworth, "Special Report of Mr. Joseph Whitworth," in *The American System of Manufactures: The Report of the Committee on the Machinery of the United States, 1855, and the Special Reports of George Wallis and Joseph Whitworth, 1854*, ed. Nathan Rosenberg (Edinburgh: Edinburgh University Press, 1969), 387–88. Despite its false assumption of persistent abundance throughout American history, Potter's *People of Plenty* correctly recognizes the transformations needed to convert natural resources into finished products. See Potter, 161–62.

10. See Hugo A. Meier, "The Technological Concept in American Social History, 1750–1850" (Ph.D. diss., University of Wisconsin, 1950), 548–50; Robert H. Bremner, *From the Depths: The Discovery of Poverty in the United States* (New York: New York University Press, 1956), pt. 1; Meier, "Technology and Democracy, 1800–1860," *Mississippi Valley Historical Review* 43 (March 1957): 618–40; Fred Somkin, *Unquiet Eagle: Memory and Desire in the Idea of American Freedom, 1815–1860* (Ithaca, N.Y.: Cornell University Press, 1967), chap. 1; and John F. Kasson, *Civilizing the Machine: Technology and Republican Values in America, 1776–1900* (New York: Grossman/Viking Press, 1976), chap. 1. Throughout *People of Plenty* the reader can sense Potter's personal unease with excessive abundance; its negative impact upon the American character is implicit in his account.

11. On Henry George's *Progress and Poverty*, see the relevant portions of John L. Thomas, *Alternative America: Henry George, Edward Bellamy, Henry Demarest Lloyd and the Adversary Tradition* (Cambridge: Harvard University Press, 1983). For evidence of similar assumptions about the likelihood of technological solutions for social problems in antebellum America, see below on the earlier technological utopians and Meier, "The Technological Concept," 556–57, 561–63.

12. In the absence of more empirical and more comprehensive evidence of American public opinion on technology in the nineteenth and early twentieth centuries, scholars have relied on impressionistic materials. Studies using such materials vary markedly in quality. The best of them include Samuel Rezneck, "The Rise and Early Development of Industrial Consciousness in the United States, 1760–1830," *Journal of Economic and Business History*, Supplement to 4 (August 1932): 784–811; Ekirch, chap. 4; Meier, "The Technological Concept"; Meier, "Technology and Democracy," 618–40; Meier, "American Technology and the Nineteenth-Century World," *American Quarterly* 10 (Summer 1958): 116–30; Ekirch, *Man and Nature in America* (New York: Columbia University Press, 1963), chap. 4; Perry Miller, *The Life of the Mind in America: From the Revolution to the Civil War* (New York: Harvest Books, 1965), bk. 3; Brooke Hindle, "The Exhilaration of Early American Technology: An Essay," in his *Technology in Early America: Needs and Opportunities for Study* (Chapel Hill: University of North Carolina Press, 1966), 3–28; Carroll W. Pursell, Jr., and Melvin Kranz-

berg, "Epilogue," in *Technology in Western Civilization*, ed. Kranzberg and Pursell (New York: Oxford University Press, 1967), 1:739–43; Boorstin, *The Americans: The Democratic Experience* (New York: Random House, 1973); Russel B. Nye, *Society and Culture in America, 1830–1860* (New York: Harper and Row, 1974), 24–31, 258–82; Welter, *Mind of America*, 113–15; Davis, Introduction and pt. 4; Alan Trachtenberg, *The Incorporation of America: Culture and Society in the Gilded Age* (New York: Hill and Wang, 1982), chap. 2. Meier's dissertation, of which his two published articles summarize only a small portion, is particularly illuminating in its breadth and in the number and variety of arguments unearthed for technology—and, to a much lesser extent, against. Meier certainly supports his principal point that the "technological concept," as he calls it, was "a vital part of earlier [late eighteenth- and early nineteenth-century] American social history" (534). Yet as Thomas Parke Hughes cautions, "There has never been a single 'American attitude' toward technology" (Introduction, *Changing Attitudes toward American Technology*, ed. Hughes [New York: Harper and Row, 1975], 1). Hughes' book is the foremost survey of those attitudes exactly because it is not a conventional social scientific survey but rather a sophisticated sampler of various attitudes in selected periods of American history. Comparable to Hughes' book in its range of primary sources is *Readings in Technology and American Life*, ed. Pursell (New York: Oxford University Press, 1969). Because Pursell's concerns are broader than Hughes', his interpretation of attitudes is comparatively more modest. A good example of an impressionistic study that ultimately fails to persuade the reader is Boorstin, *The Republic of Technology: Reflections on Our Future Community* (New York: Harper and Row, 1978). Chapters 5 and 6, in fact, have nothing to do with technology.

13. As Hughes, Introduction, *Changing Attitudes*, puts it, "Depression, wars, and vistas of unexploited land and resources have stimulated reactions or attitudes toward the uses or potential of technology. But what influences these attitudes at any one time is a complex combination of events, trends, and ideas; a war or a depression may be only the most obvious immediate cause of the crystallization of attitudes" (7–8). Attitudes toward technology and conceptions of technology are separate if clearly related issues.

14. Meier, "American Technology," 119. My own research has confirmed this statement, and for the early nineteenth century as well. Meier adds, " 'The useful arts' was perhaps the phrase most commonly used in popular writings about applied science" ("Technology and Democracy," 618).

15. On Jacob Bigelow's life, see the sketch in the *Dictionary of American Biography* (New York: Scribner, 1929), 2:257–58, and the memorial in the *Proceedings, Massachusetts Historical Society* 17 (March 1880): 383–467. On the Rumford Professorship, see Sanborn C. Brown, *Benjamin Thompson, Count Rumford* (Cambridge: MIT Press, 1979), 305–306.

16. Perry Miller contends that *Elements of Technology* "should be honored as a major document in American intellectual development; it has not been so esteemed because, of course, the technological revolution already extensively under way when Bigelow wrote has proceeded at such a pace that his little book seems rudimentary. Yet it is indeed curious that the highly industrialized society of twentieth-century America can be bullied by humanistic professors into remembering Emerson's *Nature* of 1836, or even to cherishing the candlesticks and spinning wheels of our preindustrial past, and yet will not bother to salute in Bigelow a prophet more relevant to the later economy than either Emerson or Jefferson" (289). As the subtitle of *Elements* indicated, the book derived from Bigelow's lectures at Harvard. Meier ("Technology and Democracy," 618–19) states that the term *technology* had been used in those lectures. The term does not, however, appear in Bigelow's inaugural lecture, delivered in 1816, and reprinted in the *North American Review* 4 (January 1817): 271–83. The lecture has been reprinted

in more accessible form as "A Nation of Inventors" in *The Rising Glory of America, 1760–1820*, ed. Gordon Wood (New York: Braziller, 1971), 247–52; comments regarding Count Rumford have been deleted in this reprinting, the one cited elsewhere here. Yet Bigelow does seem to be groping for a term to categorize Rumford's own contributions to what he vaguely calls "the science—of clothing, of warming, and of nourishing mankind" (*North American Review* version, 279; deleted in *The Rising Glory*).

17. See chap. 1, n. 7 above.

18. Bigelow, *Elements of Technology, Taken Chiefly from a Course of Lectures Delivered at Cambridge, on the Application of the Sciences to the Useful Arts* (Boston: Hillard, Gray, Little, and Wilkins, 1829), 1.

19. See again chap. 1, n. 7. As Raymond Williams explains, "Until C18 [the eighteenth century] most sciences were *arts*; the modern distinction between *science* and *art*, as contrasted areas of human skill and effort, with fundamentally different methods and purposes, dates effectively from mC19 [the mid-nineteenth century]" (*Keywords: A Vocabulary of Culture and Society* [New York: Oxford University Press, 1976], 34). See also Williams' complete entries on "Art" and "Science," 32–35 and 232–35. On the specifically American background of these linguistic and conceptual developments, see Bruce Sinclair, *Philadelphia's Philosopher Mechanics: A History of the Franklin Institute, 1824–1865* (Baltimore: Johns Hopkins University Press, 1974), chaps. 1, 12. In light of the changes summarized in this and preceding and succeeding paragraphs, the comments of an anonymous reviewer of *Elements of Technology* in a prestigious contemporary journal are revealing: "The word Technology gives but an imperfect idea of the contents of this volume. The end of a name would have been better answered by some title showing, that it treated of the scientific and practical principles of many of the useful, curious, and elegant arts. All the arts may safely be called useful . . ." (*North American Review* 30 [April 1830]: 337–38).

20. On these earlier writings, see Robert P. Multhauf, "Some Observations on the State of the History of Technology," *Technology and Culture* 15 (January 1974): 1–2, and Reinhard Rürup, "Reflections on the Development and Current Problems of the History of Technology," ibid. 15 (April 1974): 167–68. On related American writings of Bigelow's day, see Hindle, "A Bibliography of Early American Technology," in his *Technology in Early America*, 87–88. As that same anonymous reviewer of *Elements of Technology* concluded, "The publication of Dr. Bigelow's book will contribute to the diffusion of a better taste, by making known the essential principles of the arts, and thus preparing for the circulation of larger and more particular treatises. It will do it no less by laying before the young inquirers something like a map of the various regions of pleasant knowledge . . ." (*North American Review*, 360).

21. The book appeared in several other editions, including a two-volume edition beginning in 1840 under the title *The Useful Arts, Considered in Connexion With the Applications of Science* (Boston: Marsh, Capen, Lyon, and Webb; reprinted virtually unchanged by other publishers after 1842). *The Useful Arts* contains several new chapters, including an historical overview "of the Progress of the Arts in Ancient and Modern Times" as its first chapter, additions to a number of existing chapters, and an Appendix and Glossaries also not found in *Elements of Technology*.

22. Bigelow, *Elements of Technology*, 2.

23. Ibid., 3.

24. Ibid., 4.

25. Bigelow, "A Nation of Inventors," in *The Rising Glory*, 247. In the preceding paragraphs he distinguished individuals' transient physical and political power from mankind's permanent—and cumulative—intellectual power. And as he put it in *Elements of Technology*, "we could not return to the state of knowledge which existed even fifty or sixty years ago, without suffering both intellectual and physical degradation" (5).

26. Bigelow, *An Address on the Limits of Education* (Boston: Dutton, 1865), 3.

27. Ibid., 3–4.

28. Significantly, the "Advertisement" to the original and later editions of *Elements of Technology* had stated that the advancement of technology "*probably* . . . has contributed more than that of any *other* science, to the improved condition of the present age" (iii; my italics). As Meier observes, "By the middle of the [nineteenth] century the connotation of even the word technology was far richer than it had been in the time of its popularization by Jacob Bigelow" ("The Technological Concept," 541). On this development, see also Sinclair, chap. 12. Hindle sees in Bigelow "despite his assertions . . . an anti-intellectual and anti-scientific impulse in the desire to adapt science to 'practical men' by eliminating mathematics and as much of principles as possible" ("Bibliography," 88). The criticism may be somewhat unfair, but the important point is the maturation of technology—even if, as Hindle himself adds, most other writers on such technical subjects "were moving in the opposite direction as they sensed the need to find a mathematical basis for technology . . ." (ibid.). For Bigelow's further thoughts on the question Hindle raises, see his *An Address* and his *Remarks on Classical and Utilitarian Studies* (Boston: Little, Brown, 1867), advertised as "a more complete expression"(3) of his *An Address* in that regard. The larger and more significant issue is the actual relationship between technology and science in Bigelow's day, about which see chap. 1, n. 12 above.

29. Bigelow, *An Address*, 27–28.

30. On Timothy Walker's life, see the sketch in the *Dictionary of American Biography* (New York: Scribner, 1936), 19: 363.

31. Thomas Carlyle, "Signs of the Times," *Edinburgh Review* 49 (June 1829): 439–59, reprinted in abridged version as "The Mechanical Age" in *Nature and Industrialization: An Anthology*, ed. Alasdair Clayre (Oxford: Oxford University Press, 1977), 229–34, 229. The essay was published anonymously.

32. Carlyle in *Nature and Industrialization*, 229, 231.

33. For a fuller description of Carlyle's views on technology throughout his life, see Williams, *Culture and Society, 1780–1950* (New York: Harper Torchbooks, 1966), chap. 4, and Herbert L. Sussman, *Victorians and the Machine: The Literary Response to Technology* (Cambridge: Harvard University Press, 1968), chap. 1.

34. Walker, "Defense of Mechanical Philosophy," *North American Review* 33 (July 1831): 122–36, reprinted as "The New Age Defended," in *Readings in Technology*, 67–77, 77.

35. Bigelow, *Elements of Technology*, 4.

36. According to John C. Burnham, "It is significant that when . . . Walker . . . wanted an opponent of a mechanistic view, he could not find an American adversary but had to use an Englishman . . . (*Science in America: Historical Selections*, ed. Burnham [New York: Holt, Rinehart and Winston, 1971], 90). The same point is made by, among others, George H. Daniels, *American Science in the Age of Jackson* (New York: Columbia University Press, 1968), 61.

37. Walker in *Readings in Technology*, 68.

38. Ibid., 69, 68.

39. On changes within nineteenth-century French utopianism from a mechanical conception of the ideal social order to an organic conception, see Frank E. Manuel, *The Prophets of Paris* (New York: Harper Torchbooks, 1965), 124–129; Manuel and Fritzie P. Manuel, Introduction, *French Utopias: An Anthology of Ideal Societies*, ed. and tr. Manuel and Manuel (New York: Schocken, 1971), 8–9, 14; and Barbara Goodwin, *Social Science and Utopia: Nineteenth-Century Models of Social Harmony* (Sussex, Eng.: Harvester Press, 1978), 9–11, 167–69, 178–79.

40. See Werner Stark, *The Fundamental Forms of Social Thought* (London: Routledge and Kegan Paul, 1962), for a comprehensive study of the mechanical and organic

conceptions of the social order. Although Stark is occasionally ahistorical in his analysis, he generally recognizes the changes over time within each conception and the complex relationship between the two. He also recognizes that societies of whatever kind usually represent a mixture to some degree between the two. On this latter point, see Stark, 1, 12, 203. On the changing meanings of *mechanical* and *organic*, see Karl W. Deutsch, "Mechanism, Organism, and Society: Some Models in Natural and Social Science," *Philosophy of Science* 18 (July 1951): 230–52, and Williams' complete entries on each in *Keywords*, 167–69 and 189–92. As Williams points out, "*Mechanical* was earlier in English than *machine*, and has long had certain separable senses" (167). Moreover, *mechanical* and *organic* had, before the nineteenth century, "been very close in meaning" (168): "*Organ* first appeared in English, from C13 [the thirteenth century], to signify a musical instrument . . ." (190). On the need for historical perspective in the examination of ideas used in more than one period, see Quentin Skinner, "Meaning and Understanding in the History of Ideas," *History and Theory* 8 (1969): 3–53.

41. See Carlo M. Cipolla, *Clocks and Culture, 1300–1700* (London: Collins, 1967), and Jean Gimpel, *The Medieval Machine: The Industrial Revolution of the Middle Ages* (New York: Holt, Rinehart and Winston, 1976), chap. 7.

42. Carlyle in *Nature and Industrialization*, 232. As Sussman elaborates, "Although he [Carlyle] had spent most of his early life on the farm or at the university, he was by no means unacquainted with the mechanization of pre-railway England. . . . from the textile mills he took the specific machine figures of the essay. For example, the figure of the 'Machine of Society . . .' refers to the typical early textile mill in which each separate machine was connected by a belt to a single rotating shaft turned by either a water wheel or a stationary engine. . . . Even if the readers of the *Edinburgh Review* had never been inside a textile mill, they had almost surely seen pictures of the stationary engine and spinning machinery . . ." (16–17).

43. Walker in *Readings in Technology*, 72.

44. Carlyle in *Nature and Industrialization*, 230, 231.

45. Walker in *Readings in Technology*, 70. In part to alleviate Carlyle's concerns about the replacement of God by man, Walker invoked God's name and power. For example: "When we attempt to convey an idea of the infinite attributes of the Supreme Being, we point to the stupendous machinery of the universe" (ibid.). And similarly: "We cannot go back to the origin of mankind and trace them down to the present time, without believing it to be a part of the providence of God, that his creatures should be perpetually advancing" (ibid., 77).

46. Ibid., 72. Carlyle likewise employed mathematical language, if sadly: "Nothing is now done directly, or by hand; all is by rule and calculated contrivance" (in *Nature and Industrialization*, 229).

47. Walker in *Readings in Technology*, 70.

48. Ibid., 68.

49. Carlyle in *Nature and Industrialization*, 233.

50. See ibid., 229, and Walker in *Readings in Technology*, 70 (quoted in text above).

51. Carlyle in *Nature and Industrialization*, 234.

52. See, for example, Meier, "The Technological Concept," chap. 13; Leo Marx, *The Machine in the Garden: Technology and the Pastoral Ideal in America* (New York: Oxford University Press, 1964), chap. 4; George M. Fredrickson, *The Inner Civil War: Northern Intellectuals and the Crisis of the Union* (New York: Harper and Row, 1965); Thomas, "Romantic Reform in America, 1815–1865," *American Quarterly* 17 (Winter 1965): 656–81; Lynn L. Marshall, "The Strange Stillbirth of the Whig Party," *American Historical Review* 72 (January 1967): 445–68; John Higham, *From Boundlessness to Consolidation: The Transformation of American Culture, 1848–1860* (Ann Arbor, Mich.: Clements Library, 1969); David J. Rothman, *The Discovery of the Asylum: Social*

Order and Disorder in the New Republic (Boston: Little, Brown, 1971); Barbara G. Rosenkrantz, *Public Health and the State: Changing Views in Massachusetts, 1842–1936* (Cambridge: Harvard University Press, 1972), chaps. 1, 2; Sinclair; David B. Tyack, *The One Best System: A History of American Urban Education* (Cambridge: Harvard University Press, 1974), 41–42, 143–44; Matthew A. Crenson, *The Federal Machine: Beginnings of Bureaucracy in Jacksonian America* (Baltimore: Johns Hopkins University Press, 1975); Hughes, Introduction, Conclusion, and pt. 2, *Changing Attitudes*; Kasson, chap. 1; Henry D. Shapiro, "From Association to Community: The Organization of the A.A.A.S. and the Transformation of Scientific Society in the United States" (paper delivered at 1976 annual meeting, American Association for the Advancement of Science, Boston); Shapiro, "The Western Academy of Natural Sciences of Cincinnati and the Structure of Science in the Ohio Valley, 1810–1850," in *The Pursuit of Knowledge in the Early American Republic: American Scientific and Learned Societies from Colonial Times to the Civil War*, ed. Alexandra Oleson and Brown (Baltimore: Johns Hopkins University Press, 1976), 219–47; Theodore Dwight Bozeman, *Protestants in an Age of Science: The Baconian Ideal and Antebellum American Religious Thought* (Chapel Hill: University of North Carolina Press, 1977); Herbert Hovenkamp, *Science and Religion in America, 1800–1860* (Philadelphia: University of Pennsylvania Press, 1978); and Walters, chap. 9 and Afterword. Indicative of this new mechanical order were the contemporary writings on American mechanics and their contributions to American society. See, for example, Zachariah Allen, *The Science of Mechanics, As Applied to the Present Improvements in the Useful Arts in Europe, and in the United States of America* (Providence, R.I.: Hutchens and Cory, 1829); Henry Howe, *Memoirs of the Most Eminent American Mechanics* (New York: Blake, 1844); and Charles H. Fitch, "The Rise of a Mechanical Ideal," *Magazine of American History* 11 (June 1884): 516–27.

53. Walker in *Readings in Technology*, 76–77.

54. See the citations in n. 39 above.

55. Manuel and Manuel, Introduction, *French Utopias*, 14.

56. See Meier, "The Technological Concept," 456–95; Daniel Horowitz, "An Intellectual History of American Industrialization: Academic Economists, New England Publicists, and Engineers, 1830–1910" (manuscript, n.d., in possession of author), chap. 8; Patrick R. Brostowin, "John Adolphus Etzler: Scientific-Utopian During the 1830s and 1840s" (Ph.D. diss., New York University, 1969); Robin Linstromberg and James Ballowe, "Thoreau and Etzler: Alternative Views of Economic Reform," *Midcontinent American Studies Journal* 11 (Spring 1970): 20–29; Nydahl, Introduction, *The Collected Works of John Adolphus Etzler*, ed. Nydahl (Delmar, N.Y.: Scholars' Facsimiles and Reprints, 1977), vii–xxxi; Nelson F. Adkins, Introduction, Mary Griffith, *Three Hundred Years Hence*, ed. Adkins (Philadelphia: Prime Press, 1950), 5–20; and Arthur O. Lewis, Jr., Introduction, *American Utopias: Selected Short Fiction*, ed. Lewis (New York: Arno Press, 1971), x–xi. Only the last two citations treat Griffith's work; the rest treat Etzler's or Ewbank's or both.

57. For the principal sources of that information, see the Appendix.

58. I cannot, however, accept Hughes' further distinction between technology in the early nineteenth century as a solver of "a few specific problems" and technology in the late nineteenth and early twentieth centuries as "an abstract force applicable to a variety of problems" (333). The distinction is too fine, for technology was conceived as both an abstraction and a multiproblem solver in both periods—even though its imagery and its solutions changed.

59. Significantly, Henry Adams searched for historical continuities even as he recognized discontinuities—very like those we have just discussed. He confessed: "Yet the dynamo, next to the steam engine, was the most familiar of exhibits. For Adams's objects its value lay chiefly in its occult mechanism. Between the dynamo in the gallery

of machines and the engine-house outside, the break of continuity amounted to abysmal fracture for a historian's objects. No more relation could he discover between the steam and the electric current than between the Cross and the cathedral. . . . Historians undertake to arrange sequences—called stories, or histories—assuming in silence a relation of *cause and effect*. These assumptions, hidden in the depths of dusty libraries, have been astounding, but commonly unconscious and child-like; so much so, that if any captious critic were to drag them to light, historians would probably reply, with one voice, that they had never supposed themselves required to know what they were talking about. Adams, for one, had toiled in vain to find out what he meant. He had even published a dozen volumes of American history for no other purpose than to satisfy himself whether . . . he could fix for a familiar moment a necessary sequence of human movement. The result had satisfied him as little as at Harvard College. Where he saw sequence, other men saw something quite different, and *no one saw the same unit of measure*" ("The Dynamo and the Virgin," in *The Education of Henry Adams* [1907], reprinted in *Changing Attitudes*, 169–71; my italics). On Adams' historical outlook, see William H. Jordy, *Henry Adams: Scientific Historian* (New Haven, Conn.: Yale University Press, 1952), esp. chap. 5.

60. Meier, "The Technological Concept," 546.

61. Ibid.

62. Meier admits: "If one must crystallize a thesis, it is that there existed in American society in the period 1750 through 1860 a growing awareness that the technologist [i.e., scientist, engineer, and inventor] was playing a unique role in the making of American civilization. This role was recognized as essentially different in nature from that of the politician, the lawyer, the churchman or physician, and other long established professional groups" ("The Technological Concept," 535). The thesis must, to repeat, remain relatively imprecise. See also ibid., 540–51; Rezneck, 784–811; Marx, chap. 4; Hughes, Introduction, Conclusion, and pt. 2, *Changing Attitudes*; and Kasson, chap. 1. On the emergence of technology as a "calling" but on the emerging "technologist" as something less than a "technocrat," see Meier, "The Technological Concept," 540. The technologists Meier describes are akin to both the earlier and later technological utopians in their modest intellectual sophistication and eagerness to represent popular opinion.

63. See Rezneck, 784–811; Meier, "The Technological Concept," 544–45; Meier, "Technology and Democracy," 618–40; Meier, "American Technology," 121; Edmund S. Morgan, "The Puritan Ethic and the American Revolution," *William and Mary Quarterly*, 3d ser., 24 (January 1967): 3–43; Harry N. Scheiber, *Ohio Canal Era: A Case Study of Government and the Economy, 1820–1861* (Athens: Ohio University Press, 1969), 88–94; Wood, "Republican Technology" in *The Rising Glory*, 237–52; J. G. A. Pocock, "Virtue and Commerce in the Eighteenth Century," *Journal of Interdisciplinary History* 3 (Summer 1972): 119–34; Kasson, chaps.1–2; despite its aversion to "ideology," Boorstin, *The Republic of Technology*, chaps. 1, 4; and Michael Brewster Folsom and Steven D. Lubar, Introduction, *The Philosophy of Manufactures: Early Debates Over Industrialization in the United States*, ed. Folsom and Lubar (Cambridge: MIT Press, 1982), xix–xxxix. What Wood says about the late eighteenth century applies to the early and mid-nineteenth century as well: "The coincidence in time between the American Revolution and the beginnings of what came to be called the Industrial Revolution had momentous consequences for the way Americans came to identify technological progress with the promise of their own history. Political and physical science seemed to be providentially linked, and technology became as important as virtue in achieving America's realization of itself as a moral republic. The useful arts were designed not only to promote the physical well-being of Americans, but more important, to enhance their liberty and republicanism as well" (Wood in *The Rising Glory*, 237). Meier's "Technology and Democracy" extends these points chronologi-

cally and intellectually by demonstrating the widely perceived reciprocal relationship between technological advance and the advance of republicanism in the nineteenth century.

64. See Ekirch, *Idea of Progress*, chap. 4; Meier, "The Technological Concept," 549–55; Meier, "Technology and Democracy," 629–31; Charles L. Sanford, "The Intellectual Origins and New-Worldliness of American Industry," *Journal of Economic History* 18 (March 1958): 1–16; Clarence Mondale, "Daniel Webster and Technology," *American Quarterly* 14 (Spring 1962): 37–47; Marx, chap. 4; Marvin Fisher, *Workshops in the Wilderness: The European Response to American Industrialization, 1830–1860* (New York: Oxford University Press, 1967); and Nye, 258–82.

65. See George R. Taylor, *The Transportation Revolution, 1815–1860* (New York: Holt, Rinehart and Winston, 1951), chaps. 12, 13; Alan Dawley, *Class and Community: The Industrial Revolution in Lynn* (Cambridge: Harvard University Press, 1976), 227; Herbert G. Gutman, *Work, Culture, and Society in Industrializing America: Essays in American Working-Class and Social History* (New York: Vintage Books, 1977), 26–27; Richard A. McLeod, *Workers and Industrialization in Ante Bellum America* (St. Louis: Forum Press, 1977), 5–11; Merritt Roe Smith, *Harpers Ferry Armory and the New Technology: The Challenge of Change* (Ithaca, N.Y.: Cornell University Press, 1977); and Edward Pessen, *Jacksonian America: Society, Personality, and Politics*, 2d ed. (Homewood, Ill.: Dorsey, 1978), 115–17. Several of these authors demonstrate that strikes among American workers became a more popular tactic after the Civil War. On pre-Civil War working class violence prompted by different–if related–concerns, see David Montgomery, "The Shuttle and the Cross: Weavers and Artisans in the Kensington Riots of 1844," *Journal of Social History* 5 (Summer 1972): 411–46.

66. See Gutman, 57–60; McLeod, 12; and Pessen, 118. On the English Luddites, see E. P. Thompson, *The Making of the English Working Class* (New York: Pantheon, 1963), chap. 14; E. J. Hobsbawm, "The Machine Breakers," in his *Labouring Men: Studies in the History of Labour* (Garden City, N.Y.: Anchor Books, 1967), 7–26; and Malcolm I. Thomis, *The Luddites: Machine-Breaking in Regency England* (New York: Schocken, 1972).

67. See Meier, "The Technological Concept," 536, 548–51; Taylor, chaps. 12, 13; Meier, "Technology and Democracy," 628–29, 634–37; Montgomery, "The Working Classes of the Pre-Industrial American City, 1780–1830," *Labor History* 9 (Winter 1968): 3–22; Montgomery, "The Shuttle and the Cross," 411–46; Dawley; Kasson, chaps. 1–3; Gutman, chap. 1; McLeod; and M. R. Smith. On the accommodation of Americans in general to technological change later in the nineteenth century, see the interesting discussion in Glenn Porter, *The Rise of Big Business, 1860–1910* (Arlington Heights, Ill: AHM, 1973), 25–26 and chap. 3. On the accommodation of farmers in particular in the nineteenth and twentieth centuries, see Wayne D. Rasmussen, "The Impact of Technological Change on American Agriculture, 1862–1962," *Journal of Economic History* 22 (December 1962): 578–82; Earl W. Hayter, *The Troubled Farmer, 1850–1900: Rural Adjustment to Industrialism* (De Kalb: Northern Illinois University Press, 1968); John L. Shover, *First Majority—Last Minority: The Transforming of Rural Life in America* (De Kalb: Northern Illinois University Press, 1976), chap. 4; and David B. Danbom, *The Resisted Revolution: Urban America and the Industrialization of Agriculture, 1900–1930* (Ames: Iowa State University Press, 1979).

68. In addition to the works cited in n. 64 above, see the following writings concerning the state of American and British technology in the nineteenth century: John E. Sawyer, "The Social Basis of the American System of Manufacturing," *Journal of Economic History* 14 (Fall 1954): 361–79; Eugene S. Ferguson, "On the Origin and Development of American Mechanical 'Know-How,' " *Midcontinent American Studies Journal* 3 (Fall 1962): 3–16; H. J. Habakkuk, *American and British Technology in the Nineteenth Century: The Search for Labour-Saving Inventions* (Cambridge: Cambridge

University Press, 1962); Norman B. Wilkinson, "Brandywine Borrowings from European Technology," *Technology and Culture* 4 (Winter 1963): 1–13; Siegfried Giedion, *Mechanization Takes Command: A Contribution to Anonymous History* (New York: Norton, 1969); Rosenberg, Introduction, *The American System of Manufactures*, 1–2, esp. n. 1; and M. R. Smith, Introduction. The initial success of Humphreysville, Waltham, Lawrence, Lowell, and other early factory communities set amid rural landscapes seemed to confirm the faith of Americans and Europeans alike that America could absorb and even advance British industry and technology yet avoid Britain's crowded, diseased, and grimy city life. These and other examples of the (initial) "middle landscape" are discussed later in this chapter.

69. George Wallis, "Special Report of Mr. George Wallis," in *The American System of Manufactures*, 306–307. M. R. Smith, Introduction, comments in turn upon the British committee members' excessive optimism about Americans' adjustment to industrialization.

70. See Marx, chaps. 3–5. See also Thomas Bender, *Toward an Urban Vision: Ideas and Institutions in Nineteenth-Century America* (Lexington: University Press of Kentucky, 1975).

71. As Marx puts it, the middle landscape had become a "popular and sentimental" version of the pastoral ideal, where it had previously been an "imaginative and complex" version. Required instead were "new [cultural] symbols of possibility" (ibid., 5, 365).

72. See Segal, "Leo Marx's 'Middle Landscape': A Critique, A Revision, and An Appreciation," *Reviews in American History* 5 (March 1977): 137–50. None of the persons studied by either Marx or me used the term *middle landscape*, but our uses of it to describe their efforts do not distort history.

73. See ibid., 140–50. The middle landscape concept is applicable beyond Marx's restricted use of it, but I suggest three cautions in extending it: (1) the concept may prove *too* useful, may explain so much of American history that it explains very little specifically; as such it may not be worth pursuing; (2) the concept may apply to Europe as well and so not be as peculiarly American as Marx suggests; and (3) the concept may be imprecise insofar as Marx sees technology simply as machines rather than, as I have argued in chap. 1, as including structures as well; and imprecise again insofar as Marx sees technology as dynamic and destructive and nature as static and constructive rather than both as being partly static and partly dynamic, partly constructive and partly destructive—depending on the particular circumstances.

74. See ibid., 141–42 and nn. 11 and 12, and Cecelia Tichi, *New World, New Earth: Environmental Reform in American Literature from the Puritans through Whitman* (New Haven, Conn.: Yale University Press, 1979), 214–18. On Olmsted and his contemporaries, see also John B. Jackson, *American Space: The Centennial Years, 1865–1876* (New York: Norton, 1972). On the efforts of late nineteenth-century utopian writers to reconcile America's older agrarian values with the emerging industrial ones, see Donald C. Burt, "Utopia and the Agrarian Tradition in America, 1865–1900" (Ph.D. diss., University of New Mexico, 1973), and Burt, "The Well-manicured Landscape: Nature in Utopia," in *America as Utopia: Collected Essays*, ed. Kenneth M. Roemer (New York: Burt Franklin, 1981), 175–85. On the harmonious relationships they sought between machine and garden and city and farm, see Burt, "Utopia and the Agrarian Tradition," chaps. 3–6, and Burt, "The Well-manicured Landscape," 175–85.

75. See Tichi, esp. 55, 64, 85–86, 89–90, 103–105, 134, 165, 168, 190, 216. See also Sacvan Bercovitch, *The American Jeremiad* (Madison: University of Wisconsin Press, 1979), 111, 142, 162–63. As I indicated in chap. 4, utopianism and millenarianism rely on different principal agents of change to achieve perfection—one on man, the other on God. Neither Tichi nor Bercovitch makes this distinction, which is not technical but basic. Yet, as I indicated in n. 5 of that chapter, there are grey areas between the two,

most notably when the language of utopianism is religious or quasireligious, as in technological utopianism. On the varying attitudes of antebellum American millennialists toward technological advance, specifically as a means of hastening the millennium, see Robert K. Whalen, "Millenarianism and Millennialism in America, 1790–1880" (Ph.D. diss., State University of New York at Stony Brook, 1971), 193–203, 215–16, and Welter, *The Mind of America*, 20–21. The variations reflected differences over whether God would bring about the millennium with man's assistance (not, to be sure, whether man alone would do so). For an account of expressions of utopianism and millenarianism in late nineteenth- and early twentieth-century America that at once treats their favorable attitudes toward technology (and science) and distinguishes between them as here, see Jean B. Quandt, "Religion and Social Thought: The Secularization of Postmillennialism," *American Quarterly* 25 (October 1973): 390–409.

76. As Meier says about the early and mid-nineteenth-century exponents of technological advance, "The arguments . . . were based on the fundamental assumptions that technology . . . was [among other things] not incompatible with religion but actually an answer to the age-old prayer for fuller revelation of the divine will and goodness" ("The Technological Concept," 549). Where, as Tichi observes, we today equate environmental reform with restraint, earlier visionaries, including hers and mine, equated it with highly aggressive activity. See Tichi, viii–ix.

77. Although I know of no comprehensive studies of this subject, the summary comments of two of the foremost nineteenth-century observers of these communities are revealing. According to Charles Nordhoff (*The Communistic Societies of the United States* [1875; reprinted New York: Schocken, 1965], 389–90), "the [Oneida] Perfectionists are essentially manufacturers, using agriculture only as a subsidiary branch of business. All the other societies have agriculture as their industrial base . . . though all have some branch of manufacture." John H. Noyes (*History of American Socialisms* [1870; reprinted New York: Dover, 1966], 19), himself the founder of Oneida, confirms Nordhoff's point: "It is really ludicrous to see how uniformly an old saw-mill turns up in connection with each Association, and how zealously the brethren made much of it; but that is about all they attempted in the line of manufacturing." Consequently, "we should have advised the Phalanxes to limit their land-investments to a minimum, and put their strength as soon as possible into some form of manufacture." In his more recent analysis of American aesthetics, John A. Kouwenhoven (*The Arts in Modern American Civilization* [New York: Norton, 1967], 92) reaches similar conclusions about the Shakers: "The Shakers had no fear of the machine. Their communities actually seem to have produced more mechanics and inventors per capita than most other towns and villages of comparable size." The Shakers manufactured brooms, pails, chairs, tubs, mops, washing machines, lathes, and nails, among other items, and the Oneida perfectionists manufactured animal traps, chairs, brooms, dishwashers, traveling bags, and later, of course, silverware. On the role of technology in various utopian communities of the nineteenth and early twentieth centuries, see Arthur E. Bestor, Jr., "American Phalanxes: A Study of Fourierist Socialism in the United States, with Special Reference to the Movement in Western New York" (Ph.D. diss., Yale University, 1938), vol. 1; William A. Hinds, *American Communities* (1878; reprinted New York: Corinth Books, 1961); Nordhoff; Thomas, "Antislavery and Utopia," in *The Antislavery Vanguard: New Essays on the Abolitionists*, ed. Martin Duberman (Princeton, N.J.: Princeton University Press, 1965), 240–69; Mark Holloway, *Heavens on Earth: Utopian Communities in America, 1680–1880*, 2d ed. (New York: Dover, 1966); Noyes; Maren Lockwood Carden, *Oneida: Utopian Community to Modern Corporation* (Baltimore: Johns Hopkins University Press, 1969); John F. C. Harrison, *The Quest for the New Moral World: Robert Owen and the Owenites in Britain and America* (New York: Scribner, 1969), 38–87, 163–92, 235–54; Bestor, *Backwoods Utopias: The Sectarian Origins and the Owenite Phase of Communitarian Socialism in America: 1663–1829*, 2d ed. (Philadelphia: University of Pennsylvania Press, 1970), 74–77, 162–64; Chris-

topher Tunnard, *The City of Man*, 2d ed. (New York: Scribner, 1970), chap. 6; Michael Fellman, *The Unbounded Frame: Freedom and Community in Nineteenth Century American Utopianism* (Westport, Conn.: Greenwood Press, 1973), chaps. 1, 3; Robert S. Fogarty, "American Communes, 1865–1914," *Journal of American Studies* 9 (August 1975): 145–62; Charles Pierce LeWarne, *Utopias on Puget Sound, 1885–1915* (Seattle: University of Washington Press, 1975); Dolores Hayden, *Seven American Utopias: The Architecture of Communitarian Socialism, 1790–1975* (Cambridge: MIT Press, 1976), 9, 14, 15, 41, 323, 325–26, 328; John Egerton, *Visions of Utopia: Nashoba, Rugby, Ruskin, and the "New Communities" in Tennessee's Past* (Knoxville: University of Tennessee Press, 1977); David Lindsey, *Nineteenth Century American Utopias* (St. Louis: Forum Press, 1976), 4–15; and Walters, chaps. 2, 3. Walters says of the antebellum communities: "Communitarians, like antebellum reformers generally, were at too early a stage in the Industrial Revolution to perceive clearly that it was the force transforming the world around them. A number of communal ventures did reject urban and capitalistic society (in that, they resembled the communes of the 1960's). Others nonetheless bore a less hostile relationship to it. Many engaged in business enterprises and some were successful. Very few utopians chose to get away from it all by settling on the frontier, and even the most agricultural-minded ones usually sold crops rather than practice subsistence farming. . . . Many communitarians, moreover, admired the science, technology, and rising standard of living produced by industrialization, despite being appalled at the misery 'progress' brought with it. They did not so much oppose economic development as feel humankind had to assert rational control over it . . ." (74). On the similarly significant role of technology in English utopian communities of the same period, see Dennis Hardy, *Alternative Communities in Nineteenth-Century England* (London: Longman, 1979). Whether the American utopian communities *as communities* were ever realistic models of social change, especially in the late nineteenth and early twentieth centuries, will be discussed at the beginning of the next chapter.

Chapter 6

1. On the role of the Civil War as an agent of social, cultural, and economic change in nineteenth-century America, see the historiographical discussions and revisionist interpretations in *Economic Change in the Civil War Era*, ed. David T. Gilchrist and W. David Lewis (Greenville, Del.: Eleutherian Mills-Hagley Foundation, 1965); John L. Thomas, "Antislavery and Utopia," in *The Antislavery Vanguard: New Essays on the Abolitionists*, ed. Martin Duberman (Princeton, N.J.: Princeton University Press, 1965), 264–69; *The Economic Impact of the American Civil War*, ed. Ralph Andreano, 2d ed. (Cambridge, Mass.: Schenkman, 1967); John Higham, *From Boundlessness to Consolidation: The Transformation of American Culture, 1848–1860* (Ann Arbor, Mich.: Clements Library, 1969); Stuart Bruchey, *Growth of the Modern American Economy* (New York: Dodd, Mead, 1975), 73–83; and Richard D. Brown, *Modernization: The Transformation of American Life, 1600–1865* (New York: Hill and Wang, 1976), chap. 7. In different ways all of these works suggest that the Civil War's impact on the "modernization" of American society was less than has been thought. I am indebted to Professor Martin Jay of the University of California at Berkeley for the analogy to the revolutions of 1848 in Europe. John F. C. Harrison has, I feel, properly attributed the decline of utopian communities in Britain in the 1840s—and in America after the Civil War—to the other changes which I discuss below. He does not, however, discuss the replacement of communal experiments by writings in either country. See Harrison, *Quest for the New Moral World: Robert Owen and the Owenites in Britain and America* (New York: Scribner, 1969), 62–63, 151.

2. Edward Bellamy, "Progress of Nationalism in the United States," *North American Review* 154 (June 1892): 743, and "Concerning the Founding of Nationalist Colonies," *The New Nation* 3 (September 23, 1893): 434. These are only samples of the considerable writings by Bellamy and others on the issue of possible Nationalist communities. Thomas and Michael Fellman have noted this insight of Bellamy but have not applied it to the transformation of American utopianism in general as described here. See Thomas, Introduction, Bellamy, *Looking Backward: 2000–1887*, ed. Thomas (Cambridge: Harvard University Press, 1967), 42–43, 73, 76, and Fellman, *The Unbounded Frame: Freedom and Community in Nineteenth Century American Utopianism* (Westport, Conn.: Greenwood Press, 1973), 105–106. A slightly broader discussion of these changes is in Francine C. Cary, "Shaping the Future in the Gilded Age: A Study of Utopian Thought, 1888–1900" (Ph.D. diss., University of Wisconsin, 1975), 10–13. On the general connection between form and content in turn-of-the-century American utopian literature, see Mary J. Pfaelzer, "Utopian Fiction in America, 1880–1900: The Impact of Political Theory on Literary Form" (Ph.D. diss., University College, London, 1975), esp. chap. 2.

3. See particularly Joel Nydahl, "From Millennium to Utopia Americana" and "Early Fictional Futures: Utopia, 1798–1864," in *America as Utopia: Collected Essays*, ed. Kenneth M. Roemer (New York: Burt Franklin, 1981), 237–53 and 254–91. As I discussed in the preceding chapter, three technological utopians published works in America before the Civil War.

4. See Robert S. Fogarty, "American Communes, 1865–1914," *Journal of American Studies* 9 (August 1975): 145–62. The beginning of the article cites the major proponents of the older interpretation he rejects.

5. Charles Pierce LeWarne, *Utopias on Puget Sound, 1885–1915* (Seattle: University of Washington Press, 1975), 235. Although his pioneering studies of American utopian communities erroneously suggest that such communities virtually died out by the Civil War, Arthur E. Bestor, Jr., advances a view similar to LeWarne's: he regards any that did persist as irrelevant. See his *Backwoods Utopias: The Sectarian Origins and the Owenite Phase of Communitarian Socialism in America: 1663–1829*, 2d ed. (Philadelphia: University of Pennsylvania Press, 1970), 229, and his articles of 1953 and 1957, "Patent-Office Models of the Good Society: Some Relationships between Social Reform and Westward Expansion," and "The Transit of Communitarian Socialism to America," reprinted in ibid., 250 and 268. Fogarty does not concede this point, but he implies throughout his article that the problems that these later communities perceived and set out to solve—such as "urbanization, industrial depression, and religious change" (146)—were different from the problems confronted by their predecessors.

6. The term comes from Thomas, "Romantic Reform in America, 1815–1865," *American Quarterly* 17 (Winter 1965): 679. The notion, however, of antebellum disorder and fragmentation applies to this and other of Thomas' writings on nineteenth-century America, to Fellman's *Unbounded Frame* (the very title makes the point), to Bestor's *Backwoods Utopias*, and to other studies of antebellum utopianism and of antebellum society. The most notable of the last are David Donald, "An Excess of Democracy: The American Civil War and the Social Process," in his *Lincoln Reconsidered: Essays on the Civil War Era*, 2d ed. (New York: Vintage Books, 1961), 209–35; Stanley M. Elkins, *Slavery: A Problem in American Institutional and Intellectual Life*, 2d ed. (Chicago: University of Chicago Press, 1968), 27–34; and Rowland T. Berthoff, *An Unsettled People: Social Order and Disorder in American History* (New York: Harper and Row, 1971), pt. 2.

7. The best general account of the various strands of reform in antebellum America is Ronald G. Walters, *American Reformers, 1815–1860* (New York: Hill and Wang, 1978). No comparable work exists for the rest of the nineteenth century, but then reform

after 1860 was even less cohesive—and less pervasive—than before 1860. For more specialized studies of reform in the nineteenth and early twentieth centuries, see the numerous pertinent citations throughout this chapter.

8. As Walters puts it bluntly, "Utopian societies, if one counts active members, were among the least popular expressions of antebellum reform and radical sentiment. . . . At that, there was no other period in our history, except possibly in the late 1960s, when such a large proportion of Americans joined communal societies" (40). Even as sympathetic a student of those communities as Bestor agrees: "One must not exaggerate. The communitarian movement enlisted its thousands, not its hundreds of thousands. It was but a segment of the reform movement of the early nineteenth century, and for only a few brief years could it be considered a major segment" ("Transit of Communitarian Socialism," 27).

9. Bestor and Fogarty, among other scholars sympathetic to utopian communities, understandably argue that they are practical, lest the communities be dismissed as unrealizable. Bestor argues about the antebellum communities, "The communitarian idea was peculiarly attractive because alternative methods of social reform appeared to have reached a dead end during this particular period" (*Backwoods Utopias*, 7). And Fogarty observes, "The founders of communal societies usually thought themselves eminently practical, and some had careers as successful entrepreneurs before embarking on colony ventures. . . . Others—usually religious leaders—believed that the only practical way for an individual to proceed in the world was to ask the question: what shall I do to be saved? rather than the commonsensical one: how can I get ahead?" (*Dictionary of American Communal and Utopian History* [Westport, Conn.: Greenwood Press, 1980], xiv). Yet even these comments fail to defend the communities at any point in American history as practical vehicles for national improvement, a point I treat further below both in the text and in the notes.

10. That the vast majority of utopian communities established in America during the nineteenth and early twentieth centuries were not intended as mere escapes from the real world is confirmed by nearly all the primary and secondary works cited in this chapter and in chap. 5, n. 77.

11. Some recent communities went as far as to try to dispense with technology altogether—a naive effort because every society, as I noted in the Introduction, has had some degree of technology. The modern forms of large-scale, highly centralized technology, which other communities rejected in favor of smaller-scale and more decentralized forms, will be examined, along with their alternative, in chap. 8. Those contemporary utopian communities representing that alternative may well constitute a new, more purposeful, and more practical version of communitarianism.

12. See Frederic Cople Jaher, *Doubters and Dissenters: Cataclysmic Thought in America, 1885–1918* (New York: Free Press, 1964). See also Daniel Aaron, *Men of Good Hope: A Story of American Progressives* (New York: Oxford University Press, 1951), chap. 8; Thornton Anderson, *Brooks Adams: Constructive Conservative* (Ithaca, N.Y.: Cornell University Press, 1951); Walter B. Rideout, Introduction, Ignatius Donnelly, *Caesar's Column*, ed. Rideout (Cambridge: Harvard University Press, 1960), xv–xxxi; Alexander Saxton, "*Caesar's Column*: The Dialogue of Utopia and Catastrophe," *American Quarterly* 19 (Summer 1967): 224–38; and Gorman Beauchamp, "Jack London's Utopian Dystopia and Dystopian Utopia," in *America as Utopia*, 91–107. Beauchamp does qualify Jaher's categorization of London as simply pessimistic by demonstrating the considerable extent to which this man of many tensions and contradictions was simultaneously optimistic. Many historians have found a mixture of hopes and fears especially around the turn of the century, and I discussed this issue in chap. 3 in regard to utopians specifically. But as John Higham has pointed out, most American intellectuals of the period, unlike many of their European counterparts, "resisted pessimism" ("The Reorientation of American Culture in the 1890s," in Higham, *Writing American*

History: Essays on Modern Scholarship [Bloomington: Indiana University Press, 1970], 93). On this issue, see also Charles L. Sanford, "Technology and Culture at the End of the Nineteenth Century: The Will to Power," in *Technology in Western Civilization*, ed. Melvin Kranzberg and Carroll W. Pursell, Jr. (New York: Oxford University Press, 1967), 1: 726–39, and T. J. Jackson Lears, *No Place of Grace: Antimodernism and the Transformation of American Culture, 1880–1920* (New York: Pantheon, 1981).

13. On the nature of this vision, see the works cited in chap. 4, n. 4.

14. See Kenneth E. Boulding, *The Organizational Revolution: A Study in the Ethics of Economic Organization* (New York: Harper, 1953), chap. 11; Samuel P. Hays, *The Response to Industrialism, 1885–1914* (Chicago: University of Chicago Press, 1957), chap. 3; *Large Scale Organizations*, vol. 1, *The Emergent American Society*, ed. W. Lloyd Warner (New Haven, Conn.: Yale University Press, 1967); *Institutions in Modern America: Innovation in Structure and Process*, ed. Stephen E. Ambrose (Baltimore: Johns Hopkins University Press, 1967); Robert H. Wiebe, *The Search for Order, 1877–1920* (New York: Hill and Wang, 1967); Alfred D. Chandler, Jr., and Louis Galambos, "The Development of Large-Scale Economic Organizations in Modern America," *Journal of Economic History* 30 (March 1970): 201–17; Galambos, "The Emerging Organizational Synthesis in Modern American History," *Business History Review* 44 (Autumn 1970): 279–90; *Building the Organizational Society: Essays on Associational Activities in Modern America*, ed. Jerry Israel (New York: Free Press, 1972); Robert D. Cuff, "American Historians and the 'Organizational Factor,' " *Canadian Review of American Studies* 4 (Spring 1973): 19–31; Tom G. Hall, "Agricultural History and the 'Organizational Synthesis': A Review Essay," *Agricultural History* 48 (April 1974), 313–25; Higham, "Hanging Together: Divergent Unities in American History," *Journal of American History* 61 (June 1974): 5–28; Galambos, *The Public Image of Big Business in America, 1880–1940: A Quantitative Study in Social Change* (Baltimore: Johns Hopkins University Press, 1975), chaps. 1–9; Ellis W. Hawley, *The Great War and the Search for a Modern Order: A History of the American People and Their Institutions, 1917–1933* (New York: St. Martin, 1979); Gerald D. Nash, *The Great Depression and World War II: Organizing America, 1933–1945* (New York: St. Martin, 1979); and, more implicitly, John Whiteclay Chambers II, *The Tyranny of Change: America in the Progressive Era, 1900–1917* (New York: St. Martin, 1980). The last three works are part of a four-volume series on modern America, and their appearance—both as synthetic works in themselves and as part of a series—testifies to the growing influence of this interpretation of late nineteenth- and twentieth-century America. None of these works offers as complete a reinterpretation of American history along these lines as does Robert F. Berkhofer, "The Organizational Interpretation of American History: A New Synthesis," in *Prospects: An Annual of American Cultural Studies*, ed. Jack Salzman, 4 (New York: Burt Franklin, 1978): 611–29.

15. See Mark H. Haller, *Eugenics: Hereditarian Attitudes in American Thought* (New Brunswick, N.J.: Rutgers University Press, 1963). Haller's account covers considerably more than the "Progressive Era" yet is more reliable for that period than Donald K. Pickens' *Eugenics and the Progressives* (Nashville: Vanderbilt University Press, 1968), which surveys little more than just that period. See also Kenneth M. Ludmerer's uneven *Genetics and American Society: A Historical Appraisal* (Baltimore: Johns Hopkins University Press, 1972).

16. Haller notes that "by the middle 1930's, forty-one states had laws to prohibit marriage of the insane and feebleminded, seventeen to prohibit marriage of epileptics, and four to prohibit marriage of confirmed drunkards" (142). But many of these laws were either unenforced or else ineffective even when enforcement was attempted. Thus "The legislative battle of the eugenics movement consisted . . . of a long list of partial victories that added up to over-all defeat" (143).

17. On the assumption that perfect health can be achieved in the real world, see Barbara Gutmann Rosenkrantz, *Public Health and the State: Changing Views in Massachusetts, 1842–1936* (Cambridge: Harvard University Press, 1972), esp. Introduction.

18. See the persuasive critiques of the "organizational thesis" for insufficient attention to differences among professional groups in motivation, in alliances with business groups, in identification with technology and science, and in degree of bureaucratization in Wayne K. Hobson, "Professionals, Progressives and Bureaucratization: A Reassessment," *Historian* 39 (August 1977): 639–58; and for insufficient attention to the opposition of many farmers to greater centralization, mechanization, and organization in David B. Danbom, *The Resisted Revolution: Urban America and the Industrialization of Agriculture, 1900–1930* (Ames: Iowa State University Press, 1979). Higham's "Reorientation of American Culture in the 1890s," 88, finds ambivalence toward organizationalism in at least that period: an enthusiasm for the unorganized "strenuous life" alongside the very organization of those outdoor activities, and Morton Keller, *Affairs of State: Public Life in Late Nineteenth Century America* (Cambridge: Harvard University Press, 1977), discovers mixed feelings about the size, shape, and strength of government at all levels in the period 1865–1900. Finally, Nash's *The Great Depression* exaggerates both the extent and the consciousness of organizationalism in the 1930s and early 1940s and, in so doing, undermines his thesis that organizationalism *did* increase in that period—it did, but surely not so much as he describes. On the general limitations of the "organizational thesis" for American history, see Berkhofer, 624–26.

19. I have therefore restricted my argument and interpretation to certain tendencies within an admittedly larger, heterogeneous culture and to certain reform movements within an admittedly broader spectrum—usually termed *Progressivism* but defined variously (for that reason, I have not used the term in this study). I have consequently left out several other trends—such as nationalism and imperialism—and several other reform crusades—such as the Social Gospel and muckraking (also commonly lumped together as Progressivism). These additional trends and crusades were not necessarily antithetical to those I have included; many were even complementary. For example, the Social Gospel movement attempted to accommodate mainstream liberal Protestantism to the American industrial revolution and its aftermath. And muckraking obviously treated the same changes, if in different fashion and with somewhat different objectives. Nevertheless, they bear at most only peripherally on technological utopianism both in itself and within American culture. On pertinent other contemporary currents, see the relevant portions of Wiebe, chaps. 3, 6; R. Jackson Wilson, *In Quest of Community: Social Philosophy in the United States, 1860–1920* (New York: Wiley, 1968); Higham, "The Reorientation of American Culture," 73–102; R. Laurence Moore, "Directions of Thought in Progressive America," in *The Progressive Era*, ed. Lewis L. Gould (Syracuse, N.Y.: Syracuse University Press, 1974), 35–53; Daniel Horowitz, "An Intellectual History of American Industrialization: Academic Economists, New England Publicists, and Engineers, 1830–1910" (manuscript, n.d., in possession of author), pts. 1, 2; Horowitz, "Genteel Observers: New England Economic Writers and Industrialization," *New England Quarterly* 48 (March 1975): 65–83; and Lears. The term *Progressivism* has become encumbered with so many different meanings that Peter G. Filene has composed its "obituary": "A diffuse progressive 'era' may have occurred, but a progressive 'movement' did not. 'Progressives' there were, but of many types . . . characterized by shifting coalitions around different issues" ("An Obituary for 'The Progressive Movement,' " *American Quarterly* 22 [Spring 1970]: 33). On this issue, see also the more analytical discussions of Otis L. Graham, Jr., *The Great Campaigns: Reform and War in America, 1900–1928* (Englewood Cliffs, N.J.: Prentice-Hall, 1971), 22–23, 126–29; Thomas K. McCraw, "The Progressive Legacy," in *The Progressive Era*,

181–201; David M. Kennedy, "Overview: The Progressive Era," *Historian* 37 (May 1975): 453–68; Joan Hoff Wilson, *Herbert Hoover: Forgotten Progressive* (Boston: Little, Brown, 1975), 48–51; John D. Buenker, John C. Burnham, and Robert M. Crunden, *Progressivism* (Cambridge, Mass.: Schenkman, 1977); Chambers, Preface; and Blaine A. Brownell, "Interpretations of Twentieth-Century Urban Progressive Reform," in *Reform and Reformers in the Progressive Era*, ed. David R. Colburn and George E. Pozzetta (Westport, Conn.: Greenwood Press, 1983), 3–23. Chambers, 138, briefly discusses the organic model of society that the majority of Progressives endorsed and sought to realize.

20. On the development of conservation throughout American history, see Donald Fleming, "Roots of the New Conservation Movement," *Perspectives in American History* 6 (1972): 7–91; Roderick Nash, *Wilderness and the American Mind*, 2d ed. (New Haven, Conn.: Yale University Press, 1973); and Roderick Nash, *The American Conservation Movement* (St. Charles, Mo.: Forum Press, 1974).

21. Hays, *Conservation and the Gospel of Efficiency: The Progressive Conservation Movement, 1890–1920* (New York: Atheneum, 1969), 1. Additional opponents were resource users with immediate and particular as opposed to long-term, "systematic" objectives: sheepmen, cattlemen, big-game hunters, landowners, and others. James L. Penick, Jr., offers a similar interpretation of the conservation movement during this period. See his *Progressive Politics and Conservation: The Ballinger-Pinchot Affair* (Chicago: University of Chicago Press, 1968), Introduction and chap. 8, and "The Progressives and the Environment," in *The Progressive Era*, 115–31. Penick does, however, observe in both works that experts and bureaucracies are hardly free from politics themselves, contrary to the implications of Hays' work (despite Hays' emphasis, noted below, on political structures)—a point which I will elaborate on in chap. 7. Donald C. Swain extends Hays' overall interpretation of the conservation movement through 1933, when this study ends. See his *Federal Conservation Policy, 1921–1933* (Berkeley and Los Angeles: University of California Press, 1963). He concludes, "Contrary to widely held opinion, the national conservation program did not deteriorate in the 1920's. It expanded and matured" (170). On conservation policy for the first four decades of the twentieth century, see also Penick, "The Resource Revolution," in *Technology in Western Civilization* 2:431–48. On the role of science and government in conservation policy for the late nineteenth and early twentieth centuries, see also A. Hunter Dupree, *Science in the Federal Government: A History of Policies and Activities to 1940* (Cambridge: Harvard University Press, 1957), chap. 12.

22. Hays, *Conservation*, 265. Hays' 1969 Preface to this paperback edition of his 1959 book emphasizes that the work was intended as more a case study of these broader developments than a study of conservation policies as such. Hays makes the same point in a 1970 publication, "Conservation and the Structure of American Politics: The Progressive Era," reprinted in his collected essays, *American Political History as Social Analysis* (Knoxville: University of Tennessee Press, 1980), 233–43. A similar interpretation of American environmental planning—not conservation alone—is found in John Brinckerhoff Jackson, *American Space: The Centennial Years, 1865–1876* (New York: Norton, 1972).

23. The term *scientific management* eventually applied to four distinct if closely related phenomena: personal efficiency, or a personal attribute; mechanical efficiency (from thermodynamics), or the energy output-input ratio of a machine; commercial efficiency, or the output-input ratio of dollars; and social efficiency, or a social relationship among people. See Samuel Haber, *Efficiency and Uplift: Scientific Management in the Progressive Era, 1890–1920* (Chicago: University of Chicago Press, 1964), ix–x. See also Sumner H. Slichter, "Efficiency," in *The Encyclopaedia of the Social Sciences*, ed. Edwin R. A. Seligman (New York: Macmillan, 1931), 5:439, who condenses these four meanings into three somewhat vaguer ones. Taylor was hardly the first student of management. His originality lay in transforming an existing group of

fairly nebulous concepts into a "science" supposedly capable of universal validation. On the origins of scientific management, see Harlow Person, "Scientific Management," in *The Encyclopaedia of the Social Sciences* (1934), 13: 603–4; Haber, 18–22; Claude S. George, Jr., *The History of Management Thought* (Englewood Cliffs, N.J.: Prentice–Hall, 1972), chap. 5; Norman B. Wilkinson, "In Anticipation of Frederick W. Taylor: A Study of Work by Lammot du Pont, 1872," *Technology and Culture* 6 (Spring 1965): 208–21; and Daniel Nelson, *Frederick W. Taylor and the Rise of Scientific Management* (Madison: University of Wisconsin Press, 1980), chap 1. On Taylor's own background, see Haber, chap. 1; Frank B. Copley, *Frederick Winslow Taylor, Father of Scientific Management*, 2 vols. (New York: Harper, 1923); Sudhir Kakar, *Frederick Taylor: A Study in Personality and Innovation* (Cambridge: MIT Press, 1970); and Nelson, chaps. 2–7.

24. Haber, chap. 4.

25. Harrington Emerson, *Twelve Principles of Efficiency* (New York: Engineering Magazine, 1913), 372. On Emerson, see Haber, 55–57.

26. Concludes Haber: "Efficiency as morality, the most widespread and easily acceptable form, was quickest to evaporate. Efficiency as a series of profit-making stunts was soon discredited. Efficiency as a technique of industrial management and as a form of social control found a small but steadfast following and had more lasting effects" (74), but with the considerable alterations discussed below.

27. On Taylor's disciples, see Person, "Scientific Management," 607; Haber, chaps. 3, 9; George, chaps. 6, 7; Edwin T. Layton, Jr., *The Revolt of the Engineers: Social Responsibility and the American Engineering Profession* (Cleveland: Case Western Reserve University Press, 1971), chaps. 6, 7; and Nelson, chaps. 5–8.

28. Industrial psychology in turn proved less successful than had been expected. Its initial aim was the development of tests that would measure the aptitudes of prospective workers, who would be hired accordingly. Unfortunately, the tests were difficult to compose, to apply, and to evaluate. They gave way to a concern for environmental reforms, and from the late 1920s through the 1940s a "scientific" approach to industry meant a concern for improved working conditions. Beginning with the classic studies conducted in the Western Electric Company's Hawthorne Works outside Chicago, psychologists discovered that group perceptions of working conditions, rather than native individual abilities, most affected workers' productivity. They therefore recommended that industries expend less effort on recruiting the ablest workers available and more effort on making the conditions of all workers as pleasant as possible. On the development of industrial psychology, see Reinhard Bendix, *Work and Authority in Industry: Ideologies of Management in the Course of Industrialization* (New York: Wiley, 1956), chap. 5; Loren Baritz, *The Servants of Power: A History of the Use of Social Science in American Industry* (Middletown, Conn.: Wesleyan University Press, 1960); Haber, chap. 9; Robert H. Guest, "The Rationalization of Management," in *Technology in Western Civilization* 2:61–63; and John C. Burnham, "The New Psychology: From Narcissism to Social Control," in *Change and Continuity in Twentieth-Century America: The 1920's*, ed. John Braeman, Robert H. Bremner, and David Brody (Columbus: Ohio State University Press, 1968), 351–98, esp. 389–93. Baritz repeatedly implies that the psychological experts that he studies, like the scientific management consultants who preceded them, willingly served the interests of their employers at the expense of their supposedly scientific research. The logical outcome of environmental conditioning by experts for workers and, later, all others, was the psychology of behaviorism as initially espoused by John B. Watson in the 1910s and 1920s. Assuming that the minds of all human beings were blank slates, Watson claimed to be able to program especially babies to conform to any desired personality and outlook.

29. Not surprisingly, the relations between the Taylorites and trade unionists were particularly bitter. See Person, "Scientific Management," 607–608; Milton J. Nadworny, *Scientific Management and the Unions, 1900–1932: A Historical Analysis*

(Cambridge: Harvard University Press, 1955); Hugh G. J. Aitken, *Taylorism at Water-town Arsenal: Scientific Management in Action, 1908–1915* (Cambridge: Harvard University Press, 1960); Haber, 67–69, 127–33, 149–50; Guest, 52–64; Nelson, *Managers and Workers: Origins of the New Factory System in the United States, 1880–1920* (Madison: University of Wisconsin Press, 1975), chap. 4; and David Montgomery, *Workers' Control in America: Studies in the History of Work, Technology, and Labor Struggles* (New York: Cambridge University Press, 1979). Montgomery's essays, which cover the late nineteenth century through the present, are highly (and persuasively) critical of studies such as Aitken's and Haber's which provide a largely "technological explanation of the spread of Taylorism" (27). Such studies ignore both the persistent opposition by workers of various kinds to scientific and later forms of management and the persistent demand for worker control over work that spurred their opposition. In addition, such studies minimize the extent to which, in Guest's words, "many of Taylor's 'scientific' findings were, in fact, mere rationalizations of the class interests of management" (63). Finally, Montgomery offers compelling examples of ways in which allegedly rational forms of management had irrational underpinnings and so results. As he demonstrates, "The scientifically managed factory appeared to employers to be under rational engineering control. But to craftsmen of the prewar generation that plant resembled a bedlam: arbitrary and pretentious men in white shirts shouted orders, crept up behind workers with stopwatches, had them running incessantly back and forth to time clocks, and posted silly notices on bulletin boards" (117). Nelson nevertheless argues that "scientific management had little direct effect on the character of factory work or the lot of the worker" (*Frederick W. Taylor*, x). See ibid., chaps, 6, 8.

30. On the application of scientific management to government in general, see Leonard D. White, "Public Administration," in *The Encyclopaedia of the Social Sciences* (1930), 1: 440–50, esp. 442–43, and Dwight Waldo, *The Administrative State: A Study of the Political Theory of American Public Administration* (New York: Ronald, 1948), chap. 3. Taylorism was applied to many branches and at various levels of government, but nowhere was it applied so vigorously and self-consciously as in the State Department. There it was applied not only to the administration of the department but also to the formulation of foreign policy. See Israel, "A Diplomatic Machine: Scientific Management in the Department of State, 1906–1924," in *Building the Organizational Society*, 183–96.

31. On the application of scientific management to education in general, see Raymond E. Callahan, *Education and the Cult of Efficiency* (Chicago: University of Chicago Press, 1962), and, as a case study (using Atlanta), Philip N. Racine, "A Progressive Fights Efficiency: The Survival of Willis Sutton, School Superintendent," *South Atlantic Quarterly* 76 (Winter 1977): 103–16. On the vocational dimension of scientific management in education, see Berenice M. Fisher, *Industrial Education: American Ideals and Institutions* (Madison: University of Wisconsin Press, 1967), chap. 4, and Arthur G. Wirth, *Education in the Technological Society: The Vocational-Liberal Studies Controversy in the Early Twentieth Century* (Scranton, Pa.: Intext, 1972). David B. Tyack, *The One Best System: A History of American Urban Education* (Cambridge: Harvard University Press, 1974), pts. 4, 5, makes clear the importance of a new corporate model of institutions, to be discussed below, as a replacement for an older factory model as the principal ideal model for the public schools, and the consequently greater classification and hierarchy of ranks of staff and of students within the schools—reflections in turn of the emerging organic model of society.

32. What Hawley writes about the "search for a modern order" during the period 1917–1933 surely applies here: "It provided a framework in which the ideology of scientific management moved out from the work place to embrace whole industries, whole systems of social activity, and national economic life as a whole" (227–28).

33. Hays, Introduction, *Building the Organizational Society*, 6. But see also n. 37 below on this use of *functional*.

34. Thus in 1867 the American Society of Civil Engineers, the first professional division within engineering as a whole, was formed (after earlier attempts, in 1837 and in 1852, had largely failed). It was followed by the American Institute of Mining Engineers in 1871, the American Society of Mechanical Engineers in 1880, and the American Institute of Electrical Engineers in 1884. These four "founder societies," as they came to be called, were followed by more than a score of other, newer, and usually more specialized professional bodies in the late nineteenth and early twentieth centuries. See Monte A. Calvert, "The Search for Engineering Unity: The Professionalization of Special Interest," in *Building the Organizational Society*, 42–54. See also Layton, "Frederick Haynes Newell and the Revolt of Engineers," *Midcontinent American Studies Journal* 3 (Fall 1962): 17–26.

35. *The Emergent American Society*, 1: 280–86.

36. See Arthur M. Schlesinger, Sr., "Biography of a Nation of Joiners," in his *Paths to the Present* (New York: Macmillan, 1949), 23–50. See also Peter Dobkin Hall, *The Organization of American Culture, 1700–1900: Private Institutions, Elites, and the Origins of American Nationality* (New York: New York University Press, 1982).

37. For excellent examples of these changes, see Lynn L. Marshall, "The Strange Stillbirth of the Whig Party," *American Historical Review* 72 (January 1967): 445–68; Henry D. Shapiro, "From Association to Community: The Organization of the A.A.A.S. and the Transformation of Scientific Society in the United States" (paper delivered at annual meeting, American Association for the Advancement of Science, Boston, 1976), and Shapiro, "The Western Academy of Natural Sciences of Cincinnati and the Structure of Science in the Ohio Valley, 1810–1850," in *The Pursuit of Knowledge in the Early American Republic: American Scientific and Learned Societies from Colonial Times to the Civil War*, ed. Alexandra Oleson and Sanborn C. Brown (Baltimore: Johns Hopkins University Press, 1976), 219–47. As Hays himself admits in his Introduction to *Building the Organizational Society*, the term *functional* applies only to "people carrying out similar specialized tasks" (6) and so fits the distinctions drawn in this paragraph between types of organizations. Likewise, Hays admits too that these new functional organizations "did not bring organization to [a] society where no organization had before existed. On the contrary, they transformed [existing] patterns of social relationships . . ." (13). The qualifications made by Hobson, 639–58, regarding differences among professional groups in degree of bureaucratization, etc., should be kept in mind here as well. See n. 18 above.

38. On the development of the American corporation, see Chandler, "The Beginnings of 'Big Business' in American Industry," *Business History Review* 33 (Spring 1959): 1–31; Chandler, *Strategy and Structure: Chapters in the History of the American Industrial Enterprise* (Cambridge: MIT Press, 1962); Chandler, "The Large Industrial Corporation and the Making of the Modern American Economy," in *Institutions in Modern America*, 71–101; Chandler and Stephen Salsbury, *Pierre S. DuPont and the Making of the Modern Corporation* (New York: Harper and Row, 1971); Thomas C. Cochran, *Business in American Life: A History* (New York: McGraw-Hill, 1972), chaps. 9, 16; Glenn Porter, *The Rise of Big Business, 1860–1910* (Arlington Heights, Ill.: AHM, 1973); Chandler, *The Visible Hand: The Managerial Revolution in American Business* (Cambridge: Harvard University Press, 1977); and Cuff, "From Market to Manager," *Canadian Review of American Studies* 10 (Spring 1979): 47–54. The last is an excellent summary and review of these and other, older or briefer, studies by Chandler unnecessary to cite here. On the origins of corporate administration in general—which took place in association with the origins of the industrial revolution that later transformed the United States—see Sidney Pollard, *The Genesis of Modern Management: A Study of the Industrial Revolution in Great Britain* (London: Arnold, 1965). Pollard, 270–72, attributes to management a far greater share of the credit for Britain's industrial revolution than have previous historians.

39. Chandler, *Strategy and Structure*, 8. On the continued growth of corporate

administration and of administrative centralization in the 1930s, when the corporate structures Chandler describes had already achieved their basic form, see Graham, "The Planning Ideal and American Reality: The 1930's," in *The Hofstadter Aegis: A Memorial*, ed. Elkins and Eric McKitrick (New York: Knopf, 1974), 274–77. But on the continued growth of small businesses until the present, see *Small Business in American Life*, ed. Stuart W. Bruchey (New York: Columbia University Press, 1980).

40. For a provocative study of Carnegie's own major contributions to modern corporate management, see Harold C. Livesay, *Andrew Carnegie and the Rise of Big Business* (Boston: Little, Brown, 1975).

41. On the public relations campaigns of big businesses and big businessmen to enhance their corporate images and on their treatment by the media, see Sigmund Diamond, *The Reputation of the American Businessman* (Cambridge: Harvard University Press, 1955), esp. 178–79; Morrell Heald, "Management's Responsibility to Society: The Growth of an Idea," *Business History Review* 31 (Winter 1957): 375–84; Alan R. Raucher, *Public Relations and Business, 1900–1929* (Baltimore: Johns Hopkins University Press, 1968); Heald, *The Social Responsibilities of Business: Company and Community, 1900–1960* (Cleveland: Case Western Reserve University Press, 1970); and Galambos, *The Public Image of Big Business*, chap. 9. The ideal corporate image described by all four historians is one reflecting the pervasive organic model of society, not least in its preference for organization men and so for groups over individual entrepreneurs and executives.

42. See Hays, "The Social Analysis of American Political History, 1880–1920," *Political Science Quarterly* 80 (September 1965): 373–94, esp. 383–84, 391–92, and Hays, "Political Parties and the Community-Society Continuum," in *The American Party System: Stages of Political Development*, ed. William Nisbet Chambers and Walter Dean Burnham (New York: Oxford University Press, 1967), 152–81, esp. 170–71.

43. See Peter Woll, *American Bureaucracy* (New York: Norton, 1963), 174. On the development of bureaucracy in American government, see Woll, chap. 2, and *The Emergent American Society* 1:5–6. The figures come from *The Emergent American Society* 1:5, and John Chambers, 241. We should not, however, exaggerate the degree of development in the civil service, the hallmark of government bureaucracy. On its limited growth before the Civil War, see Matthew A. Crenson, *The Federal Machine: Beginnings of Bureaucracy in Jacksonian America* (Baltimore: Johns Hopkins University Press, 1975). On its greater but hardly smoother growth after the Civil War, see Ari Hoogenboom, *Outlawing the Spoils: A History of the Civil Service Reform Movement, 1865–1883* (Urbana: University of Illinois Press, 1961). As Hoogenboom says of the situation in 1865, "The civil service lacked system" (2).

44. Sidney Fine, *Laissez-Faire and the General-Welfare State: A Study of Conflict in American Thought, 1865–1901* (Ann Arbor: University of Michigan Press, 1956), vii. But Keller cautions about assuming that this development came about as easily and as uniformly as Hays, Fine, and others suggest. As indicated in n. 18 above, he demonstrates Americans' pervasive and persistent ambivalence toward "big" government at all levels.

45. Certainly there were other kinds of local and state reform campaigns—which Melvin G. Holli calls "social" rather than "structural"—that pursued not the centralization and bureaucratization of government but rather such social measures as the expansion of the franchise; the introduction of primaries, initiatives, referendums, and recalls; the direct election of U.S. Senators; the lowering of tax, utility, and transit rates for the poor; the passage of social justice legislation regulating child labor, hours and wages, and factory working conditions; and the redistribution of the local or state wealth. See Holli, *Reform in Detroit: Hazen S. Pingree and Urban Politics* (New York: Oxford University Press, 1969), chap. 8. On municipal and state reform drives and the

distinctions between types of reform crusades, see Benjamin Parke DeWitt, *The Progressive Movement* (New York: Macmillan, 1915), pts. 3, 4; George E. Mowry, *The Era of Theodore Roosevelt and the Birth of Modern America, 1900–1912* (New York: Harper, 1958), chap. 4; Hays, "The Politics of Reform in Municipal Government in the Progressive Era," *Pacific Northwest Quarterly* 55 (October 1964): 157–69; Holli, *Reform in Detroit*; Roy Lubove, *Twentieth-Century Pittsburgh: Government, Business, and Environmental Change* (New York: Wiley, 1969), chaps. 2, 3; Stanley P. Caine, *The Myth of a Progressive Reform: Railroad Regulation in Wisconsin, 1903–1910* (Madison: State Historical Society of Wisconsin, 1970); Graham, *The Great Campaigns*, pts. 1, 3, esp. pp. 27–33; David P. Thelen, *The New Citizenship: Origins of Progressivism in Wisconsin, 1885–1900* (Columbia: University of Missouri Press, 1972), esp. 1–3, 311–12; Holli, "Urban Reform in the Progressive Era," in *The Progressive Era*, 133–51; McCraw, 181–201; and Martin J. Schiesl, *The Politics of Efficiency: Municipal Administration and Reform in America, 1800–1920* (Berkeley and Los Angeles: University of California Press, 1977). The social reforms were clearly less elitist, less purely administrative, and more democratic than the structural reforms. Placing the two together points up some of the many tensions within "Progressivism" and so provides further reason not to invoke the term here. Yet many times the same reformers sought both varieties of reform, either simultaneously or alternately, and they sometimes saw the two types as complementary. Holli's own study of Detroit Mayor and later Michigan Governor Hazen Pingree offers an excellent example of such integration. Still, it is the structural reforms that had the greater influence and success and bear the closer relationship to technological utopianism.

46. By 1917, at its peak of popularity, nearly 500 cities had adopted the commission form of urban government. By 1930 well over 300 cities had adopted the city manager form. And by 1930 over 300 cities had chosen the manager-commission variety. On city commission, city manager, and manager-commission types of government, see Hays, "The Politics of Reform," 159–65; Weinstein, chap. 4; Graham, *The Great Campaigns*, 132; Holli, "Urban Reform in the Progressive Era," 147–48; Richard J. Stillman II, *The Rise of the City Manager: A Public Professional in Local Government* (Albuquerque: University of New Mexico Press, 1974); Richard M. Bernard and Bradley R. Rice, "Political Environment and the Adoption of Progressive Municipal Reform," *Journal of Urban History* 1 (February 1975): 149–74; Rice, *Progressive Cities: The Commission Government Movement in America, 1901–1920* (Austin: University of Texas Press, 1977); Schiesl, chaps. 7, 9, 10; and Peter R. Gluck and Richard J. Meister, *Cities in Transition: Social Changes and Institutional Responses in Urban Development* (New York: Franklin Watts, 1979), chaps. 5, 6. Rice does find multiple motives for instituting the commission form of government: besides efficiency, he uncovers businessmen's sheer greed, their desire for continued or expanded political power, their and others' civic boosterism, and their and others' simple faddishness. He even finds selected support for the reform from among labor leaders and socialists and selected opposition to it from among businessmen. Whatever their motives, the decision makers saw their choices as promoting efficiency, a validating quality in their minds. Yet DeWitt, 319–20, back in 1915, cautioned municipal reformers about excessive reliance on efficiency, on treating it as an end in itself rather than as a means to other reforms.

47. On municipal research bureaus, see Waldo, 31–33; Hays, "The Politics of Reform," 159; Jane S. Dahlberg, *The New York Bureau of Municipal Research: Pioneer in Government Administration* (New York: New York University Press, 1966); and Holli, "Urban Reform in the Progressive Era," 144–46. On similar management agencies at the state level, see Graham, *The Great Campaigns*, 133.

48. See Holli, "Urban Reform in the Progressive Era," 148–51. The traditional histories of this period usually portrayed sharp divisions between big businessmen and their allies (professionals and the well-to-do)—few of whom were deemed true "Pro-

gressives"—and political bosses, trade unionists, ethnic and immigrant leaders, working-class radicals, and other poor and uneducated political groups—many of whom, aside from the bosses, were deemed true "Progressives," along with middle class reformers. Revisionist histories, reflected in this and other portions of the present chapter, have now shown that such simplistic divisions did not often occur. Rice's conclusion about the commission reform crusade applies to these others as well: "While business elites almost invariably favored and laboring classes usually opposed the adoption of the commission . . . the story involved much more than a conflict between a single-minded business community and a unified lower-class opposition. Neither side was monolithic" (110). Nevertheless, the social reforms, when enacted, did produce more serious change in city and state government than the structural reforms—but not the full-fledged, egalitarian democracy that was sought. On the backgrounds and achievements of the municipal and state structural and social reformers, see Hays, "The Politics of Reform," 168; Hays, "The Social Analysis of American Political History," 387–88; Hays, "Political Parties and the Community-Society Continuum," 172; Weinstein, chap. 4; Caine, chap. 10; Graham, The Great Campaigns, 27–33, 131–32, 136; Lyle W. Dorsett, "The City Boss and the Reformer: A Reappraisal," Pacific Northwest Quarterly 63 (October 1972): 150–54; Brownell and Warren E. Stickle, Introduction, pt. 3, Bosses and Reformers: Urban Politics in America, 1880–1920, ed. Brownell and Stickle (Boston: Houghton Mifflin, 1973), 164–66; Holli, "Urban Reform in the Progressive Era," 148–51; McCraw, 185–91; Bernard and Rice, 149–74; Howard P. Chudacoff, The Evolution of American Urban Society (Englewood Cliffs, N.J.: Prentice-Hall, 1975), 153–55; Charles N. Glaab and A. Theodore Brown, A History of Urban America, 2d ed. (New York: Macmillan, 1976), 196–99; Rice; Schiesl; John Chambers, 238–39; and Hays, Introduction, American Political History as Social Analysis, 23–26.

49. Graham's representative comment on the chronology of developments may be chronologically correct but is causally insufficient; it ignores the influence of both the quest for efficiency and the organic model of society: "Even before progressivism at the state level had reached its apogee, reformers began to think of Washington" (The Great Campaigns, 33). His comment on the relationship between local and state reform drives (ibid., 30) is similar.

50. Thus Robert Wiebe and Gabriel Kolko, for all their disagreements over the worth of the reformers' efforts and over the degree of unity among them, agree that businessmen and their political supporters were "the most important single factor—or set of factors—in the development of economic regulation" (Wiebe, Businessmen and Reform: A Study of the Progressive Movement [Cambridge: Harvard University Press, 1962], 217). See ibid., and Kolko, The Triumph of Conservatism: A Reinterpretation of American History, 1900–1915 (Glencoe, Ill.: Free Press, 1963). Kolko argues, "The basic fact of the Progressive Era was the large area of consensus and unity among key business leaders and most political factions on the role of the federal government in the economy. There were disagreements, of course, but not on fundamentals" (208). Even so traditional an account of the relationship between business and reform as that of Edward D. Kirkland, Dream and Thought in the Business Community, 1860–1900 (Ithaca, N.Y.: Cornell University Press, 1956), chap. 5, acknowledges closer ties between them after 1900. By contrast, Hoogenboom, ix-x, 196–97, finds little support by businessmen for civil service reform at any level in the period 1865–1883. And Keller, esp. chap. 11, uncovers ambivalence in many sectors, including business, for government regulation of the economy in the period 1865–1900.

51. Whether these agencies were originally established as genuine regulatory bodies, as Wiebe believes, or were from the outset tools of business, as Kolko contends, the result, both agree, was that later generations of businessmen "benefitted most" from the very organs of government that, at least for Wiebe, "their predecessors once feared

would destroy them" (Wiebe, *Businessmen and Reform*, 224). Kolko calls these developments part of the "political rationalization" (2) of modern American society: not merely "the improvement of efficiency, output, or internal organization of a company" but, more broadly, "the organization of the economy and the larger political and social spheres in a manner that will allow corporations to function in a predictable and secure environment permitting reasonable profits over the long run" (3). On the issue of government regulation in this period and its beneficiaries, see also, besides Wiebe, *Businessmen and Reform*, chap. 9, and Kolko, chaps. 9, 10, and Conclusion, the basically complementary interpretations, via case studies, of Gordon M. Jensen, "The National Civic Federation: American Business in an Age of Social Change and Social Reform, 1900–1910" (Ph.D. diss., Princeton University, 1956); Daniel Levine, *Varieties of Reform Thought* (Madison: State Historical Society of Wisconsin, 1964), chap. 3 (on the Civic Federation of Chicago, out of which grew the National Civic Federation); Kolko, *Railroads and Regulation, 1877–1916* (Princeton, N.J.: Princeton University Press, 1965); Melvin I. Urofsky, *Big Steel and the Wilson Administration: A Study in Business-Government Relations* (Columbus: Ohio State University Press, 1969); Weinstein, chaps. 1, 3 (on the National Civic Federation and the Federal Trade Commission); Caine (whose study of railroad regulation in Wisconsin is intended to apply to the national level); Bruno Ramirez, *When Workers Fight: The Politics of Industrial Relations in the Progressive Era, 1898–1916* (Westport, Conn.: Greenwood Press, 1978); and Anne Kusener Nelson, "Policy Formulation and Implementation: The Case of the U.S. Industrial Commission [of 1898–1902]," in *Retrospective Technology Assessment— 1976*, ed. Joel A. Tarr (San Francisco: San Francisco Press, 1977), 149–63. See as well the largely antithetical interpretations of government regulation, also via case studies, of Robert W. Harbeson, "Railroads and Regulation, 1877–1916: Conspiracy or Public Interest?" *Journal of Economic History* 27 (June 1967): 230–42; Gerald Kurland, *Seth Low: The Reformer in an Urban and Industrial Age* (New York: Twayne, 1971), chap. 16 (on the National Civic Federation); and Albro Martin, *Enterprise Denied: Origins of the Decline of American Railroads, 1897–1917* (New York: Columbia University Press, 1971). Finally, see the more general, more balanced, and more persuasive assessments of government regulation in this period of Marver H. Bernstein, *Regulating Business By Independent Commission* (Princeton, N.J.: Princeton University Press, 1955); Milton Derber, *The American Idea of Industrial Democracy, 1865–1965* (Urbana: University of Illinois Press, 1970), pts. 2, 3; Graham, *The Great Campaigns*, 33–49, 135–38; Cochran, 215–29, 240–41; McCraw, 187–91, 196–99; Gould, *Reform and Regulation: American Politics, 1900–1916* (New York: Wiley, 1978); John Chambers, 40–42, chaps. 4, 8 (esp. pp. 242–44); and *Regulation in Perspective: Historical Essays*, ed. McCraw (Cambridge: Harvard University Press, 1981). These last studies recognize the growing complexity of American society, particularly in the early twentieth century, and the consequent need for unprecedented government regulation at all levels of the economy and so of the businesses that were shaping, or reshaping, it. They view the impact as mixed.

52. On these agencies, see Gerald Nash, "Experiments in Industrial Mobilization: WIB and NRA," *Mid-America* 45 (July 1963): 157–74; William E. Leuchtenburg, "The New Deal and the Analogue of War," in *Change and Continuity in Twentieth-Century America*, ed. Braeman, Bremner, and Everett Walters (Columbus: Ohio State University Press, 1964), 81–143; Daniel R. Beaver, *Newton D. Baker and the American War Effort, 1917–1919* (Lincoln: University of Nebraska Press, 1966); Paul A. C. Koistinen, "The 'Industrial-Military Complex' in Historical Perspective: World War I," *Business History Review* 41 (Winter 1967): 378–403; Weinstein, chap. 8; Urofsky, chaps. 3–8; Graham, *The Great Campaigns*, 103–107; Cuff, "The Cooperative Impulse and War: The Origins of the Council of National Defense and Advisory Commission," in *Building the Organizational Society*, 233–46; Pursell, Introduction, *The Military-Industrial Complex*, ed. Pursell (New York: Harper and Row, 1972), 2–4; Cuff, *The War Industries*

Board: Business-Government Relations During World War I (Baltimore: Johns Hopkins University Press, 1973); Cuff, "We Band of Brothers—Woodrow Wilson's War Managers," Canadian Review of American Studies 5 (Fall 1974): 135–48; Kennedy, "The Political Economy of World War I," Reviews in American History 2 (March 1974): 102–107; Graham, Toward a Planned Society: From Roosevelt to Nixon (New York: Oxford University Press, 1976), 9–13; Hawley, chap. 2; and Kennedy, Over Here: The First World War and American Society (New York: Oxford University Press, 1980), chap. 2. These authors do not concur on the precise degree of business-government ties in the early twentieth century. One of the most sophisticated among them, Cuff, concludes that "Insofar as the WIB experience is any guide, the keynotes of business-government relations during the war are complexity, hesitancy, and ambiguity" (The War Industries Board, 7). Moreover, "What the WIB experience suggests . . . is that the phrase 'business-government relations' embraces a whole series of processes which are shaped at any one time by a complex interplay of fluctuating variables" (ibid., 275). Nevertheless see ibid., 271–75, for Cuff's partial integration of the other major interpretations of business-government relations with his own. See also his "We Band of Brothers," 143, for a description of Woodrow Wilson's "organic, historical, evolutionary view of society and social change." Graham comments on the consequences of those relations, however varying and ambiguous: "the perversion of the idea of a regulatory state [occurred] as business groups came to dominate governmental policy" (The Great Campaigns, 103).

53. So lasting was their impact that two of the earliest and most important New Deal agencies, the National Recovery Administration and the Agricultural Adjustment Administration, were modeled on these First World War agencies and were headed by veterans of them. On the impact of those agencies on Herbert Hoover's career, see Hawley, 226–27. On the growing post–First World War relations between big business and the military, see Koistinen, "The 'Industrial-Military Complex' in Historical Perspective: The Interwar Years," Journal of American History 56 (March 1970): 819–39; Burton I. Kaufman, Efficiency and Expansion: Foreign Trade Organization in the Wilson Administration, 1913–1921 (Westport, Conn.: Greenwood Press, 1974), chaps. 8–11; Roger W. Lotchin, "The City and the Sword: San Francisco and the Rise of the Metropolitan-Military Complex, 1919–1941," Journal of American History 65 (March 1979): 996–1020; Lotchin, "The Metropolitan-Military Complex in Comparative Perspective: San Francisco, Los Angeles, and San Diego, 1919–1941," Journal of the West 18 (July 1979): 19–30; Nash, The Great Depression, chaps. 9, 11; and Hawley, chaps. 3, 4, 6, 10. On the growing post–First World War relations between big business and the federal government in civilian matters, see Galambos, Competition and Cooperation: The Emergence of a National Trade Association (Baltimore: Johns Hopkins University Press, 1966); Graham, "The Planning Ideal and American Reality," 280–84; Wilson, 70–73; and Robert F. Himmelberg, The Origins of the National Recovery Administration: Business, Government, and the Trade Association Issue, 1921–1933 (New York: Fordham University Press, 1976). These works show continued business efforts in favor of greater regulation (and so organization) and consequently reduced competition and increased monopolization and profits. Yet they also demonstrate differences within the business community over specific policies. So, too, do some of the works already cited on business-military relations and several others on business involvement in foreign policy: for example, Carl P. Parrini, Heir to Empire: U.S. Economic Diplomacy, 1916–1923 (Pittsburgh: University of Pittsburgh Press, 1969); and Wilson, American Business and Foreign Policy, 1920–1933 (Lexington: University Press of Kentucky, 1971). On the vacillating and improvised policies of the New Deal regarding regulation by the federal government, see Hawley, The New Deal and the Problem of Monopoly: A Study in Economic Ambivalence (Princeton, N.J.: Princeton University Press, 1966).

54. Between 1865 and 1895 Congress had sponsored four investigations of government management, but none of their recommendations had been implemented. Nor had the proposals of Presidents Theodore Roosevelt through Coolidge ever been adopted. In each case, Richard Polenberg observes, the objective was merely "to promote economy through the introduction of improved working methods, and the underlying assumption was the responsibility of the administrative system to the Legislature" (*Reorganizing Roosevelt's Government: The Controversy over Executive Reorganization, 1936–1939* [Cambridge: Harvard University Press, 1966], 3). On these earlier congressional and presidential reorganization efforts, see Polenberg, 3–5; Herbert Emmerich, *Federal Organization and Administrative Management* (University: University of Alabama Press, 1971), 29–32, 38–43; and William E. Pemberton, *Bureaucratic Politics: Executive Reorganization during the Truman Administration* (Columbia: University of Missouri Press, 1979), chap. 2. The proposals of Hoover and Franklin Roosevelt were far broader in scope and were urged far more vigorously. Efforts during this same period on the federal level, and on the municipal and state levels as well, to develop a genuine civil service were only moderately more successful. See Hoogenboom for the years 1865–1883, and, for the whole late nineteenth and early twentieth centuries, Paul Van Riper, "American Civil Service Reform," in *Bureaucracy in Historical Perspective*, ed. Michael T. Dalby and Michael S. Werthman (Glenview, Ill.: Scott, Foresman, 1971), 126–34.

55. To facilitate achieving those goals, Roosevelt's commission recommended the following: (1) appointing six executive assistants to the President; (2) strengthening the federal civil service; (3) improving fiscal management by budget planning and independent auditing; (4) establishing a permanent National Resources Planning Board to coordinate all government programs; and (5) creating two new cabinet posts and thereafter placing every executive agency, including the hitherto (relatively) independent regulatory commissions, under one of the twelve cabinet departments. Although approved by the Senate, the Executive Reorganization Bill failed in the House, where Roosevelt's opponents, with considerable public support, mounted an effective campaign against it in the name of liberty. As Polenberg suggests, "perhaps no other [New Deal] proposal was so little understood by the common man—or as angrily denounced by him" (193). The revised bill, passed by both houses the following year (1939), bore little resemblance to the original: it allowed Roosevelt to appoint his six proposed administrative assistants but dropped the other major provisions, forcing him to rearrange his administration within the existing institutional framework. See Polenberg, 184–88. It did, however, allow for the formal establishment of the Executive Office of the President. On the backgrounds and views of the members of Roosevelt's Committee on Administrative Management, see Emmerich, chap. 3, and Barry D. Karl, *Executive Reorganization and Reform in the New Deal: The Genesis of Administrative Management, 1900–1939* (Cambridge: Harvard University Press, 1963), chaps. 2–4. In addition, there persisted until 1943 a National Resources Planning Board, which had been founded a decade earlier—that is, before the establishment of Roosevelt's committee—but which never enjoyed the powers envisioned for it by the committee members. It became part of the Executive Office of the President in 1939, as did the Bureau of the Budget. On the board's history, see Charles E. Merriam, "The National Resources Planning Board: A Chapter in American Planning Experience," *American Political Science Review* 38 (December 1944): 1075–88; Philip W. Warken, "A History of the National Resources Planning Board, 1933–1943" (Ph.D. diss., Ohio State University, 1969); Karl, *Charles E. Merriam and the Study of Politics* (Chicago: University of Chicago Press, 1974), chaps. 12, 13; Arlene Inouye and Charles Susskind, " 'Technological Trends and National Policy,' 1937: The First Modern Technology Assessment," *Technology and Culture* 18 (October 1977): 593–621; and Marion Clawson, *New Deal*

Planning: The National Resources Planning Board (Baltimore: Johns Hopkins University Press, 1981).

56. They would not, however, eliminate the ongoing centralization and bureaucratization of government. Rather, they would help create unique forms of government administration. Their watchwords, besides of course efficiency, would be volunteerism, cooperation, service, flexibility, and integrity. Toward that end, and over the course of his career, Hoover arranged more than 3,000 conferences, commissions, and trade association meetings of various kinds attended largely by experts from those different realms. In Hoover's mind the governmental process should consist of an ongoing debate, temporarily suspended at critical junctures, among an "intellectual, extra-governmental community which considered itself professional [i.e., objective and apolitical] in the study of society, the President and his Cabinet as managers of government, and Congress with its responsibility to represent and serve a constituency defined as an electorate or an interest" (Karl, "Presidential Planning and Social Science Research: Mr. Hoover's Experts," *Perspectives in American History* 3 [1969]: 396). In this light Polenberg surely exaggerates the originality of Roosevelt's objectives. He claims that Roosevelt's predecessors, Hoover included, had been concerned *only* with "the reduction of expenditures," where Roosevelt alone saw in reorganization the means to make the Presidency "more responsive to the national interest and better able to serve that interest" (Polenberg, 7). It is nevertheless questionable whether, as Karl contends, Hoover "was aiming at a rational revolution designed to create a Bellamy-like utopia" (Karl, "Presidential Planning," 363). A "pragmatic utopianism that defied standard economic and political classifications," as Wilson (*Herbert Hoover*, 56) puts it, might be more accurate. On Hoover's conception and style of government, see also Emmerich, pp. 43–45 and chaps. 5, 6; Craig Lloyd, *Aggressive Introvert: A Study of Herbert Hoover and Public Relations Management, 1912–1932* (Columbus: Ohio State University Press, 1972); Carolyn Grin, "The Unemployment Conference of 1921: An Experiment in National Cooperative Planning," *Mid-America* 55 (April 1973): 83–107; David Burner and Thomas R. West, "A Technocrat's Morality: Conservatism and Hoover the Engineer," in *The Hofstadter Aegis*, 235–56; Hawley, "Herbert Hoover, the Commerce Secretariat, and the Vision of an 'Associative State,' 1921–1928," *Journal of American History* 61 (June 1974): 116–40; Wilson, *Herbert Hoover*, chaps. 3–7, 9; Burner, *Herbert Hoover: A Public Life* (New York: Knopf, 1979), chaps. 9–12; Hawley, *The Great War*, chaps. 4, 6, 10–12; and *Herbert Hoover as Secretary of Commerce: Studies in New Era Thought and Practice*, ed. Hawley (Iowa City: University of Iowa Press, 1981). On government reorganization as, contrary to Hoover, a nonscientific process, see Waldo, 38; Polenberg, 192; Karl, "Presidential Planning," 388, 408–409; and Emmerich, 8, 16–17. On the backgrounds and views of the members of Hoover's Committee on Recent Social Trends, see Karl, "Presidential Planning," 347–409, and Karl, *Charles E. Merriam*, chap. 11.

57. On these different uses of political and social scientists—and sometimes the very same experts—by Hoover and Roosevelt, see Karl, "Presidential Planning," 408–409. While reorganization would, according to Roosevelt's committee, benefit all Americans through greater governmental efficiency, it would especially benefit the underprivileged. See Polenberg, 26.

58. Despite these differences in both their views and styles of government, Hoover and Roosevelt were not so far apart in their ultimate objectives as traditionally portrayed. As more recent interpretations of Hoover's career—such as those cited above—have made clear, Hoover was far from a reactionary, and Roosevelt was far from a radical. Moreover, both were committed to reorganizing the Presidency, if not the whole federal government, to save the existing economic and social as well as political order. As Hawley argues, with considerable persuasiveness, "the 1920's are best understood not as the Indian summer of an outmoded order or even as the seedtime of the

reforms of the New Deal but rather as the premature spring of the kind of modern capitalism that would take shape in the America of the 1940's and 1950's" (*The Great War*, vi). On the connections between the two Presidencies, see also ibid., 213–14. In a negative sense, the two administrations were also tied together by their hesitant commitment to national planning of some kind. As the discussion below of national planning will show, the commitment in other quarters of the country during this same period was, at best, equally half-hearted. Finally, on Hoover's later efforts, also through commissions (and with himself as chief expert), to streamline the executive branch of the federal government along the lines of his earlier schemes, see Wilson, *Herbert Hoover*, 224–28, and Burner, Epilogue. Pemberton nevertheless concludes, "The basic problems of society could not be solved by administrative reform." Rather, "Meaningful change in administrative organization could only occur with fundamental change in the [liberal democratic capitalist] economic system" (7).

59. On the evolution of the social science techniques refined by these researchers, see Karl, "Presidential Planning," 348–51. "The optimism of the social science of the 1920's [and in this case, at least, of the preceding decade as well] rested on a faith in the natural, rational limitations of the social process. It assumed the availability of finite bodies of data to which precise instruments of analysis could be applied, once those instruments had been perfected. Such analysis would yield programs of social and political behavior which would insure progress, at a sane and sedate pace, toward the ultimate human happiness" (ibid., 368). Inouye and Susskind, 593–621, discuss the considerable persistence of such optimism into the 1930s and early 1940s despite the effects of the Great Depression on social scientific analyses of the time. See also Gene M. Lyons, *The Uneasy Partnership: Social Science and the Federal Government in the Twentieth Century* (New York: Russell Sage Foundation, 1969), chaps. 1–3.

60. David W. Eakins argues, in the course of criticizing these centers for their "corporate liberalism," social stability ". . . was increasingly defined in terms of efficiency, of greater control, of greater centralization, of closer cooperation between businessmen and a rationalizing government. It became, then, more and more logical to create experts who understood that version of the science and society. . . . But their 'efficiency' and their effectiveness was acquired at the expense of democratic decision-making" (Eakins, "The Origins of Corporate Liberal Policy Research, 1916–1922: The Political-Economic Expert and the Decline of Public Debate," in *Building the Organizational Society*, 179). On the ideology of "corporate liberalism" no doubt espoused, at least, in part, by these research centers, see, besides ibid., 163–79, Eakins, "The Development of Corporate Liberal Policy Research in the United States, 1885–1965" (Ph.D. diss., University of Wisconsin, 1966); Weinstein, Introduction and chap. 1; and Eakins, "Policy Planning for the Establishment," in *A New History of Leviathan: Essays on the Rise of the American Corporate State*, ed. Ronald Radosh and Murray N. Rothbard (New York: Dutton, 1972), 188–205.

61. On the origins of these centers, see Eakins, "The Development of Corporate Liberal Policy Research," and "The Origins of Corporate Liberal Policy Research," 163–79. On the transformation of the previously mentioned National Civic Federation, founded in 1900, into a similar kind of research organ by the First World War, see Jensen, and Weinstein, chap. 1. A good example of the traditional, uncritical view of such centers as wholly objective analysts of policy issues is Arthur F. Burns, *Wesley Mitchell and the National Bureau* (New York: National Bureau of Economic Research, 1949), published as pp. 3–55 of the Bureau's Twenty-ninth Annual Report. But see Eakins' critique of the National Bureau in his "Policy-Planning for the Establishment," 193, 199–205.

62. On the development of city planning, see John W. Reps, *The Making of Urban America: A History of City Planning in the United States* (Princeton, N.J.: Princeton University Press, 1965); John L. Hancock, "Planners in the Changing American City,

1900–1940," *Journal of the American Institute of Planners* 33 (September 1967): 290–304; Jon A. Peterson, "The Origins of the Comprehensive City Planning Ideal in the United States, 1840–1911" (Ph.D. diss., Harvard University, 1967); Mel Scott, *American City Planning since 1890* (Berkeley and Los Angeles: University of California Press, 1969); Albert Fein, "The American City: The Ideal and the Real," in *The Rise of an American Architecture*, ed. Edgar Kaufmann, Jr. (New York: Praeger, 1970), 51–112; Thomas S. Hines, *Burnham of Chicago: Architect and Planner* (New York: Oxford University Press, 1974); Peterson, "The City Beautiful Movement: Forgotten Origins and Lost Meanings," *Journal of Urban History* 2 (August 1976): 415–34; and Stanley K. Schultz and Clay McShane, "To Engineer the Metropolis: Sewers, Sanitation, and City Planning in Late Nineteenth-Century America," *Journal of American History* 65 (September 1978): 389–411. Fein observes about the slow development of the profession of city planning, "The designers failed to create a single profession that might have unified the practitioners of architecture, engineering, and landscape architecture, who possessed a common concern for the environment. Instead, there developed separate professional schools and mutually conflicting objectives" (104). Yet he, Peterson, and especially Schultz and McShane argue that the origins of comprehensive city planning lie as much in social and cultural developments as in economic, political, and technological ones—a view which would support the significance of the organic model as a partial explanation for the assumption that cities could indeed be systematically planned.

63. The exceptions to this generalization were the considerable efforts in economic planning at the state and national levels of government throughout the nineteenth century. See George Rogers Taylor, *The Transportation Revolution, 1815–1860* (New York: Holt, Rinehart and Winston, 1951), chaps. 16, 17; Douglas C. North, *Growth and Welfare in the American Past: A New Economic History* (Englewood Cliffs, N.J.: Prentice-Hall, 1966), esp. chap. 8; and Harry N. Scheiber, *Ohio Canal Era: A Case Study of Government and the Economy, 1820–1861* (Athens: Ohio University Press, 1969), Preface and Conclusion.

64. On these individual "systems," see Peterson, "The Origins of the Comprehensive City Planning Ideal," chaps. 1, 2. The absence of genuinely comprehensive planning during the early (and often later) growth of most American cities exemplifies what Sam Bass Warner, Jr. has properly called the triumph of "privatism": "By and large the productivity and social order of the metropolis flowed from private institutions and individual adjustments. So did its weaknesses. Privatism left the metropolis helpless to guarantee its citizens a satisfactory standard of living. Privatism encouraged the building of vast new sections of the city in a manner well below contemporary standards of good layout and construction. Privatism suffered and abetted a system of politics which was so weak it could not deal effectively with the economic, physical, and social events that determined the quality of life within the city" (*The Private City: Philadelphia in Three Periods of Its Growth* [Philadelphia: University of Pennsylvania Press, 1968], 202). Warner's *The Urban Wilderness: A History of the American City* (New York: Harper and Row, 1972) discusses privatism in American cities in general and calls for not just more and better city planning but more "democratic national and regional planning" (276) as well.

65. On the "City Beautiful" movement, see Peterson, "The Origins of the Comprehensive City Planning Ideal," chaps. 3, 4; Scott, chap. 2; Hines, 141–42; and Peterson, "The City Beautiful Movement," 415–34. European cities were their model, but, as Reps observes, American planners rarely matched European standards. The results were further examples of piecemeal city planning. See Reps, *Town Planning in Frontier America* (Princeton, N.J.: Princeton University Press, 1969), chap. 12.

66. On Olmsted as scientific planner of both the natural and man-made environments, see chap. 5, p. 96, and n. 74 there. On Olmsted's career, see the works cited there,

esp. Fein, *Frederick Law Olmsted and the American Environmental Tradition* (New York: Braziller, 1972), and Laura W. Roper, *FLO: A Biography of Frederick Law Olmsted* (Baltimore: Johns Hopkins University Press, 1973). Of contemporary landscape architects—and Olmsted was more than a landscape architect in any case—only Horace William Shaler Cleveland, Robert Morris Copeland, and Charles Eliot approached his achievement. Although his work influenced Olmsted's, Andrew Jackson Downing was not urban enough in his orientation to warrant inclusion here. On landscape architecture in America, see Lubove's excellent review essay, "Social History and the History of Landscape Architecture," *Journal of Social History* 9 (Winter 1975): 268–75.

67. See Peterson, "The Origins of the Comprehensive City Planning Ideal," chaps. 3, 4. Peterson does rightly credit the fair with promoting the idea of comprehensive city planning but not with directly inspiring or reflecting it. See ibid., pp. 183–89. On the fair alone, see the conventional interpretations by Maurice Neufeld, "The White City: The Beginnings of a Planned Civilization in America," *Journal of the Illinois State Historical Society* 27 (April 1934): 71–93, a summary of his "The Contributions of the World's Columbian Exposition of 1893 to the Idea of a Planned Society in the United States" (Ph.D. diss., University of Wisconsin, 1935); William P. Shaw, "The World's Columbian Exposition: Its Revelations and Influences" (Master's thesis, Clark University, 1935); Lubove, *The Progressives and the Slums: Tenement House Reform in New York, 1890–1917* (Pittsburgh: University of Pittsburgh Press, 1962), 217–19; David H. Crook, "Louis Sullivan, the World's Columbian Exposition, and American Life" (Ph.D. diss., Harvard University, 1963); Michael T. Klare, "The Architecture of Imperial America," *Science and Society* 33 (Summer-Fall 1969): 257–84; and David F. Burg, *Chicago's White City of 1893* (Lexington: University Press of Kentucky, 1976); and the revisionist interpretations by Lowell Tozer, "American Attitudes toward Machine Technology, 1893–1933" (Ph.D. diss., University of Minnesota, 1953); John G. Cawelti, "America on Display: The World's Fairs of 1876, 1893, and 1933," in *The Age of Industrialism in America: Essays in Social Structure and Cultural Values,* ed. Jaher (New York: Free Press, 1968), 317–63; and Alan Trachtenberg, *The Incorporation of America: Culture and Society in the Gilded Age* (New York: Hill and Wang, 1982), chap. 7. Scott, 31–37, esp. 44, downplays the influence of the "White City" on city planning. Hines, 74, 123–24, 138, 141–42, 218, is more passive. On the McMillan Commission alone, see Scott, 47–57, and Hines, chap. 7. Despite these qualifications concerning its influence on comprehensive city planning, the fair did publicize the triumph of electricity, a key technological component in technological utopianism as in the real world. In addition, the fair promoted—but, contrary to those earlier interpretations, did not originate—the notion that "a city is an organism of interrelated, interdependent parts whose efficiency depends upon planned and orderly [i.e., evolutionary] growth" (Lubove, *The Progressives and the Slums,* 218).

68. The dates and figures in the paragraph are taken from Hancock, 294. On the conference and the institute, see Scott, 95–100, 163–64. Though called by their designers "comprehensive plans," the McMillan Commission and most other pre–First World War schemes were nevertheless fragmentary. Critical items such as sewers, shops, and homes were often left out, even in the most famous prewar design, Daniel Burnham and Edward Bennett's Plan of Chicago (1909). Only the Pittsburgh Survey (1907–1914), which collected and interpreted data on that city's population, traffic, taxation, health, housing, labor, education, and other problems, met the requirements for genuine comprehensiveness that future planners would almost take for granted. Yet the survey itself was more an analysis of existing conditions than a blueprint for improving them. On the limitations of city plans in general, see Peterson, "The Origins of the Comprehensive City Planning Ideal," chap. 6. On the limitations of the Plan of Chicago, see Hancock, 295; Scott, 37–39, 101–109; and Hines, chap. 14. On the Pittsburgh Survey,

see Bremner, *From the Depths: The Discovery of Poverty in the United States* (New York: New York University Press, 1956), 154–57; Lubove, *Twentieth-Century Pittsburgh*, chaps. 1, 2; Scott, 93–95; Clarke Chambers, *Paul U. Kellogg and the Survey: Voices for Social Welfare and Social Justice* (Minneapolis: University of Minnesota Press, 1971), 32–40; John F. McClymer, "The Pittsburgh Survey, 1907–1914: Forging an Ideology in the Steel District," *Pennsylvania History* 41 (April 1974): 169–86; and McClymer, *War and Welfare: Social Engineering in America, 1890–1925* (Westport, Conn.: Greenwood Press, 1980), pt. 1. Curiously, where Bremner and Chambers view the survey as objective social science at its pioneering—and humanitarian—best, McClymer sees it as a primary example of elitist, technocratic social engineering. The more balanced perspectives of Lubove and Scott are more persuasive.

69. Lubove, Introduction, *The Urban Community: Housing and Planning in the Progressive Era*, ed. Lubove (Englewood Cliffs, N.J.: Prentice-Hall, 1967), 14. As Lubove here also observes, the new concern with efficiency and professionalism often had the negative effect of limiting their creativity.

70. See Scott, 72, and Glaab and Brown, 223.

71. See chap. 5, pp. 95–96, and nn. 70–74.

72. The electrified street railways were the logical successors to the omnibuses of the 1820s and the horse-drawn street railways of the 1850s; neither of the latter two could travel as quickly, as comfortably, or as far as growing numbers of disenchanted city dwellers wished. On the electrified street railways, see Glen E. Holt, "The Changing Perception of Urban Pathology: An Essay on the Development of Mass Transit in the United States," in *Cities in American History*, ed. Kenneth T. Jackson and Schultz (New York: Knopf, 1972), 324–43, and Tarr, "From City to Suburb: The 'Moral' Influence of Transportation Technology," in *American Urban History: An Interpretive Reader with Commentaries*, ed. Alexander B. Callow, Jr., 2d ed. (New York: Oxford University Press, 1973), 202–12. For their impact on the growth of Boston, see Warner, *Streetcar Suburbs: The Process of Growth in Boston, 1870–1900* (Cambridge: Harvard University Press, 1962); on Milwaukee, McShane, *Technology and Reform: Street Railways and the Growth of Milwaukee, 1887–1900* (Madison: State Historical Society of Wisconsin, 1974); and on Norfolk and elsewhere, Glaab and Brown, 149.

73. On Shaker Heights, see Glaab and Brown, 256–57. On the Country Club District, see ibid., 266–68, and Norman T. Newton, *Design on the Land: The Development of Landscape Architecture* (Cambridge: Harvard University Press, 1971), 471–74.

74. On Pullman, see Graham R. Taylor, *Satellite Cities: A Study of Industrial Suburbs* (New York: Appleton, 1915), and Stanley Buder, *Pullman: An Experiment in Industrial Order and Community Planning, 1880–1930* (New York: Oxford University Press, 1967). Taylor's second chapter was sarcastically entitled "Rediscovering An Employer's Utopia." On Gary, see Graham Taylor, chaps. 6, 7; Raymond A. Mohl and Neil Betten, "The Failure of Industrial City Planning: Gary, Indiana, 1906–1910," *Journal of the American Institute of Planners* 38 (July 1972): 203–14; Anthony Brook, "Gary, Indiana: Steeltown Extraordinary," *Journal of American Studies* 9 (April 1975): 35–53; and James B. Lane, *"City of the Century": A History of Gary, Indiana* (Bloomington: Indiana University Press, 1978). In light of Gary's swift decline, an article by Henry B. Fuller in *Harper's Weekly* (51 [October 12, 1907]: 1482–83, 1495) is especially ironic: "An Industrial Utopia: Building Gary, Indiana, to Order." See also John K. Mumford, "This Land of Opportunity: Gary, the City that Rose from a Sandy Waste," *Harper's Weekly* 52 (July 4, 1908): 22–23, 29. A rare exception to this commercial orientation was N. O. Nelson, a wealthy manufacturer who in 1890 established the model town of Leclaire in Illinois and who insisted on both progressive architecture and profit sharing for his employees. See John S. Garner, "Leclaire, Illinois: A Model Company Town (1890–1934)," *Journal of the Society of Architectural Historians* 30 (October 1971): 219–27. On other model company towns, see Graham Taylor, chaps. 1, 4, 5, 8, 9; George

H. Miller, "Fairfield, a Town with a Purpose," *The American City* 9 (September 1913): 213–19; A. T. Luce, "Kincaid, Illinois: A Model Mining Town," *The American City* 13 (July 1915): 10–13; Ida M. Tarbell, *New Ideals in Business: An Account of Their Practice and Their Effects Upon Men and Profits* (New York: Macmillan, 1917), 22–23, 118–20, 146–55, 172–73; and Lubove, *Twentieth-Century Pittsburgh*, 14–19. As indicated in chap. 1, n. 28, Graham Taylor used *system* explicitly, and in the ways described there. As Buder notes, so did George Pullman, and no less broadly. See Buder, Preface.

75. The book was reprinted in 1904 as *Garden Cities of Tomorrow* and became more prominent under that title.

76. According to Peter Batchelor, Howard's contribution to city planning lay not in any specific ideas, for all of his concepts derived from existing sources, but in the synthesis he made of them. See Batchelor, "The Origin of the Garden City Concept of Urban Form," *Journal of the Society of Architectural Historians* 28 (October 1969): 184–200. See also Walter L. Creese, *The Search for Environment: The Garden City: Before and After* (New Haven, Conn.: Yale University Press, 1966); Buder, "Ebenezer Howard: The Genesis of a Town Planning Movement," *Journal of the American Institute of Planners* 35 (November 1969): 390–98; and Robert Fishman, *Urban Utopias in the Twentieth Century: Ebenezer Howard, Frank Lloyd Wright, and LeCorbusier* (New York: Basic Books, 1977), pt. 1. Howard's many American admirers incorporated some but never all of his principles, even in the Garden City Association of America, formed in New York City in 1906. On the association, see Paul K. Conkin, *Tomorrow a New World: The New Deal Community Program* (Ithaca, N.Y.: Cornell University Press, 1959), 65–66, and Scott, 90.

77. Forest Hills Gardens proved so expensive to construct that it became a predominantly residential community, rather than a commercial as well as residential one, and a sanctuary for the prosperous rather than a home for wage earners of all kinds. Sunnyside Gardens and especially Radburn—which was a more rural community designed for the automobile age—were initially successful but later failed because of inadequate financial support. In both, high costs precluded a greenbelt, "privatism" precluded merely leasing rather than selling land, and incompleteness precluded industry. The two eventually became predominantly residential suburbs. On Forest Hills Gardens, see Scott, 90–91, and Newton, 474–78. On Sunnyside Gardens and Radburn, see Clarence Stein, *Toward New Towns for America*, 2d ed. (New York: Reinhold, 1957), chaps. 1, 2; Scott, 188, 259–60; Newton, 489–95; and Daniel Schaffer, *Garden Cities for America: The Radburn Experience* (Philadelphia: Temple University Press, 1982). On the New Deal communities, see Stein, chap. 8; Conkin, chap. 14; Scott, 338–41; Joseph L. Arnold, *The New Deal in the Suburbs: A History of the Greenbelt Town Program, 1935–1954* (Columbus: Ohio State University Press, 1971); and Newton, 502–507. Forest Hills Gardens was designed by Frederick L. Olmsted, Jr. (son of the landscape architect) and Grosvenor Atterbury and was sponsored by the Russell Sage Foundation Homes Company. Both Sunnyside Gardens and Radburn were designed by Clarence Stein and Henry Wright and sponsored by the City Housing Corporation of New York City, headed by Alexander Bing. Plans for additional greenbelt communities in New Jersey and Missouri were drawn up but later scrapped because of local opposition. As Conkin writes about the hundred New Deal communities he treats, "These were to be examples of a new organic society, with new values and institutions"(6).

78. On the RPAA, see Lubove, *Community Planning in the 1920's: The Contribution of the Regional Planning Association of America* (Pittsburgh: University of Pittsburgh Press, 1963), and Park Dixon Goist, *From Main Street to State Street: Town, City, and Community in America* (Port Washington, N.Y.: Kennikat, 1977), chap. 11. Leading RPAA members included Lewis Mumford, the planner and social critic; Stuart Chase, an economist and social critic; Stein and Wright, the designers of Sunnyside Gardens and of Radburn; Frederick Ackerman, an architect who assisted them on those

two projects; Charles Whitaker, editor of the *Journal of the American Institute of Architects*; Benton MacKaye, a forester and conservationist; and Bing and Robert Kohn, housing experts and government officials.

79. Lubove, *Community Planning in the 1920's*, observes: "In reviewing the American past, the RPAA uncovered few precedents for its community planning synthesis. A number of partial expressions had appeared—the New England colonial farm village, the romantic pedestrian-scale suburbs of Frederick Law Olmsted at Riverside, Illinois, and Roland Park near Baltimore, the superior examples of industrial town planning like Pullman, Illinois, and Kingsport, Tennessee, and the federal housing program of World War I. Members . . . found nothing at all to commend in the dominant nineteenth- and early twentieth-century tradition of housing betterment through minimum standards legislation" (2–3).

80. See Lubove, Introduction, *The Urban Community*, 16, 18. On the specific contributions of the RPAA to planning, see Lubove, *Community Planning in the 1920's*, 122–27.

81. In addition, through the establishment of (1) employee hiring and training programs; (2) homes, schools, libraries, small local cooperative industries, and recreational areas for them and their families; and (3) the model town of Norris, Tennessee, the TVA did attain a modest measure of those non-economic reforms sought by Morgan. On the TVA, see Roy Talbert, "The Human Engineer: Arthur E. Morgan and the Launching of the TVA" (Master's thesis, Vanderbilt University, 1967); Scott, 311–16; Talbert, "Arthur E. Morgan's Social Philosophy and the Tennessee Valley Authority," *East Tennessee Historical Society Publications* 41 (1969): 86–99; McCraw, *Morgan vs. Lilienthal: The Feud within the TVA* (Chicago: Loyola University Press, 1970); Robert E. Barde, "Arthur E. Morgan, First Chairman of TVA," *Tennessee Historical Quarterly* 30 (Fall 1971): 299–314; McCraw, *TVA and the Power Fight, 1933–1939* (Philadelphia: Lippincott, 1971); Talbert, "Beyond Pragmatism: The Story of Arthur E. Morgan" (Ph.D. diss., Vanderbilt University, 1971); and Howard P. Segal, "Arthur E. Morgan's Conception of a Humane Technological Society: The Tennessee Valley Authority as a Case Study" (paper delivered at annual meeting, Southern Historical Association, New Orleans, November 1977). Norris, built near the site of Norris Dam, was to provide a model for other small communities throughout the region but failed to produce more than a few offspring; instead, it became a quite ordinary, if quite handsome, suburban community, with a greenbelt surrounding it. On the failure of a plan for regional power generation in the northeastern United States, see Terry Kay Rockefeller, "The Failure of Planning for Electrical Power Supply: The Case of the Electrical Engineers and 'Superpower,' 1915–1924," in *Retrospective Technology Assessment—1976*, 191–215.

82. George Soule, "Prospects of General Economic Planning," in *Planned Society: Yesterday, Today, Tomorrow*, ed. Findlay MacKenzie (New York: Prentice-Hall, 1937), 915. Graham observes, as a (further) contrast with national planning, "The planning ethos has worked its way deep into American life via two institutions in particular, the city planning aspect of urban government, and the large corporation" (*Toward a Planned Society*, xiii–xiv). For Soule's own plans, see his *A Planned Society* (New York: Macmillan, 1932), chaps. 9, 10, and R. Alan Lawson, *The Failure of Independent Liberalism, 1930–1941* (New York: Capricorn Books, 1972), pt. 1., chap. 5, and pp. 234–37.

83. Mumford, Foreword, *Planned Society*, x. On the desired organic model of society, see ibid., viii; Richard H. Pells, *Radical Visions and American Dreams: Culture and Social Thought in the Depression Years* (New York: Harper and Row, 1973), 75, 101, 102; and Graham, "The Planning Ideal and American Reality," 289.

84. A prominent exception to this generalization was Chase, who, following participation in regional planning through the RPAA, became a leading advocate of national planning. For Chase's vision of a planned society, see his *A New Deal* (New York:

Macmillan, 1932), chaps. 9–12, and Lawson, pt. 1, chap. 6, and pp. 226–33. Mumford was far less active in national planning than in regional planning but clearly was an advocate of both.

85. NESPA leaders included David Coyle of an engineering consulting firm; Marion Hedges of the International Brotherhood of Electrical Workers; Ford Hinrichs of the Department of Labor; Lewis Lorwin of the Brookings Institution; Harlow Person, a disciple of Frederick Taylor (and the author of "Waste," cited in chap. 1, n. 26, and of "Scientific Management," cited in n. 23 above), of the Society for the Advancement of Management; and Soule of the *New Republic*. Lorwin became the NESPA's first director but resigned within a year and was succeeded by Soule. On the NESPA and its intellectual and cultural milieu, see Soule, "NESPA: December 1934–December 1940," *Plan Age* 6 (November-December 1940): 289–94; Lawson, pt. 1, chap. 4; and Pells, pt. 2, chaps. 4, 8, and pt. 3, chaps. 1, 6.

86. See Graham, *Toward a Planned Society*, 65–68. On the role of national planning in modern American history, see *Planned Society*; Merriam, "The National Resources Planning Board," 1075–88; Soule, *Planning U.S.A.* (New York: Viking, 1967); Graham, "The Planning Ideal and American Reality," 257–99; Karl, *Charles E. Merriam*, chaps. 11–13; Graham, *Toward a Planned Society*; Inouye and Susskind, 593–621; and Clawson, 253–54, 269–73. Clawson misreads Graham as arguing for the persistence of actual national planning in twentieth-century America. On the association in the 1930s (and early 1940s) of national planning with socialism and other supposed evils, see Theodore Rosenof, "Freedom, Planning, and Totalitarianism: The Reception of F. A. Hayek's *Road to Serfdom*," *Canadian Review of American Studies* 5 (Fall 1974): 149–65. Like Graham, Rosenof, 156–57, perceives that, after the early 1930s, even liberal politicians, academics, and civil servants rejected national planning as undemocratic or unnecessary or both. In a provocative essay, Martin Meyerson argues that utopian writers have generally failed to take serious account of the physical dimensions of their particular schemes (for cities and, by extension, regions and whole societies) while utopian city (and, again, by extension, regional and national) planners have generally failed to do likewise concerning social, cultural, political, and economic life. See Meyerson, "Utopian Traditions and the Planning of Cities," in *The Future Metropolis*, ed. Lloyd Rodwin (New York: Braziller, 1961), 233–50. The distinction, however, is not entirely valid, for each group has, to varying degrees, certainly considered both physical and nonphysical dimensions of utopian cities. Clearly the technological utopians treat both at length. On this same alleged dichotomy, see the less explicit discussions in Reps, "Ideal Cities" (Master of Regional Planning thesis, Cornell University, 1947), chap. 12, and in S. Lang, "The Ideal City: From Plato to Howard," *Architectural Review* 112 (August 1952): 91–101. And on the tensions within at least American planning between utopian ideals and models and real world values and practices, see Michael Lee Vasu, *Politics and Planning: A National Study of American Planners* (Chapel Hill: University of North Carolina Press, 1979), chap. 2.

87. On Veblen's association with future Technocrats, see Joseph Dorfman, *Thorstein Veblen and His America* (New York: Viking, 1934), 453–55, 459–63, 510–15; Layton, "Veblen and the Engineers," *American Quarterly* 14 (Spring 1962): 64–72; Henry Elsner, Jr., *The Technocrats: Prophets of Automation* (Syracuse, N.Y.: Syracuse University Press, 1967), chap. 2; Layton, *The Revolt of the Engineers*, 145–47, 226–27; and William E. Akin, *Technocracy and the American Dream: The Technocrat Movement, 1900–1941* (Berkeley and Los Angeles: University of California Press, 1977), chaps. 1, 2. Not surprisingly, in view of persistent allegations of Veblen's decisive influence on him, Howard Scott later claimed that Veblen had influenced him hardly at all; that Bellamy Nationalism and scientific management had not influenced him either; and that Technocracy—and, for that matter, the Technical Alliance, its precursor (about which more below)—were consequently based primarily on his own ideas as

well as his own organizational skills. See Scott's two letters of 1964 to Professor J. Kaye Faulkner, reprinted in the *Northwest Technocrat* 28 (July 1965): 6, 9–11, 21–22. See also "A Note on the Work of Thorstein Veblen," in Scott et al., *Introduction to Technocracy* (New York: Day, 1933), 59–61, which more briefly disclaims Veblen's influence. Given the inaccuracies concerning Scott's background and the confusion over Technocracy's statistics (discussed below in notes 98 and 94 respectively), it is difficult to accept Scott's dogmatic assertions. If anything, their rigidity—and defensiveness—point to a more complex story. The issue of precise influence is not, however, of principal concern here.

88. The book reprinted a series of magazine articles published two years earlier in the *Dial*. The book itself was reprinted in 1963 (New York: Harcourt, Brace and World) with a fine introduction by Daniel Bell placing the work in broader historical perspective. Bell's introduction appeared first as "Reappraisals: Veblen and the New Class," in the *American Scholar* 32 (Autumn 1963): 616–38. It is, however, openly based on several of the other relevant works cited here.

89. Veblen, *The Engineers and the Price System* (New York: Viking, 1933), 135. On Veblen's gradual disillusionment with engineers, see Layton, "Veblen and the Engineers," 64–72; Haber, 143–45; and Layton, *The Revolt of the Engineers*, 227. In his correspondence with Faulkner cited in n. 87 above, Scott (9) himself opposed the "Soviet of Technicians" as being too radical; yet in the same publication (22), he paradoxically criticized Veblen for being too conservative. Rockefeller's study of the "Superpower" plan is intended in part as a confirmation of Veblen's disillusionment with American engineers as excessively passive and apolitical.

90. On the Technical Alliance's membership, see Dorfman, 459–60; Elsner, 24–25; and Akin, 34–35. Four of its leaders—Frederick Ackerman, Stuart Chase, Benton MacKaye, and Charles Whitaker—were also leaders of the RPAA. See n. 78 above. In his correspondence with Faulkner cited in n. 87 above, Scott (6) claimed that the Alliance began in 1918—shortly before he met Veblen. No professional historian of the Alliance or of Technocracy confirms this earlier date. What several do confirm is the establishment, in 1916, of another organization, the New Machine. The latter was founded by Henry Gantt, a one-time disciple of Frederick Taylor who thought the master indifferent to social issues. Its platform included modification of the profit system in the direction Veblen desired, replacement of absentee owners of big business and of politicians with engineers, "public service banks" for the honest poor, and job placement bureaus for the unemployed and occupationally discontent. Unfortunately, the First World War and Gantt's death in 1919 stopped the New Machine in its tracks and ended its influence on the nascent Technocracy movement. Certainly the New Machine was, at the time of its demise, potentially a much more radical organization than either the Technical Alliance or Technocracy ever became. On the New Machine, see Layton, "Veblen and the Engineers," 68–69; Bell, "Reappraisals," 628; Layton, *The Revolt of the Engineers*, 145–47; and Akin, 11–12. Scott, in that same correspondence with Faulkner (10–11), characteristically dismissed the New Machine as an influence on Technocracy.

91. *Technocracy*, the new movement reluctantly acknowledged, was not a new term but dated back at least to 1882. But even here Scott took credit for an original definition of the term and so declared again the absence of outside influence on his ideas. See his correspondence with Faulkner in the July 1965 *Northwest Technocrat*, 19–20.

92. Other categories were "expenditure of energy per product, employment and working hours, volume and rate of growth of production, and total installed horsepower for each industry" (Elsner, 2). As Scott acknowledged—for once—there were outside influences on his choice of categories: "We derive most of our concepts of *thermodynamics* and energy determinants from the works of J. Willard Gibbs" (corre-

spondence with Faulkner, *Northwest Technocrat*, July 1965, 9; my italics). On Technocracy's categories, see also *Introduction to Technocracy*, 7–9, 48, and Akin, 42–43, 143–44. In an otherwise favorable summary of Technocracy's basic platform, Chase criticized its narrow conception of economic activity: "Economic activity includes many valuable services not susceptible to measurement in terms of energy—the work of teachers, doctors, artists, professional baseball players, traffic officers, research workers. In 1930 roughly half as many people were employed in the 'service' trades as in the production and distribution of physical goods. The ratio has been growing rapidly in recent years, especially since 1920" (*Technocracy: An Interpretation* [New York: Day, 1933], 31).

93. *Introduction to Technocracy*, 48. As Scott made remarkably clear in his illuminating correspondence with Faulkner in that same July 1965 *Northwest Technocrat*, he and his fellow Technocrats eschewed genuine prophecy and, for that matter, serious social—as opposed to economic—change. Scott dismissed Bellamy's *Looking Backward* and *Equality* as naive "idealistic projection" (10) on a par with Jules Verne's *Twenty Thousand Leagues Under the Sea*. Significantly, Scott lumped Harold Loeb's *Life in a Technocracy*, an authentic technological utopia, with Bellamy's two similar works and ridiculed them all as "imaginary fiction" (21). On Loeb's attitude in the book toward Scott, see n. 100 below. Scott then boasted that Technocracy had "never advocated social change. We have pointed out the factors that would create it and have come pretty close to predicting its arrival, but that is an entirely different thing than advocating social change per se, [supposedly] for social change's sake" (15). See also ibid., 31. In the very face of Technocracy's aversion to utopian schemes, Akin (133–34, chap. 8), tries to piece one together from among Technocracy's various components. This vision, which he calls a "technological utopia" in itself, reflected the Technocrats' "faith in positivistic science; their mechanistic view of man with his essentially animal-like rationality, his desire for security, abundance, and tranquility; the organizational imperative caused by natural inequality; and the dominance of technology" (148). Yet the vision, as Akin constructs it, remains incomplete and is usefully supplemented by Loeb's book, about which more in chap. 8 below. Moreover, the Technocratic publications which do try to portray life in the better future are themselves limited in scope. See *Technocracy Briefs* (Seattle: Technocracy Inc., n.d.); *The Energy Certificate* (Rushland, Pa.: Technocracy Inc., 1938); and *Technocracy: Technological Social Design* (Savannah, Ohio: Technocracy Inc., 1975). The second work, esp. 7, uses *system* in the broad sense described above in chap.1 but fails to specify its many manifestations under Technocracy. Finally, the specific technological advances that Technocracy did predict were not only few but also unoriginal for the 1930s and early 1940s: namely, streamlined automobiles and flying-wing aircraft. See Elsner, 95–97, 110, 152, regarding them. Elsner's book is thus missubtitled as *Prophets of Automation*. The Technocrats were, as we have seen, anything but prophets. Leon Ardzrooni, a disciple of Veblen, perceived at the height of Technocracy's popularity that the movement was fundamentally conservative—much more so, as I have indicated, than even technological utopianism. See his "Veblen and Technocracy," *Living Age* 344 (March 1933): 39–42.

94. Quoted from Ralph Chaplin, Foreword, Howard Scott, *Science versus Chaos* (New York: Technocracy Inc., 1933), reprinted in *Northwest Technocrat* (July 1965), 28. On the still more ironic dispute in 1932 and 1933 over Technocracy's use of statistics, presumably the "bottom line" of scientific objectivity, see Elsner, 8–9, and Akin, 153–56.

95. From Chaplin, quoted also in *Northwest Technocrat* (July 1965), 28.

96. "Technocracy—Boom, Blight, or Bunk?" *Literary Digest* 114 (December 31, 1932): 5.

97. On the address, see Elsner, 11–13, and Akin, 87–88. Chase, more charitably,

described Scott as a speaker as characteristically "delivering an amazing flow of technical information and discovery, sidewise, out of a wry mouth" (*Technocracy: An Interpretation*, 7).

98. On Loeb's cosmopolitan background, see Elsner, 56–57, and Akin, 29, 100, 118–19. The minor scandal that simultaneously broke over Scott's own background—the disclosure that he had in fact had no certified technical training or formal experience—represented two kinds of failure, undermining further as it did both Scott's leadership and Technocracy's credibility as a scientific organization. Scott boasted of having received a first-class engineering education in Europe and of having worked on engineering projects throughout the world. His real background, exposed in the *New York Herald Tribune* in December 1932, was considerably more modest: he had been foreman of a cement-pouring gang and had been fired for incompetence and, ironically, inefficiency; he had then become part owner of a floor-wax manufacturing firm. His engineering expertise, such as it was, had been acquired entirely on his own rather than in any schools. On the scandal, see Elsner, 5–9, and Akin, 28–29. Significantly, from the outset of their meetings, Veblen had been suspicious of Scott's alleged technical expertise. See Dorfman, 454. Three decades after the scandal, in the correspondence with Faulkner in the July 1965 *Northwest Technocrat*, 6, Scott unhesitatingly called himself the "Chief Engineer" of the Technical Alliance and so, presumably, an engineer still. Even the 1975 brochure *Technocracy*, cited in n. 93 above, described the then-deceased Scott as an "outstanding consulting and industrial engineer" (55). In addition, these controversies prompted Columbia University—whose Department of Industrial Engineering had provided free rooms for Technocracy—to evict the organization's members from that space, lest the university's reputation be further damaged.

99. On the CCT leadership, see Elsner, 10–11, and Akin, 99–100, 117–18. On the Technocracy Inc. leadership, see Elsner, 108–12, and Akin, 100–101.

100. Loeb's *Life in a Technocracy: What It Might Be Like* (New York: Viking, 1933), written before his final break with Scott, nevertheless indicates in its Foreword the extent of their disagreement: "The following essay is based on the ideas of Howard Scott. The credit for whatever originality it may possess belongs to him. The interpretation and development of the ideas, however, are mine, and much of it would probably not meet with Scott's approval. He believes in sticking closely to the engineering aspects" (vi). Among the areas covered by Loeb but not by Scott, all of them discussed above in chap. 2, are government, education, religion, art, and recreation. As I will make clear in chap. 8, the differences between them concerning the role of technology under a Technocracy were deeper yet. On Scott's attitude toward the book following Loeb's defection, see n. 93 above. On his even more negative attitude toward Loeb himself, see his correspondence with Faulkner in the July 1965 *Northwest Technocrat*, 20–21.

101. To be sure, Technocracy steadfastly maintained a nonpartisan stance toward all political and ideological questions. Yet, like technological utopianism, it manifested at least covert political and ideological leanings, as just indicated (and as we will see further in chap. 7 regarding technocratic schemes in general). Hence the hollowness of its declarations of absolute neutrality—as, for example, in Scott's correspondence with Faulkner in the July 1965 *Northwest Technocrat*, 12–13, 29. In fact, that same issue (8–9, 33–39) first discussed and then reprinted two articles Scott had written back in 1920, as head of the Technical Alliance, for the International Workers of the World (the IWW), the radical labor union. On Scott's activities for the IWW as its "Research Director," see Akin, 37–42. Akin perceives how Scott's technocratic conservatism even then could fit in with the IWW's syndicalism. In this regard see Arthur Lipow, *Authoritarian Socialism in America: Edward Bellamy and the Nationalist Movement* (Berkeley and Los Angeles: University of California Press, 1982). On the adoption in Europe in the 1920s and early 1930s by left-wing and right-wing political groups alike

of scientific management and other technocratic schemes, see Charles S. Maier, "Between Taylorism and Technocracy: European Ideologies and the Vision of Industrial Productivity in the 1920's," *Journal of Contemporary History* 5(1970): 27–61. On scientific management in Europe in general, see Judith A. Merkle, *Management and Ideology: The Legacy of the International Scientific Management Movement* (Berkeley and Los Angeles: University of California Press, 1980), pt. 2.

102. On this streak of anti-Catholicism, see Elsner, 148–49, 164–66, 181–82.

103. On this recruitment campaign, see Elsner, chap. 8. On the attractiveness of Technocracy Inc. as a mass movement, see Elsner, 194–95; Akin, chap. 6; and esp. Elsner, "Messianic Scientism: Technocracy, 1919–1960" (Ph.D. diss., University of Michigan, 1963), chap. 11. This dissertation chapter, eliminated from the book, is based on a questionnaire that the author distributed to many former Technocrats. Scott's comments, in his correspondence with Faulkner in the July 1965 *Northwest Technocrat*, that he and his fellow Technocrats "were not concerned with wasting our time trying to capture the entire scientific and engineering profession of this Continent as members" (31) may well reflect—or rationalize—the failure of earlier efforts to do just that. So may his comments in ibid., 24–25, minimizing the failure of his campaign to become "Director General of Defense." On Scott's disavowal both of violence and of politics, see ibid., 13, 15. On the likely contribution of the latter stance to the overall decline of Technocracy as a mass movement, see Akin, 110–11.

104. Scott remained Director-in-Chief of Technocracy Inc. until his death at age 79 in 1970. At least three official Technocracy Inc. publications survive: *The Technocrat, Technocracy Digest*, and the *Northwest Technocrat*. All are published quarterly on the West Coast of the United States and Canada but are rotated so that a different one appears each month. On the general fate of Technocracy Inc. after the Second World War, see Eugene R. Wutke, "Technocracy: It Failed to Save the Nation" (Ph.D. diss., University of Missouri at Kansas City, 1964), chap. 4, and Elsner, *The Technocrats*, chaps. 9, 10.

105. *Introduction to Technocracy*, 46.

106. On the New Machine and the Technical Alliance, see n. 90 above. On the Utopian Society of America, see Newton Van Dalsem, *History of the Utopian Society of America: An Authentic Account of its Origin and Development up to 1942* (Los Angeles: Utopian Society, 1942); Elizabeth Sadler, "One Book's Influence: Edward Bellamy's *Looking Backward*," *New England Quarterly* 17 (December 1944): 547–48; Arthur M. Schlesinger, Jr., *The Age of Roosevelt*, vol. 3, *The Politics of Upheaval* (Boston: Houghton Mifflin, 1960), 110–11; and Elsner, 37, 67. The Utopian Society originated on the West Coast in 1932 and, as Van Dalsem's pamphlet made clear, was influenced by both Technocracy and Bellamy Nationalism. The society achieved its greatest popularity in 1934 and declined rapidly thereafter. The pamphlet was its de facto epitaph.

107. On world's fairs between the 1851 Crystal Palace Exhibition and the 1939–1940 World of Tomorrow in New York discussed below, see the surveys and comparisons by Merle Curti, "America at the World Fairs, 1851–1893," *American Historical Review* 55 (July 1950): 833–56; Tozer; Calvert, "American Technology at World Fairs, 1851–1876" (Master's thesis, University of Delaware, 1962); Paul F. Norton, "World's Fairs in the 1930's," *Journal of the Society of Architectural Historians* 24 (March 1965): 27–30; Eugene S. Ferguson, "Expositions of Technology, 1851–1900," in *Technology in Western Civilization* 1: 706–26; Cawelti, 317–63; and Ada Louise Huxtable, "You Can't Go Home to Those Fairs [of the 1930s] Again," *New York Times*, Arts and Leisure sect., October 28, 1973.

108. On these world's fairs of the 1930s in the United States, see, besides Norton, Cawelti, and Huxtable cited in the preceding note, Tozer, "A Century of Progress,

1833–1933: Technology's Triumph over Man," *American Quarterly* 4 (Spring 1952): 78–81, and Helen A. Harrison et al., *Dawn of a New Day: The New York World's Fair, 1939–40* (New York: The Queens Museum /New York University Press, 1980). Tozer perceptively analyzes the significance of the official motto of the Century of Progress Exposition—"Science Finds, Industry Applies, Man Conforms"—in relation both to the fair itself and to broader American attitudes toward science and technology. None of this is to deny that all of these fairs of the 1930s, like their predecessors and their successors, were designed in part simply to entertain their patrons.

109. On these industrial designers and their contributions to the World of Tomorrow, see Donald J. Bush, *The Streamlined Decade* (New York: Braziller, 1975), chap. 8; Bush, "Futurama: World's Fair as Utopia," *Alternative Futures* 2 (Fall 1979): 3–20; and Jeffrey L. Meikle, *Twentieth Century Limited: Industrial Design in America, 1925–1939* (Philadelphia: Temple University Press, 1979), chap. 9.

110. Van Dalsem, 8.

111. Schlesinger, *The Age of Roosevelt*, vol. 1, *The Crisis of the Old Order, 1919–1933* (Boston: Houghton Mifflin, 1957), 464. On the legacy of the Technocracy movement for the so-called managerial revolution of the post-1940 era, see Elsner, 213–19, and Akin, 169–70. As the poet Archibald MacLeish perceived as early as 1933, the questions raised by Technocracy about America's future were crucial ones, even if its own answers were inadequate. See his "Machines and the Future," *Nation* 36 (February 8, 1933): 140–42.

112. Scott, *Northwest Technocrat* (July 1965), 13. As Akin, 134–35, 143–45, nearly concedes, this very fact could work against Technocracy, as elaborated below.

113. Where the streamlined style of the World of Tomorrow was intended by its promoters to remain in vogue long after most of the fair's machines and structures had met their planned demolishment, the more eclectic styles of the earlier fairs of the 1930s were conceived without such pretensions of permanence (as were most of their machines and structures as well). Thus the clever title of Meikle's illuminating book, *Twentieth Century Limited*, a double-entendre referring to the premature end of both the streamlined style and the legendary New York–Chicago luxury train which embodied it, might apply further to the centuries-old dream of technological utopia itself, which has clearly found its partial fulfillment but perhaps its ultimate limitations in twentieth-century America.

114. Failure to predict the "real" future is hardly peculiar to these particular prophets. See the many examples of partial or wholesale failures during the same period in George Wise, "Technological Prediction, 1890–1940" (Ph.D. diss., Boston University, 1976). Wise's list of "the fifty most frequent predictors" (318) includes, as indicated in chap. 3, n. 13 above, utopians Bellamy and Robert Thurston. Moreover, among the many technological achievements largely unanticipated by those fifty and other prophets was the computer, the keystone of the revolution. See ibid., 237. Finally, predictions by these prophets about technological advances, however seriously flawed, were nevertheless much more accurate than their predictions about the effects of those advances. Hence the technological utopians, Technocrats, and industrial designers should not be singled out for criticism. Partial summaries of Wise's study are his "The Accuracy of Technological Forecasts," *Futures* 8 (October 1976): 411–19, and his "Past Efforts at Technology Assessment and Prediction: 1890–1940," in *Retrospective Technology Assessment—1976*, 245–64.

115. See Segal, "Are Fairs Obsolete?" Op-Ed Page, *New York Times*, June 3, 1981, 25.

116. Still, I would not yet compose an obituary for world's fairs. Just as there have been periodic revivals of utopian communities, most recently in the 1960s, and of utopian writings, as in the work of Ursula Le Guin, so fairs may revive as well—revive, I

mean, as important social and cultural events, unlike the 1982 Energy Expo in Knoxville and, I fear, the several other world's fairs now planned through the year 2000. People may find it necessary after all to view technological developments in the flesh and not just on television or computer screens. And ongoing developments in robotics and genetic engineering may provide further reasons for world's fairs. For their part, world's fairs might relegate their traditional entertainment functions to video arcades and theme amusement parks. The fairs might be smaller, more decentralized, and even mobile, perhaps traveling from one site to another. And they might look for sponsorship from multinational corporations rather than states and nations.

117. As Akin writes, somewhat unclearly, about Technocracy, "despite the ridicule heaped upon it at the time, [it] was not an aberration of American thought but was instead a close approximation of a modern utopia" (131).

Chapter 7

1. George Kateb, *Utopia and Its Enemies* (New York: Free Press, 1963), 14–15.

2. See Sebastian de Grazia, *Of Time, Work, and Leisure* (Garden City, N.Y.: Doubleday Anchor Books, 1964), 62, 79–80, 83, 89–90, and Hannah Arendt, *The Human Condition* (Garden City, N.Y.: Doubleday Anchor Books, 1959), 340–41, n. 85. On work in present-day society, see also Georges Friedmann, *Industrial Society: The Emergence of the Human Problems of Automation*, ed. Harold L. Sheppard (Glencoe, Ill.: Free Press, 1955); Daniel Bell, *Work and Its Discontents* (Boston: Beacon, 1956); Nels Anderson, *Dimensions of Work: The Sociology of a Work Culture* (New York: McKay, 1964); *The Human Shape of Work: Studies in the Sociology of Occupations*, ed. Peter L. Berger (Chicago: Regnery, 1964); Emma Rothschild, *Paradise Lost: The Decline of the Auto-Industrial Age* (New York: Random House, 1973); Edward Shorter, Introduction, *Work and Community in the West*, ed. Shorter (New York: Harper and Row, 1973), 1–33; *Work in America: Report of a Special Task Force to the Secretary of Health, Education, and Welfare* (Cambridge: MIT Press, 1973); and Studs Terkel, *Working: People Talk about What They Do All Day and How They Feel about What They Do* (New York: Pantheon, 1974). Shorter provides an excellent overview of the differences not just in hours but also in conditions and in mores among pre-industrial, industrial, and post-industrial workers.

3. See de Grazia, 128–30. "The things [the "average" American] now wants," he observes, "cost money, money costs work, work costs time" (211).

4. The notion of increasing division between workers and work is taken, of course, from Karl Marx's general theory of alienation in modern capitalist society. The application here, however, is more general. Nevertheless, on the nature and persistent relevance of Marx's formulation, see, for example, David McLellan, *The Thought of Karl Marx: An Introduction* (New York: Harper and Row, 1971), pt. 2, chap. 1; Bertell Ollman, *Alienation: Marx's Conception of Man in Capitalist Society* (Cambridge: Cambridge University Press, 1971); and István Mészáros, *Marx's Theory of Alienation* (New York: Harper Torchbooks, 1972).

5. The idea of a diminished sense of "calling" to work in modern capitalist society derives from Max Weber and in particular from *The Protestant Ethic and the Spirit of Capitalism*, tr. Talcott Parsons (New York: Scribner, 1958), chap. 5. For the American background to this and the other changes in work and leisure discussed here, see the complementary studies by James B. Gilbert, *Work without Salvation: America's Intellectuals and Industrial Alienation, 1830–1910* (Baltimore: Johns Hopkins University Press, 1977), 1, and Daniel T. Rodgers, *The Work Ethic in Industrial America, 1850–1920* (Chicago: University of Chicago Press, 1978).

6. Arendt, 5.

7. On the classical conception of leisure, see Arendt, 15–18; de Grazia, chap. 1; and Paul Weiss, "A Philosophical Definition of Leisure," in *Leisure in America: Blessing or Curse*, ed. James C. Charlesworth (Philadelphia: American Academy of Political and Social Science, 1964), 21–29. This is not, of course, to suggest that the majority of citizens in any Western society since classical times either enjoyed or sought leisure as such. It is rather to describe the de facto utopian ideal of leisure held by many intellectuals and other cultural leaders from classical times to the present. And by *utopian* here I mean "impossible" or "improbable," given the absence of sufficient means to realize that ideal. This ideal, moreover, was apparently not shared by the majority of farmers and craftsmen in pre-industrial Europe. As several historians have now shown, persons whose occupations were at least minimally rewarding economically and psychologically generally integrated leisure—in other, "impure" forms— with work and saw no clear distinction between them and no need to draw any clear distinction. Yet unlike the majority of citizens in contemporary Western society, they did not allow work to absorb leisure. Indeed, they did not, as we saw above, work as long as do most citizens today. Thus their integration of work and leisure stands, in effect, midway between the classical dream and the current reality. See Keith Thomas, "Work and Leisure in Pre-Industrial Society" and "Discussion [of Thomas' paper]," *Past and Present* 29 (December 1964): 50–62 and 63–66; "Work and Leisure in Industrial Society," ibid. 30 (April 1965): 96–103; E. P. Thompson, "Time, Work-Discipline, and Industrial Capitalism," *Past and Present* 38 (December 1967): 56–97; Shorter, 6–12; Alaisdair Clayre, *Work and Play: Ideas and Experience of Work and Leisure* (New York: Harper and Row, 1974), pt. 2; Michael R. Marrus, Introduction, *The Emergence of Leisure*, ed. Marrus (New York: Harper and Row, 1974), 1–7; and Marrus, *The Rise of Leisure in Industrial Society* (St. Charles, Mo.: Forum Press, 1974), 2–6. As Clayre cautions, however, "Work done for its own sake, enjoyed for its own intrinsic satisfaction, seemed . . . less widespread in the past than in the imaginations of those philosophers who wrote about work in the nineteenth century . . ." (183). A separate if related qualification is made by Marrus, who argues that leisure as a mass activity arose exactly because of the industrial revolution and the untangling of the older integration of work and leisure. See both his Introduction to *The Emergence of Leisure*, 8–9, and his *The Rise of Leisure*, 6–12. Yet this form of leisure is far different from the classical variety described above and is tied up with work, primarily as a conscious escape from work.

8. De Grazia, 35–36.

9. Arendt, pts. 1, 4–6. On the rise of the hobby in modern work-centered society, see Clement Greenberg, "Work and Leisure under Industrialism," in *Mass Leisure*, ed. Eric Larrabee and Rolf Meyersohn (Glencoe, Ill.: Free Press, 1958), 38–43, esp. 41.

10. See Josef Pieper, *Leisure: The Basis of Culture*, tr. Alexander Dru (New York: Mentor–Omega Books, 1963), esp. chap. 2. Joffre Dumazedier (*Toward a Society of Leisure*, tr. Stewart McClure [New York: Free Press, 1967]) and Stanley Parker (*The Future of Work and Leisure* [London: MacGibbon and Kee, 1971]) each attempt to integrate rather than separate work and leisure in the hope of revitalizing both activities, yet in the process, they still subordinate leisure to work.

11. See Johan Huizinga, *Homo Ludens: A Study of the Play-Element in Culture* (New York: Roy, 1950), esp. chap. 1. See also the reconsiderations and partial critiques of the book in Anthony Giddens, "Notes on the Concepts of Play and Leisure," *Sociological Review*, n.s., 12 (March 1964): 73–89; Jacques Ehrmann, "Homo Ludens Revisited," *Yale French Studies*, no. 41: 31–57; and E. H. Gombrich, "Huizinga and 'Homo Ludens,' " *Times Literary Supplement* 3787 (October 4, 1974): 1083–89.

12. On these developments, see Huizinga, chaps. 11–12.

13. Just as some social theorists would revive leisure by integrating it with work, so others would revive play by integrating it with work—with similar dubious consequences for traditional understandings of play. See William R. Torbert and Malcolm P. Rogers, *Being for the Most Part Puppets: Interactions among Men's Labor, Leisure, and Politics* (Cambridge, Mass.: Schenkman, 1973), and to a lesser extent, Giddens, 86–87. As in the case of leisure, however, so in that of play, the ideal described here was not one necessarily endorsed by the majority of citizens at any place or time. And like leisure, play in other, "impure" forms was frequently mixed with work.

14. On the unprecedented role of technical experts, see Jean Meynaud, *Technocracy*, tr. Paul Barnes (London: Faber and Faber, 1964); John McDermott, "Knowledge Is Power," *Nation* 208 (April 14, 1969): 458–62; Norman Birnbaum, "The Problem of a Knowledge Elite," in his *Toward a Critical Sociology* (New York: Oxford University Press, 1971), 416–41; and Bell, *The Coming of Post-Industrial Society: A Venture in Social Forecasting* (New York: Basic Books, 1973), chaps. 3, 6.

15. On the contemporary knowledge explosion, see Fritz Machlup, *The Production and Distribution of Knowledge in the United States* (Princeton, N.J.: Princeton University Press, 1962), and Robert E. Lane, "The Decline of Politics and Ideology in a Knowledgeable Society," *American Sociological Review* 31 (October 1966): 649–62, esp. 650.

16. These critics nevertheless do argue that technical knowledge has been repeatedly used as propaganda for various political, economic, and cultural purposes. Some go as far as to contend that knowledge per se, not just technical knowledge, is intrinsically political. See Barrington Moore, Jr., "Totalitarian Elements in Pre-Industrial Societies," in his *Political Power and Social Theory: Seven Studies* (Cambridge: Harvard University Press, 1958), 85–87; Herbert Marcuse, *One-Dimensional Man: Studies in the Ideology of Advanced Industrial Society* (Boston: Beacon, 1964), chaps. 5–7; Jacques Ellul, *Propaganda: The Formation of Men's Attitudes*, tr. Konrad Kellen and Jean Lerner (New York: Knopf, 1966); Marcuse, "Industrialization and Capitalism in the Work of Max Weber," in his *Negations: Essays in Critical Theory*, tr. Shapiro (Boston: Beacon, 1968), 201–26; McDermott, "Knowledge Is Power," 458–62; McDermott, "Technology: The Opiate of the Intellectuals," *New York Review of Books* 13 (July 31, 1969): 25–35; Jurgen Habermas, *Toward a Rational Society: Student Protest, Science, and Politics*, tr. Jeremy Shapiro (Boston: Beacon, 1970), chaps. 4–6; and Habermas, *Knowledge and Human Interests*, tr. Shapiro (Boston: Beacon, 1971), esp. appendix.

17. On the kind of knowledge these theorists seek to cultivate and to utilize, see Marcuse, "A Note on Dialectic," new Preface to his *Reason and Revolution: Hegel and the Rise of Social Theory* (Boston: Beacon, 1960), vii–xiv; Marcuse, *One-Dimensional Man*, chaps. 8–10; Marcuse, "Philosophy and Critical Theory," in *Negations*, 134–58; Habermas, *Knowledge and Human Interests*, Appendix; and Max Horkheimer, "The Social Function of Philosophy," in his *Critical Theory: Selected Essays*, tr. Matthew J. O'Connell et al. (New York: Herder and Herder, 1972), 253–72.

18. See Lane, "The Politics of Consensus in an Age of Affluence," *American Political Science Review* 59 (December 1965): 874–95; Lane, "The Decline of Politics and Ideology," 649–62; the essays by Bell, Seymour Martin Lipset, and Edward Shils, among others, in *The End of Ideology Debate*, ed. Chaim I. Waxman (New York: Clarion Books, 1969); Zbigniew Brzezinski, *Between Two Ages: America's Role in the Technetronic Era* (New York: Viking, 1970); and Bell, *The Coming of Post-Industrial Society*, chap. 6.

19. Says Meynaud: "The advent of industrial civilization is far from being the cause of the disappearance or even the decline of political activity in human societies. The thesis that politics must inevitably be absorbed into technics [i.e., technology] cannot

really be supported," given the "impossibility of totally mechanizing human decisions" (293). In his essay, "Approaches to the Study of Political Power," Franz Neumann uses the case of the "laissez faire night-watchman" state, the ideal of eighteenth- and nineteenth-century classical liberalism, to demonstrate the persistence of politics in societies officially hostile to politics. The night-watchman state seemed to be unregulated and thereby apolitical—there being no government to regulate it. It seemed to function automatically, just as the governments of technological utopia and, to a lesser extent, of our own society seem or at least aspire to function automatically. In actuality, says Neumann, "No society in recorded history has ever been able to dispense with political power. This is as true of liberalism as of absolutism, as true of laissez faire as of an interventionist state. No greater disservice has been rendered to political science than the statement that the liberal state was a 'weak' state. It was precisely as strong as it needed to be in the circumstances. It acquired substantial colonial empires, waged wars, held down internal disorders, and stabilized itself over long periods of time" (The Democratic and the Authoritarian State: Essays in Political and Legal Theory, ed. Marcuse [New York: Free Press, 1964], 8). The same is true of the governments of technological utopia and of real world America. As Neumann perceives, the claim to be apolitical masks the interests actually served and so serves those interests all the more (6–7, pt. 4). In fact, the centralization of government resulting from technological advance means that "The higher the state of technological development, the greater the concentration of political power" (10). A useful case study confirming these points is Michael Lee Vasu, Politics and Planning: A National Study of American Planners (Chapel Hill: University of North Carolina Press, 1979).

20. On the fate of politics in modern times, see Arendt, pts. 2, 5; Sheldon S. Wolin, Politics and Vision: Continuity and Innovation in Western Political Thought (Boston: Little, Brown, 1960), chap. 10; Wolin, "Paradigms and Political Theories," in Politics and Experience: Essays Presented to Michael Oakeshott, ed. Preston King and B. C. Parekh (Cambridge: Cambridge University Press, 1968), 125–52; Wolin, Hobbes and the Epic Tradition of Political Theory (Los Angeles: Clark Memorial Library, University of California, Los Angeles, 1970); George McKenna, "On Hannah Arendt: Politics: As It Is, Was, Might Be," in The Legacy of the German Refugee Intellectuals, ed. Robert Boyers (New York: Schocken, 1972), 104–22; Margaret Canovan, The Political Thought of Hannah Arendt (New York: Harcourt Brace Jovanovich, 1974); Wolin, "Political Theory as a Vocation," in Machiavelli and the Nature of Political Thought, ed. Martin Fleisher (New York: Atheneum, 1972), 23–75; and Wolin, "Political Theory and Political Commentary," in Political Theory and Political Education, ed. Melvin Richter (Princeton, N.J.: Princeton University Press, 1980), 190–203. The arguments of the following two paragraphs derive from these works as well.

21. In defense of contemporary political analysis, see David Easton, "An Approach to the Analysis of Political Systems," World Politics 9 (April 1957): 383–400; Robert A. Dahl, "The Behavioral Approach in Political Science: Epitaph for a Monument to a Successful Protest," American Political Science Review 55 (December 1961): 763–72; and Heinz Eulau, "The Behavioral Movement in Political Science: A Personal Document," Social Research 35 (Spring 1968): 1–29. But see also Gerald De Maio and Bernard F. Lynch, "The Recrudescence of Political Philosophy," Centennial Review 21 (Spring 1977): 159–75, and Richter, Introduction, Political Theory and Political Education, 3–56.

22. This is largely because they have not, to my knowledge, written explicitly on this issue in recent years. A notable exception, however, is Bell, whose sequel to The Coming of Post-Industrial Society, The Cultural Contradictions of Capitalism (New York: Basic Books, 1976), is considerably more pessimistic than the earlier volume on this and other issues. For an illuminating comparison of the two works, see Thomas E. Jones, "Daniel Bell's Evolving Vision of the Post-Industrial Society," World Future

Society Bulletin 13 (January-February 1979), 7–23. Historian W. Warren Wagar has recently argued convincingly for the growth of technocracy in general as the dominant *covert*—and utopian—ideology in the contemporary world, capitalist and noncapitalist segments alike. See his "The Steel-Gray Saviour: Technocracy as Utopia and Ideology," *Alternative Futures* 2 (Spring 1979): 38–54. See also Judith A. Merkle, *Management and Ideology: The Legacy of the International Scientific Management Movement* (Berkeley and Los Angeles: University of California Press, 1980).

23. Victor C. Ferkiss, "Bureaucracy," in *Technology and Change*, ed. John G. Burke and Marshall C. Eakin (San Francisco: Boyd and Fraser, 1979), 86–87. See also Alvin W. Gouldner, "Metaphysical Pathos and the Theory of Bureaucracy," *American Political Science Review* 49 (June 1955): 496–507.

24. See Bell, *The Cultural Contradictions of Capitalism*, 34–35, 39, 53, 100, 103. See also James Ackerman, "The Demise of the *Avant Garde*: Notes on the Sociology of Recent American Art," *Comparative Studies in Society and History* 2 (October 1969): 371–84, and Jacques Barzun, *The Use and Abuse of Art* (Princeton, N.J.: Princeton University Press, 1974).

25. On these changes within utopianism concerning the role of art and artists, see Robert C. Elliott, "Literature and the Good Life: A Dilemma," *Yale Review* 65 (Autumn 1975): 24–37; Jon E. Thiem, "The Artist in the Ideal State: A Study of the Troubled Relations between Arts and Society in Utopian Fiction" (Ph.D. diss., Indiana University, 1975); and Gorman Beauchamp, "Art in Utopia," *Southern Quarterly* 15 (July 1977): 319–33. An exception to the generalizations about art and artists in technological utopia made in this and the preceding two paragraphs is Harold Loeb's *Life in a Technocracy*, as I will discuss at the beginning of the next chapter.

26. See Marcuse, *Eros and Civilization: A Philosophical Inquiry into Freud* (New York: Vintage Books, 1962).

27. David Bleich ("Eros and Bellamy," *American Quarterly* 16 [Fall 1964]: 445–59) interprets Bellamy's *Looking Backward* as a fictional fulfillment of Marcuse's own utopian vision in *Eros and Civilization*, yet ignores other writings of Marcuse that criticize societies like that depicted in *Looking Backward* for simultaneously restricting human development.

28. See Nicholas Rescher, "Technological Progress and Human Happiness," in his *Unpopular Essays on Technological Progress* (Pittsburgh: University of Pittsburgh Press, 1980), 3–22.

29. Ibid., 19. Kateb has analyzed traditional and recent views of what, aside from technological and scientific advances, constitutes the "good life" in leading utopian and non-utopian reform schemes. As he lists them in ascending order of importance, they are "the good life as laissez-faire, as the greatest amount of pleasure, as play, as craft, as political action, and, finally, as the life of the mind [i.e., contemplation]" ("Utopia and the Good Life," in *Utopias and Utopian Thought: A Timely Appraisal*, ed. Frank E. Manuel [Boston: Beacon, 1967] 240). How popular any of these would be on a mass scale can be gauged by the preceding discussion, which touches on most of them. Significantly, all assume the "leisure and abundance" (ibid.) that, we have noted repeatedly, invariably require technological progress of some kind.

Chapter 8

1. According to Harold Loeb, as recorded by Henry Elsner, Jr., the manuscript had been composed in 1930—following Loeb's second meeting with Howard Scott, mentioned below—and "unsuccessfully offered . . . to various publishers. When the publicity struck, the manuscript was suddenly in demand and was finally published in January,1933" (*The Technocrats: Prophets of Automation* [Syracuse, N.Y.: Syracuse

University Press, 1967], 57). According to Scott, however, as recorded in his lengthy correspondence with Professor J. Kaye Faulkner cited in chap. 6 above, "Back at the time that . . . Loeb was seeking a publisher for his book, three publishing firms turned down the publication because Technocracy refused to approve the manuscript in any way, shape or form. . . . Loeb was never a member of Technocracy" (*Northwest Technocrat* 28 [July 1965]: 20). Since Technocracy itself began only in 1932, it is once again difficult to accept Scott's facts—if, at this point, no less difficult to confirm Loeb's. In any case, the seeds of their split and, more important, the intellectual bases for it, are clear from Loeb's book—whatever the reason(s) for its delayed publication. On the relationship between the two men, see also chap. 6, nn. 93, 100. The latter note includes the relevant passage from Loeb's Foreword.

2. Loeb, *Life in a Technocracy: What It Might Be Like* (New York: Viking, 1933), v.

3. Byron Brooks, *Earth Revisited* (Boston: Arena, 1893), 45. The same quotation appears in chap. 2.

4. Loeb, 60. The same quotation appears in chap. 2, n. 31.

5. Ibid., 45.

6. Ibid., 106.

7. Ibid., 42.

8. Ibid., 114. The first part of the quotation appears in chap. 2, n. 36.

9. Ibid., 127.

10. Ibid., 130.

11. Ibid., 141–42.

12. Ibid., 163.

13. Ibid., 178.

14. Ibid., 123.

15. Ibid., 126.

16. For a clear contrast between Loeb's expansive vision and Scott's narrow one, compare the above summary of the former with the summary of the latter in William E. Akin, *Technocracy and the American Dream: The Technocrat Movement, 1900–1941* (Berkeley and Los Angeles: University of California Press, 1977), chap. 8. See also chap. 6, n. 93 above, regarding Akin's attempt to piece together what he calls Scott's vision of "technological utopia."

17. On Loeb's changing views, see Akin, chap. 7.

18. See Jean Gimpel, *The Medieval Machine: The Industrial Revolution of the Middle Ages* (New York: Holt, Rinehart and Winston, 1976). Gimpel does concede that several of the principal negative—and often unanticipated—consequences of the second industrial revolution have their parallels in the first: most notably, destruction of much of the natural environment, especially deforestation and coal mining; and atmospheric, water, and noise pollution, particularly near cities and towns. See ibid., chap. 4. As Gimpel readily acknowledges, his study not only is based on but also is a logical extension of the pioneering investigations of medieval technology in Lynn White, Jr., *Medieval Technology and Social Change* (New York: Oxford University Press, 1962).

19. Gimpel, xi.

20. The term "third" industrial revolution comes from Norbert Wiener, *The Human Use of Human Beings: Cybernetics and Society* (Boston: Houghton Mifflin, 1950), chap. 9. Wiener also coined the term "cybernetics" and was a pioneer in the field. His book was a popularized, revised version of his landmark *Cybernetics; Or, Control and Communication in the Animal and the Machine* (Cambridge, Mass.: Technology Press, 1948).

21. Gimpel, 249.

22. See, in this connection, the editorial in the *New York Times* of September 28, 1979, "The Concorde's Destination," regarding the decision of the British and the

French to stop their joint production of the Concorde supersonic airplane. See also the more general discussion of the technological imperative in Daniel J. Boorstin, *The Americans: The Democratic Experience* (New York: Random House, 1973), pt. 9 and epilogue. Boorstin does not use the term, but his implicit criticism of the phenomenon—surprising, in view of his other writings on America's past and future progress—is perceptive and persuasive.

23. See Noel Perrin, *Giving Up the Gun: Japan's Reversion to the Sword, 1543–1879* (Boulder, Col.: Shambhala, 1980).

24. On this polarity of views regarding technology, see Charles L. Sanford, "Technology and Culture at the End of the Nineteenth Century: The Will to Power," in *Technology in Western Civilization*, ed. Melvin Kranzberg and Carroll W. Pursell, Jr. (New York: Oxford University Press, 1967), 1: 726–39.

25. The foremost biography of Simone Weil is Simone Pétrement, *Simone Weil: A Life*, tr. Raymond Rosenthal (New York: Pantheon, 1976), of which chap. 8 details her work in factories and its effect on her life and thought. But see also the earlier, briefer discussions of those episodes in E. W. F. Tomlin, *Simone Weil* (New Haven, Conn.: Yale University Press, 1954), 13–14, 18–20, 54–58; David Anderson, *Simone Weil* (London: SCM Press, 1971), chap. 3 and pp. 47–49; and George Lichtheim, "Simone Weil," in his *Collected Essays* (New York: Viking, 1973), 458–76.

26. Weil, *Gravity and Grace*, tr. Arthur Wills (New York: Putnam, 1952), 235. She said the same in, among other places, *The Need for Roots: Prelude to a Declaration of Duties toward Mankind*, tr. Wills (New York: Harper Colophon Books, 1971), 300.

27. Weil, *Gravity and Grace*, 233, 236. See also her *Need for Roots*, 94–97, 295–96, 300–302.

28. Weil, *Oppression and Liberty*, tr. Wills and John Petrie (Amherst: University of Massachusetts Press, 1973), 13.

29. Weil, *Need for Roots*, 72–73.

30. Ibid., 74. For more specific information about these desired machines see ibid., 56–58.

31. Ibid., 55.

32. Ibid., 60.

33. Ibid., 96. The best study of Weil's thoughts on work and spirituality is Fred Rosen, "Labour and Liberty: Simone Weil and the Human Condition," *Theoria to Theory* 7 (October 1973): 33–47.

34. Weil, *Oppression and Liberty*, 19.

35. On these exponents of liberation, see, for example, the excellent comprehensive study by Richard King, *The Party of Eros: Radical Social Thought and the Realm of Freedom* (Chapel Hill: University of North Carolina Press, 1972).

36. See E. F. Schumacher, *Small Is Beautiful: Economics as if People Mattered* (New York: Harper and Row, 1973); *Appropriate Visions: Technology, the Environment, and the Individual*, ed. Richard C. Dorf and Yvonne L. Hunter (San Francisco: Boyd and Fraser, 1978); Schumacher, *Good Work* (New York: Harper and Row, 1979); and George McRobie, *Small Is Possible* (New York: Harper and Row, 1981).

37. On these more philosophical exponents of a technological plateau, see, for instance, on Lewis Mumford, his own *Art and Technics* (New York: Columbia University Press, 1952), and *Technics and Civilization*, with a new Introduction (New York: Harcourt, Brace, and World, 1963); and Allan Temko, "Which Guide to the Promised Land: Fuller or Mumford?" *Horizon* 10 (Summer 1968): 25–30; William Kuhns, *The Post-Industrial Prophets: Interpretations of Technology* (New York: Weybright and Talley, 1971), chap. 3; and René Dubos, "The Despairing Optimist," *American Scholar* 42 (Summer 1973): 378–82. On Dubos, see, for example, his own *So Human an Animal* (New York: Scribner, 1968), and "The Despairing Optimist," *American Scholar* 40 (Winter 1970–1971): 16–20; and Richard Kostelanetz, "The Five

Careers of René Dubos," *Michigan Quarterly Review* 19 (Spring 1980): 194–202. On John Passmore, see his own *Man's Responsibility for Nature: Ecological Problems and Western Traditions* (New York: Scribner, 1974), esp. chap. 7. As Dubos argues in his 1970–1971 essay, there is currently a need for sober, technically-versed humanist critics of technology. Passmore is such a critic. This work goes beyond C. P. Snow's famous plea for closer connections between scientists (and engineers) and literary intellectuals in *The Two Cultures; And A Second Look* (New York: Cambridge University Press, 1964).

38. Mumford, *The Story of Utopias* (New York: Viking Compass Books, 1962), 306–7.

39. Ernest Callenbach, *Ecotopia: The Notebooks and Reports of William Weston* (New York: Bantam Books, 1977), 60.

40. On the issue of freedom and constraint within the space colonies, see Gerard K. O'Neill, *The High Frontier: Human Colonies in Space* (Garden City, N.Y.: Doubleday Anchor Books, 1982), 234–35.

41. On O'Neill's view of human nature, see ibid., 272–73. See also his *2081: A Hopeful View of the Human Future* (New York: Simon and Schuster, 1981), 61–75.

42. See Marshall McLuhan and Quentin Fiore, *War and Peace in the Global Village* (New York: Bantam Books, 1968). See also the critique of this concept in Jonathan Miller, *Marshall McLuhan* (New York: Viking, 1971), 115–19.

43. See the works cited in n. 36 above and *Alternative Sources of Energy: Practical Technology and Philosophy for a Decentralized Society*, ed. Sandy Eccli et al. (New York: Seabury, 1974). In this case, as with the next two notes, the citations given are illustrative of a much larger, and ever growing, body of literature on these various subjects.

44. See Emma Rothschild, *Paradise Lost: The Decline of the Auto-Industrial Age* (New York: Random House, 1973); Robert B. Goldman, *A Work Experiment: Six Americans in a Swedish Plant* (New York: Ford Foundation, 1976); David F. Noble, "Social Choice in Machine Design: The Case of Automatically Controlled Machine Tools, and a Challenge for Labor," *Politics and Society* 8 (1978): 313–47; and Seymour Melman, "Alternative Criteria for the Design of Means of Production" (paper delivered at annual meeting, Organization of American Historians, New York, April 1978).

45. See Harvey Brooks, "Technology Assessment in Retrospect," in *Technology and Change*, ed. John G. Burke and Marshall C. Eakin (San Francisco: Boyd and Fraser, 1979), 465–76, and *Technology Assessment: Creative Futures*, ed. Mark A. Boroush, Kan Chen, and Alexander N. Christakis (New York: Elsevier North Holland, 1980).

46. On this greater sophistication and balance among Americans regarding technological progress and social progress, see the illuminating case study by Jon D. Miller, "The Impact of Two Decades of Space Exploration on the Development of American Attitudes toward Science and Technology," in *Retrospective Technology Assessment—1976*, ed. Joel A. Tarr (San Francisco: San Francisco Press, 1977), 265–90. Space exploration is particularly interesting in that it has often engendered dogmatic attitudes, pro and con, among Americans and others. Miller concludes, "There is [still] a base of public support for science and technology, but the case will have to be made for each new venture" (288). See also the essays by various hands—from historians to artists to economists—in *Growth in America*, ed. Chester L. Cooper (Westport, Conn.: Greenwood Press, 1976), and Boorstin, 590–600.

47. See Herman Kahn et al., *The Next 200 Years: A Scenario for America and the World* (New York: Morrow, 1976); Kahn, *The Coming Boom: Economic, Political, and Social* (New York: Simon and Schuster, 1982); Donella H. Meadows et al., *The Limits to Growth*, 2d ed. (New York: Signet Books, 1974); Julian L. Simon, *The Ultimate Resource* (Princeton, N.J.: Princeton University Press, 1981); and Gerald O. Barney et al., *The*

Global 2000 Report (Washington, D.C.: Council on Environmental Quality and United States Department of State, 1980).

48. On this wholesale reversal by the Club of Rome, see Simon's damning critique, 286–87.

49. See Mumford, *The Story of Utopias*, 1962 Preface, 2, 6–7. By contrast, even so enlightened a visionary as O'Neill is guilty not only of misreading fundamental points in *Looking Backward* on the degrees of nationalism, socialism, and technological development that are desirable, but also of the following erroneous generalization: "most prophets overestimated how much the world would be transformed by social and political change and underestimated the forces of technological change" (*2081*, 20). More often than not the situation has been the reverse. On *Looking Backward*, see ibid., 28–29.

50. The terms *scientific* and *utopian* appear in several writings of Marx and Engels but most explicitly in Engels' famous *Socialism: Utopian and Scientific* (1880). The points noted here will be discussed further in chap. 9.

51. Harrison, "Robert Owen's Quest for the New Moral World in America," in *Robert Owen's American Legacy: Proceedings of the Robert Owen Bicentennial Conference*, ed. Donald E. Pitzer (Indianapolis: Indiana Historical Society, 1972), 30. See also Michael Fellman, "Anachronism, Context, and Progress in Nineteenth-Century American Communitarianism," *Canadian Review of American Studies* 15 (Spring 1984), 35–48.

Chapter 9

1. For convenient summaries of many of those charges and countercharges, see Chad Walsh, *From Utopia to Nightmare* (New York: Harper and Row, 1962), chap. 12; George Kateb, *Utopia and Its Enemies* (New York: Free Press, 1963); Anthony Arblaster and Steven Lukes, Introduction, *The Good Society: A Book of Readings*, ed. Arblaster and Lukes (London: Methuen, 1971), 1–21; and Kateb, Introduction, *Utopia*, ed. Kateb (New York: Atherton, 1971), 1–26.

2. But see the discussion of the *Republic* in this regard in chap. 4, n. 2.

3. Regarding the *Republic* see ibid.; regarding *Utopia* see chap. 4, pp. 56–57, and nn. 2–8. As chap. 4 makes clear, other classical utopias likewise held out the prospect of their own realization. The scholars and writings referred to in the paragraph are Judith Shklar, "The Political Theory of Utopia: From Melancholy to Nostalgia," and Adam Ulam, "Socialism and Utopia," in *Utopias and Utopian Thought: A Timely Appraisal*, ed. Frank E. Manuel (Boston: Beacon, 1967), 101–15 and 116–34; and Elisabeth Hansot, *Perfection and Progress: Two Modes of Utopian Thought* (Cambridge: MIT Press, 1974). See also the briefer discussions of these points in Shklar, "Rousseau's Two Models: Sparta and the Age of Gold," *Political Science Quarterly* 81 (March 1966): 25–27, and *Men and Citizens: A Study of Rousseau's Social Theory* (Cambridge: Cambridge University Press, 1969), 1–3; and Hansot, "Reflections on War, Utopias, and Temporary Systems," in *Small Comforts for Hard Times: Humanists on Public Policy*, ed. Michael Mooney and Florian Stuber (New York: Columbia University Press, 1977), 246–59. For Hansot, the seventeenth century marks the dividing line between classical and modern utopias; for Shklar, the eighteenth century does; and for Ulam, the nineteenth century does. Shklar refuses outright to label modern utopias genuine utopias; Hansot and Ulam do so reluctantly. Despite their aversion to modern practical proposals for actually effecting perfection, all three do favor utopias that, like the classical ones, induce reflection on the proper nature of existing society—and the (improper) nature of its citizens. See also Lyman Tower Sargent's related critique of these interpretations of

utopian traditions in the Introduction to his *British and American Utopian Literature,
1516–1975: An Annotated Bibliography* (Boston: G .K. Hall, 1979), xiii–xv.

4. As Kateb, *Utopia and Its Enemies*, succinctly puts it, genuine utopias provide
"an image of what life in its net effect would be like, if life (social life) were perfect" (7).
And as Frank and Fritzie P. Manuel observe, these utopias are, on the one hand,
extraordinarily, "unrealistically" clear visions of future society; and, on the other, more
"realistic" than existing society in a Hegelian-like sense of the ideal being real. See
Manuel and Manuel, Introduction, *French Utopias: An Anthology of Ideal Societies*,
ed. and tr. Manuel and Manuel (New York: Schocken, 1971), 2, 3, 5, and Manuel and
Manuel, *Utopian Thought in the Western World* (Cambridge: Harvard University Press,
1979), 27–28. On this point, see also Hansot, *Perfection and Progress*, 4–5, 17, 198–201,
who there calls all such utopias, whether classical or modern, "thought experiments."
Hansot, ibid., 201–203, also likens those utopias to play in Johan Huizinga's sense, as
described in chap. 7 above. The Dutch futurist Frederick L. Polak (b. 1907) has contributed more than any other contemporary defender of utopianism to applying this point
to actual social issues. See the review of his writings by Elise Boulding, "Retrospective:
Remembering the Future: Reflections on the Work of Fred Polak," *Alternative Futures* 2
(Summer 1979): 96–105.

5. See Kenneth M. Roemer, *Build Your Own Utopia: An Interdisciplinary Course in
Utopian Speculation* (Washington, D.C.: University Press of America, 1981). See also
Roemer, "Using Utopia to Teach the 80's: A Case for Guided Design," *World Future
Society Bulletin* 14 (July-August 1980), 1–5. Less explicit but similarly oriented are the
defenses of "practical" utopianism in Daniel Aaron, *Men of Good Hope: A Story of
American Progressives* (New York: Oxford University Press, 1951), xii; David Riesman,
"Some Observations on Community Plans and Utopia," in his *Individualism Recon-
sidered and Other Essays* (Glencoe, Ill.: Free Press, 1954), 70–98; Paul Goodman,
"Utopian Thinking," in his *Utopian Essays and Practical Proposals* (New York: Ran-
dom House, 1962), 3–22; François Bloch-Lainé, "The Utility of Utopias for Reformers,"
and Bertrand de Jouvenel, "Utopia for Practical Purposes," in *Utopias and Utopian
Thought*, 201–18 and 219–35; and Jost Hermand, "The Necessity of Utopian Thinking,"
Soundings 58 (Spring 1975): 97–111. As Hermand aptly states, "Utopias must be
concrete and must include a design for their enactment" (111). As Roy Pierce shows,
Simone Weil had a comparable attitude toward utopianism. See Pierce, "Sociology and
Utopia: The Early Writings of Simone Weil," *Political Science Quarterly* 77 (December
1962): 505–25, and its elaboration in Pierce, *Contemporary French Political Thought*
(New York: Oxford University Press, 1966), chap. 4.

6. On these schemes, see chap. 2, n. 63. The most illuminating, in this context, of
the studies cited there, Barbara Goodwin, *Social Science and Utopia: Nineteenth-
Century Models of Social Harmony* (Sussex, Eng.: Harvester Press, 1978), also treats
utopianism as an ideology alternative to existing society's own, but less explicitly and
less systematically than I do here. See Goodwin, chap. 8. As Robert A. Nisbet argues,
"At first thought, utopianism and a genuine social science may seem to be incompatible. But they are not. Utopianism is compatible with every thing but determinism, and
it can as easily be the over-all context of social science as can any other creative vision"
(*The Quest for Community*, 2d ed. [New York: Oxford University Press, 1969], xviii). As
I discuss in this chapter, however, the equation between utopianism and social science
models ordinarily presuming realizable blueprints ought not be taken too far. Utopianism, unlike social science generally, must maintain a certain distance from the real
world in order to criticize it and to try to change it. Still, technological utopianism, as
suggested in chap. 2, n. 63, surely is as social scientific in its objectives as the earlier
expressions of utopianism Goodwin and the other scholars cited there examine.

7. Karl Mannheim, *Ideology and Utopia: An Introduction to the Sociology of
Knowledge*, tr. Louis Wirth and Edward Shils (New York: Harcourt, Brace, 1936), 262.

This first English translation of the book was also an expanded version of the 1929 German original. On Mannheim's actual treatment of ideologies and utopias as outlined in this paragraph, see ibid., esp. chap. 2, pts. 1 and 10, and chap. 4, pt. 1. As he confessed, "To determine concretely . . . what in a given case is ideological and what utopian is extremely difficult. We are confronted here with the application of a concept involving values and standards" (196). Hence Kurt H. Wolff's comment that what Mannheim really advocated was ideology *and* utopia. See Wolff, Introduction, *From Karl Mannheim*, ed. Wolff (New York: Oxford University Press, 1971), lxv. And hence the appropriateness of W. Warren Wagar's argument that technological advance, in the form of technocracy, has become the dominant utopian ideology in the contemporary real world. See Wagar, "The Steel-Gray Saviour: Technocracy as Utopia and Ideology," *Alternative Futures* 2 (Spring 1979): 38–54. See also Judith A. Merkle, *Management and Ideology: The Legacy of the International Scientific Management Movement* (Berkeley and Los Angeles: University of California Press, 1980).

8. Friedrich Engels, *Socialism: Utopian and Scientific* (1880) in *The Marx-Engels Reader*, ed. Robert C. Tucker, 2d ed. (New York: Norton, 1978), 687.

9. On the pervasive inability of the leading late nineteenth- and early twentieth-century American prophets of technological change to predict the "real" future, see chap. 6, n. 114. As the student of the prophets cited there, George Wise, concludes, "As so often happens, the actual future [i.e., the mid- and late twentieth century] was not one of several perceived alternative futures, but rather a mixture of elements of the various alternatives" ("Technological Prediction, 1890–1940" [Ph.D. diss., Boston University, 1976], 74). To be sure, Marx and Engels repeatedly emphasized the various stages of development of Western society—as from capitalism to socialism—as each growing out of its predecessor. The inextricable connections they tried to draw between existing capitalist and emerging socialist society were, in part, what made their theories more "scientific" than those of the "utopians." For the "utopians" posited a more complete disjunction between the two stages of development—if still within a bourgeois ideological framework—as well as the possibility of shaping the latter stage to their particular specifications.

10. This notion of theoretical activity, to be discussed next, is part of Sheldon S. Wolin's defense of traditional political theory against contemporary behavioral political science, as summarized above in chap. 7, pp. 135–36, nn. 20, 21. On the complex evolution of "theory" and its relationship to "practice," see Nicholas Lobkowicz, *Theory and Practice: History of a Concept from Aristotle to Marx* (Notre Dame, Ind.: University of Notre Dame Press, 1967), and, more briefly, the complete entry, "Theory," in Raymond Williams, *Keywords: A Vocabulary of Culture and Society* (New York: Oxford University Press, 1976), 266–68.

11. Wolin, *Hobbes and the Epic Tradition of Political Theory* (Los Angeles: William Andrews Clark Memorial Library, 1970), 4, 5 (his italics), 8. See also Wolin, "Paradigms and Political Theories," in *Politics and Experience: Essays Presented to Professor Michael Oakeshott on the Occasion of His Retirement*, ed. Preston King and B. C. Parekh (Cambridge: Cambridge University Press, 1968), 125–52; Wolin, "Political Theory as a Vocation," in *Machiavelli and the Nature of Political Thought*, ed. Martin Fleisher (New York: Atheneum, 1972), 23–75; Wolin, "Hannah Arendt and the Ordinance of Time," *Social Research* 44 (Spring 1977): 91–105; Wolin, "Political Theory and Political Commentary," in *Political Theory and Political Education*, ed. Melvin Richter (Princeton, N.J.: Princeton University Press, 1980), 190–203; and Wolin, "Max Weber: Legitimation, Method, and the Politics of Theory," *Political Theory* 9 (August 1981): 401–24.

12. Wolin, "Paradigms and Political Theories," 148. The objective of such political theorists, he continues, "has been to change society itself: not simply to alter the way men look at the world, but to alter the world" (ibid., 144). See also Wolin, "Political

Theory as a Vocation," 58, and Wolin, "Political Theory and Political Commentary," 196–203.

13. Wolin, "Paradigms and Political Theories," 139–52, elaborates on this last point in regard to epic political theory and, moreover, actual political societies.

14. The conception of utopian theorizing outlined in this paragraph is applied, less explicitly, to Saint-Simon, Owen, Fourier, and William Godwin by Goodwin, chaps. 1, 7, 8; and to Comte by Gertrud Lenzer, Foreword and Introduction, *Auguste Comte and Positivism: The Essential Writings*, ed. Lenzer (New York: Harper Torchbooks, 1975), xiii–lxviii. Hansot, *Perfection and Progress*, 4–5, 17, 198–201, explicitly compares utopian theories with political theories but, as indicated, without the concern for trying to improve the real world emphasized by Wolin. See also Wolin's brief discussion of the "utopian element" in political theorizing in his "Political Theory and Political Commentary," 203.

15. On the historical and intellectual development of the "Frankfurt School," see Martin Jay, *The Dialectical Imagination: A History of the Frankfurt School and the Institute of Social Research, 1923–1950* (Boston: Little, Brown, 1973).

16. Max Horkheimer, "The Social Function of Philosophy," in his *Critical Theory: Selected Essays*, tr. Matthew J. O'Connell et al. (New York: Herder and Herder, 1972), 264–65. As historian Hayden White has observed, however, Hegel and other late eighteenth- and early nineteenth-century German idealists "constituted the last school of *self-confident* utopians in Western thought" ("Review Essay: George Armstrong Kelly, *Idealism, Politics and History: Sources of Hegelian Thought*," *History and Theory* 9 [1970]: 346; his italics).

17. See Herbert Marcuse, *One-Dimensional Man: Studies in the Ideology of Advanced Industrial Society* (Boston: Beacon, 1964).

18. On these points, see, besides the works cited in chap. 7, nn. 24 and 25, Roemer, " 'Utopia Made Practical': Compulsive Realism," *American Literary Realism* 7 (Summer 1974): 273–76; Stuart Ewen, *Captains of Consciousness: Advertising and the Social Roots of the Consumer Culture* (New York: McGraw-Hill, 1976); D. H. Meyer, "American Intellectuals and the Victorian Crisis of Faith," in *Victorian America*, ed. Daniel Walker Howe (Philadelphia: University of Pennsylvania Press, 1976), 65–66; Ann Douglas, *The Feminization of American Culture* (New York: Knopf, 1977); and Howard P. Segal, "Vonnegut's *Player Piano*: An Ambiguous Technological Dystopia," in *No Place Else: Explorations in Utopian and Dystopian Fiction*, ed. Eric S. Rabkin et al. (Carbondale and Edwardsville: Southern Illinois University Press, 1983), 162–81.

19. As Wolin argues about even epic political theory, "The giant, routinized structures defy fundamental alteration and, at the same time, display an unchallengeable legitimacy, for the rational, scientific, and technological principles on which they are based seem to be in perfect accord with an age committed to science, rationalism, and technology. Above all, it is a world which appears to have rendered epic theory superfluous" ("Political Theory as a Vocation," 73). A similar assessment of Western society as early as the 1910s was among the principal reasons that the philosopher Ernst Bloch (1885–1977) persistently emphasized "traces" of utopianism in existing society as opposed to full-scale blueprints for the future. In numerous writings over many years, Bloch stressed the potentials in the past and the present, "the memories," to link them to the "not yet," as he called the future. Like Marx and Engels, he was skeptical of rigid guidelines for new worlds and preferred more fragmentary utopian schemes. Indeed, he relished, as they did not, a plurality of utopian fragments to reflect the desired plurality of everyday experience in the ideal as well as real worlds. He feared uniformity and even authoritarianism in more concrete utopian plans. A good example of Bloch's writing on utopianism is *A Philosophy of the Future*, tr. John Cumming (New York: Herder and Herder, 1970), esp. 84–144. On Bloch, see, for example, Fredric

Jameson, *Marxism and Form: Twentieth-Century Dialectical Theories of Literature* (Princeton, N.J.: Princeton University Press, 1971), 116–59, and Jurgen Habermas, "Ernst Bloch—A Marxist Romantic," in *The Legacy of the German Refugee Intellectuals,* ed. Robert Boyers (New York: Schocken, 1972), 292–306.

20. This is not, however, to deny Frank Manuel's blunt statements: "to attack utopias is about as meaningful as to denounce dreaming"; and "Perhaps our danger lies elsewhere, in the possibility that the utopian quest may become all too matter-of-fact" ("Toward a Psychological History of Utopias," in *Utopias and Utopian Thought,* 95) and so insufficiently dreamlike. In this regard see also Manuel and Manuel, *Utopian Thought in the Western World,* 813–14.

21. On this attitude toward existing American society, see Arblaster and Lukes, 13, and Wolin, "Political Theory and Political Commentary," 192–99. Yet as Williams properly points out at the end of his entry, "Ideology," the term "is still mainly used in the [negative] sense given by Napoleon. Sensible people rely on *experience,* or have a *philosophy;* silly people rely on ideology" (130). Confirmation of this contemporary (ideological) stance on ideology can be found in Lewis S. Feuer, *Ideology and the Ideologists* (New York: Harper and Row, 1975), and in Melvin J. Lasky, *Utopia and Revolution* (Chicago: University of Chicago Press, 1976). Hence the considerable difficulties faced by George Lichtheim, "The Concept of Ideology," *History and Theory* 4 (1965): 164–95, and Emmet Kennedy, *A Philosophe in the Age of Revolution: Destutt de Tracy and the Origins of "Ideology"* (Philadelphia: American Philosophical Society, 1978), among others, in trying to restore and preserve the concept's original meaning and purpose—an effort closely akin to that of Wolin and others regarding political theory in general. See the Introduction, pp. 4–5 and nn. 7–9 above.

22. On this prospect, see, for example, Wilbert E. Moore, "The Utility of Utopias," *American Sociological Review* 31 (December 1966): 765–72; Robert L. Heilbroner, "Economic Problems of a 'Postindustrial' Society," *Dissent* 20 (Spring 1973): 176; Hazel Henderson, *Creating Alternative Futures: The End of Economics* (New York: Berkley Windhover Books, 1978); and Alexandra Aldridge, "Imagining Alternative Futures: The Polarities of Contemporary Utopian Thought," *JGE: The Journal of General Education* 33 (Spring 1981): 80–89.

23. Marcuse, "The End of Utopia," in his *Five Lectures: Psychoanalysis, Politics, and Utopia,* tr. Jeremy J. Shapiro and Shierry M. Weber (Boston: Beacon, 1970), 62. Here as elsewhere Marcuse attempts to connect utopianism with Marxism despite the traditional divergence between them. So, too, more explicitly and more optimistically, does Bloch. Their positive reading of the "end of utopia"—utopia as the "impossible"—contrasts with the negative reading of that same development, if for earlier periods, by Shklar, Hansot, and Ulam. Shklar declares sadly and even sarcastically, "The end of [classical] utopian literature did not mark the end of hope; on the contrary, it coincided with the birth of historical optimism" ("The Political Theory of Utopianism," 107).

24. Ironically, the Manuels, harsh critics of Marcuse in chap. 34 of their *Utopian Thought in the Western World,* nevertheless say in their epilogue, lamenting the contemporary absence of "significant utopian thought" (813): "Just when magnificent new scientific powers have become available to us, we are faced with a paucity of invention in utopian modalities. . . . What distresses a critical historian today is the discrepancy between the piling up of technological and scientific instrumentalities for making all things possible, and the pitiable poverty of goals" (811).

Bibliography

The Bibliography includes none of the primary and few of the secondary works cited in the Appendix. It includes nearly all the other primary works and secondary works cited in the notes plus some other works not cited in them. Pertinent cross-references have been provided.

Aaron, Daniel. *Men of Good Hope: A Story of American Progressives.* New York: Oxford University Press, 1951.

Ackerman, James. "The Demise of the *Avant Garde:* Notes on the Sociology of Recent American Art." *Comparative Studies in Society and History* 2 (October 1969): 371–84.

Adams, Robert P. "The Social Responsibilities of Science in *Utopia, New Atlantis,* and After." *Journal of the History of Ideas* 10 (June 1949): 374–98.

Aitken, Hugh G. J. *Taylorism at Watertown Arsenal: Scientific Management in Action, 1908–1915.* Cambridge: Harvard University Press, 1960.

Akin, William E. *Technocracy and the American Dream: The Technocrat Movement, 1900–1941.* Berkeley and Los Angeles: University of California Press, 1977.

Aldridge, Alexandra. "Imagining Alternative Futures: The Polarities of Contemporary Utopian Thought." *JGE: The Journal of General Education* 33 (Spring 1981): 80–89.

Allain, Mathé, ed. *France and North America: Utopias and Utopians.* Lafayette: Center for Louisiana Studies, University of Southwestern Louisiana, 1978.

Allen, Zachariah. *The Science of Mechanics, As Applied to the Present Improvements in the Useful Arts in Europe, and in the United States of America.* Providence, R.I.: Hutchens and Cory, 1829.

Almond, Gabriel A., Marvin Chodorow, and Roy Harvey Pearce, eds. *Progress and Its Discontents.* Berkeley and Los Angeles: University of California Press, 1982.

Ambrose, Stephen E., ed. *Institutions in Modern America: Innovation in Structure and Process.* Baltimore: Johns Hopkins University Press, 1967.

Anderson, David. *Simone Weil.* London: SCM Press, 1971.

Anderson, Nels. *Dimensions of Work: The Sociology of a Work Culture.* New York: McKay, 1964.

Anderson, Thornton. *Brooks Adams: Constructive Conservative.* Ithaca, N.Y.: Cornell University Press, 1951.

Andreae, Johann Valentin. *Christianopolis.* Translated by Felix Held. In Held, *Johann Valentin Andreae's* Christianopolis: *An Ideal State of the Seventeenth Century,* 131–280. Urbana: Graduate School, University of Illinois, 1914.

Andreano, Ralph, ed. *The Economic Impact of the American Civil War.* 2d ed. Cambridge, Mass.: Schenkman, 1967.

Arblaster, Anthony, and Steven Lukes, eds. *The Good Society: A Book of Readings.* London: Methuen, 1971.

Ardzrooni, Leon. "Veblen and Technocracy." *Living Age* 344 (March 1933): 39–42.

Arendt, Hannah. *The Human Condition.* Garden City, N.Y.: Doubleday Anchor Books, 1959.

Armytage, Walter H. G. *The Rise of the Technocrats: A Social History.* London: Routledge and Kegan Paul, 1965.

———. *Yesterday's Tomorrows: A Historical Survey of Future Societies.* London: Routledge and Kegan Paul, 1968.

Arnold, Joseph L. *The New Deal in the Suburbs: A History of the Greenbelt Town Program, 1935–1954.* Columbus: Ohio State University Press, 1971.

Arnold, Matthew. *Culture and Anarchy.* Edited by J. Dover Wilson. Cambridge: Cambridge University Press, 1960.

Artz, Frederick B. *The Development of Technical Education in France, 1500–1800.* Cambridge: MIT Press, 1966.

Ashton, T. S. *The Industrial Revolution: 1760–1830.* New York: Oxford University Press, 1964.

Avineri, Shlomo. "Marx's Vision of Future Society and the Problem of Utopianism." *Dissent* 20 (Summer 1973): 323–31.

Axelos, Kostas. *Alienation, Praxis, and Techne in the Thought of Karl Marx.* Translated by Ronald Bruzina. Austin: University of Texas Press, 1976.

Bacon, Francis. *The Great Instauration and New Atlantis.* Edited by J. Weinberger. Arlington Heights, Ill.: AHM, 1980.

———. *The New Atlantis.* In *Ideal Commonwealths,* edited by Henry Morley, 171–213. London: Routledge, 1896.

Baczko, Bronislaw. *Lumières de l'utopie (Critique de la Politique).* Paris: Payot, 1978.

Bailey, J. O. *Pilgrims through Space and Time: Trends and Patterns in Scientific and Utopian Fiction.* New York: Argus Books, 1947.

Baker, Keith Michael. *Condorcet: From Natural Philosophy to Social Mathematics.* Chicago: University of Chicago Press, 1975.

Barde, Robert E. "Arthur E. Morgan, First Chairman of TVA." *Tennessee Historical Quarterly* 30 (Fall 1971): 299–314.

Baritz, Loren. "The Idea of the West." *American Historical Review* 66 (April 1961): 618–40.

———. *The Servants of Power: A History of the Use of Social Science in American Industry.* Middletown, Conn.: Wesleyan University Press, 1960.

Barney, Gerald O., et al. *The Global 2000 Report.* Washington, D.C.: Council on Environmental Quality and U.S. Department of State, 1980.

Barzun, Jacques. *The Use and Abuse of Art.* Princeton, N.J.: Princeton University Press, 1974.

Basalla, George. "Museums and Technological Utopianism." In *Technological Innovation and the Decorative Arts,* edited by Ian M. G. Quimby and Polly Anne Earl, 355–73. Charlottesville: University Press of Virginia, 1974.

Batchelor, Peter. "The Origin of the Garden City Concept of Urban Form." *Journal of the Society of Architectural Historians* 28 (October 1969): 184–200.

Beard, Charles A. "Is Western Civilization in Peril?" *Harper's* 157 (August 1928): 265–73.

————, ed. *Toward Civilization*. New York: Longmans, Green, 1930.

Beauchamp, Gorman. "Art in Utopia." *Southern Quarterly* 15 (July 1977): 319–33.

————. "The Dream of Cockaigne: Some Motives for the Utopias of Escape." *Centennial Review* 24 (Fall 1981): 345–62.

————. "New Maps of Utopia." Review essay. *Michigan Quarterly Review* 19 (Spring 1980): 261–69.

————. "Utopia and Its Discontents." *Midwest Quarterly* 16 (Winter 1975): 161–74.

Beaver, Daniel R. *Newton D. Baker and the American War Effort, 1917–1919*. Lincoln: University of Nebraska Press, 1966.

Beck, Evelyn Torton. "Sexism, Racism and Class Bias in German Utopias of the Twentieth Century." *Soundings* 58 (Spring 1975): 112–29.

Bell, Daniel. "Charles Fourier: Prophet of Eupsychia." *American Scholar* 38 (Winter 1968–69): 41–58.

————. *The Coming of Post-Industrial Society: A Venture in Social Forecasting*. New York: Basic Books, 1973.

————. *The Cultural Contradictions of Capitalism*. New York: Basic Books, 1976.

————. *The End of Ideology: On the Exhaustion of Political Ideas in the Fifties*. 2d ed. New York: Free Press, 1965.

————. "Reappraisals: Veblen and the New Class." *American Scholar* 32 (Autumn 1963): 616–38.

————. *Work and Its Discontents*. Boston: Beacon, 1956.

Bellamy, Edward. *Edward Bellamy Speaks Again! Articles, Public Addresses, Letters*. Chicago: Peerage Press, 1937.

Bender, Thomas. *Community and Social Change in America*. New Brunswick, N.J.: Rutgers University Press, 1978.

————. *Toward an Urban Vision: Ideas and Institutions in Nineteenth-Century America*. Lexington: University Press of Kentucky, 1975.

Bendix, Reinhard. *Work and Authority in Industry: Ideologies of Management in the Course of Industrialization*. New York: Wiley, 1956.

Bercovitch, Sacvan. *The American Jeremiad*. Madison: University of Wisconsin Press, 1979.

Berger, Peter L., ed. *The Human Shape of Work: Studies in the Sociology of Occupations*. Chicago: Regnery, 1964.

Berkhofer, Robert F. "The New or the Old Social History?" *Reviews in American History* 1 (March 1973): 21–28.

————. "The Organizational Interpretation of American History: A New Synthesis." In *Prospects: An Annual of American Cultural Studies*, edited by Jack Salzman, 4: 611–29. New York: Burt Franklin, 1978.

Berman, Marshall. *The Politics of Authenticity: Radical Individualism and the Emergence of Modern Society*. New York: Atheneum, 1972.

Bernard, Richard M., and Bradley R. Rice. "Political Environment and the Adoption of Progressive Municipal Reform." *Journal of Urban History* 1 (February 1975): 149–74. (See also Rice, Bradley R.)

Berneri, Marie L. *Journey Through Utopia*. New York: Schocken, 1971.

Bernstein, Marver H. *Regulating Business by Independent Commission*. Princeton, N.J.: Princeton University Press, 1955.

Berthoff, Rowland T. *An Unsettled People: Social Order and Disorder in American History*. New York: Harper and Row, 1971.

Bestor, Arthur E., Jr. "American Phalanxes: A Study of Fourierist Socialism in the United States, with Special Reference to the Movement in Western New York." Ph.D. diss., Yale University, 1938.

————. *Backwoods Utopias: The Sectarian Origins and the Owenite Phase of Communitarian Socialism in America, 1663–1829*. 2d ed. Philadelphia: University of Pennsylvania Press, 1970.

Bienvenu, Richard. "Reflections: European Utopias and Society"; and Michael Fellman, comment on Bienvenu's paper. American Historical Association annual meeting, 1978. (See also Fourier, Charles.)

Bierman, Judah. "Science and Society in *The New Atlantis* and Other Renaissance Utopias." *Publications of the Modern Language Association* 78 (December 1963): 492–500.

Bigelow, Jacob. *An Address on the Limits of Education.* Boston: Dutton, 1865.

———. *Elements of Technology, Taken Chiefly from a Course of Lectures Delivered at Cambridge, on the Application of the Sciences to the Useful Arts.* Boston: Hillard, Gray, Little, and Wilkins, 1829.

———. *Remarks on Classical and Utilitarian Studies.* Boston: Little, Brown, 1867.

———. *The Useful Arts, Considered in Connexion with the Applications of Science.* Boston: Marsh, Capen, Lyon, and Webb, 1840.

Billington, David P. "Structures and Machines: The Two Sides of Technology." *Soundings* 57 (Fall 1974): 275–88.

———. "Technology and the Structuring of Cities." In *Small Comforts for Hard Times: Humanists on Public Policy,* edited by Michael Mooney and Florian Stuber, 182–98. New York: Columbia University Press, 1977. (See also Salvadori, Mario G.)

Birnbaum, Norman. "The Problem of a Knowledge Elite." In his *Toward a Critical Sociology.* New York: Oxford University Press, 1971, 416–41.

Bleich, David. "Eros and Bellamy." *American Quarterly* 16 (Fall 1964): 445–59.

———. "Utopia: The Psychology of a Cultural Fantasy." Ph.D. diss., New York University, 1968.

Bloch, Ernst. *A Philosophy of the Future.* Translated by John Cumming. New York: Herder and Herder, 1970.

Boettner, Loraine. *The Millennium.* Philadelphia: Presbyterian and Reformed Church, 1958.

Boggs, W. Arthur. "Looking Backward at the Utopian Novel, 1888–1900." *Bulletin of the New York Public Library* 64 (June 1960): 329–36.

Boguslaw, Robert. *The New Utopians: A Study of System Design and Social Change.* Englewood Cliffs, N.J.: Prentice-Hall, 1965.

Bonansea, Bernardino M. *Tommaso Campanella: Renaissance Pioneer of Modern Thought.* Washington, D.C.: Catholic University of America Press, 1969.

Boorstin, Daniel J. *The Americans: The Democratic Experience.* New York: Random House, 1973.

———. *The Genius of American Politics.* Chicago: University of Chicago Press, 1953.

———. *The Republic of Technology: Reflections on Our Future Community.* New York: Harper and Row, 1978.

Boroush, Mark A., Kan Chen, and Alexander N. Christakis, eds. *Technology Assessment: Creative Futures.* New York: Elsevier North Holland, 1980.

Boulding, Elise. "Retrospective: Remembering the Future: Reflections on the Work of Fred Polak." *Alternative Futures* 2 (Summer 1979): 96–105.

Boulding, Kenneth E. *The Organizational Revolution: A Study in the Ethics of Economic Organization.* New York: Harper, 1953.

———. Michael Kammen, and Seymour Martin Lipset. *From Abundance to Scarcity: Implications for the American Tradition.* Columbus: Ohio State University Press, 1978.

Boyer, Paul. *Urban Masses and Moral Order in America, 1820–1920.* Cambridge: Harvard University Press, 1978.

Bozeman, Theodore Dwight. *Protestants in an Age of Science: The Baconian Ideal and Antebellum American Religious Thought.* Chapel Hill: University of North Carolina Press, 1977.

Braeman, John, Robert H. Bremner, and David Brody, eds. *Change and Continuity in Twentiety-Century America: The 1920's.* Columbus: Ohio State University Press, 1968.

————, and Everett Walters, eds. *Change and Continuity in Twentieth-Century America*. Columbus: Ohio State University Press, 1964. (See also Bremner, Robert H.)

Brann, Henry Walter. "Max Weber and the United States." *Southwestern Social Science Quarterly* 25 (June 1944): 18–30.

Braunthal, Alfred. *Salvation and the Perfect Society: The Eternal Quest*. Amherst: University of Massachusetts Press, 1979.

Bremner, Robert H. *From the Depths: The Discovery of Poverty in the United States*. New York: New York University Press, 1965. (See also Braeman, John.)

Brook, Anthony. "Gary, Indiana: Steeltown Extraordinary." *Journal of American Studies* 9 (April 1975): 35–53.

Brooks, Harvey. "Technology Assessment in Retrospect." In *Technology and Change*, edited by John G. Burke and Marshall C. Eakin, 465–76. San Francisco: Boyd and Fraser, 1979.

Brown, Laurie B. *Ideology*. Baltimore: Penguin Books, 1973.

Brown, Richard D. *Modernization: The Transformation of American Life, 1600–1865*. New York: Hill and Wang, 1976.

Brown, Sanborn C. *Benjamin Thompson, Count Rumford*. Cambridge: MIT Press, 1979.

Brownell, Blaine A., and Warren E. Stickle, eds. *Bosses and Reformers: Urban Politics in America, 1880–1920*. Boston: Houghton Mifflin, 1973.

Bruchey, Stuart. *Growth of the Modern American Economy*. New York: Dodd, Mead, 1975.

————, ed. *Small Business in American Life*. New York: Columbia University Press, 1980.

Brzezinski, Zbigniew. *Between Two Ages: America's Role in the Technetronic Era*. New York: Viking, 1970.

Buber, Martin. *Paths in Utopia*. Translated by R. F. C. Hull. Boston: Beacon, 1958.

Buder, Stanley. "Ebenezer Howard: The Genesis of a Town Planning Movement." *Journal of the American Institute of Planners* 35 (November 1969): 390–98.

————. *Pullman: An Experiment in Industrial Order and Community Planning, 1880–1930*. New York: Oxford University Press, 1967.

Buenker, John D., John C. Burnham, and Robert M. Crunden. *Progressivism*. Cambridge, Mass: Schenkman, 1977. (See also Burnham, John C., and Crunden, Robert M.)

Bugliarello, George, and Dean B. Doner, eds. *The History and Philosophy of Technology*. Urbana: University of Illinois Press, 1979.

Burchard, John, and Albert Bush-Brown. *The Architecture of America: A Social and Cultural History*. Boston: Atlantic Monthly Books, 1966.

Burg, David F. *Chicago's White City of 1893*. Lexington: University Press of Kentucky, 1976.

Burner, David. *Herbert Hoover: A Public Life*. New York: Knopf, 1979.

————, and Thomas R. West. "A Technocrat's Morality: Conservatism and Hoover the Engineer." In *The Hofstadter Aegis: A Memorial*, edited by Stanley Elkins and Eric McKitrick, 235–56. New York: Knopf, 1974. (See also West, Thomas R.)

Burnham, John C., ed. *Science in America: Historical Selections*. New York: Holt, Rinehart and Winston, 1971. (See also Buenker, John D.)

Burns, Arthur F. *Wesley Mitchell and the National Bureau*. New York: National Bureau of Economic Research, 1949.

Burt, Donald C. "Utopia and the Agrarian Tradition in America, 1865–1900." Ph.D. diss., University of New Mexico, 1973.

Bury, J. B. *The Idea of Progress: An Inquiry into its Origin and Growth*. New York: Macmillan, 1932.

Bush, Donald J. "Futurama: World's Fair as Utopia." *Alternative Futures* 2 (Fall 1979): 3–20.

————. *The Streamlined Decade*. New York: Braziller, 1975.

Butt, John, ed. *Robert Owen: Prince of Cotton Spinners: A Symposium*. Newton Abbot, Eng.: David and Charles, 1971.

Cabet, Etienne. *Voyage en Icarie*. 2d ed. Paris: Mallet, 1842.

Caine, Stanley P. *The Myth of a Progressive Reform: Railroad Regulation in Wisconsin, 1903–1910*. Madison: State Historical Society of Wisconsin, 1970.

Callahan, Raymond E. *Education and the Cult of Efficiency*. Chicago: University of Chicago Press, 1962.

Callenbach, Ernest. *Ecotopia: The Notebooks and Reports of William Weston*. New York: Bantam Books, 1977.

Calvert, Monte A. "American Technology at World Fairs, 1851–1876." Master's thesis, University of Delaware, 1962.

Campanella, Tommaso. *The City of the Sun*. Translated by Thomas W. Halliday. In *Peaceable Kingdoms: An Anthology of Utopian Writings*, edited by Robert L. Chianese, 8–41. New York: Harcourt Brace Jovanovich, 1971.

Canovan, Margaret. *The Political Thought of Hannah Arendt*. New York: Harcourt Brace Jovanovich, 1974.

Carden, Maren Lockwood. *Oneida: Utopian Community to Modern Corporation*. Baltimore: Johns Hopkins University Press, 1969.

Carey, James W., and John J. Quirk. "The Mythos of the Electronic Revolution." *American Scholar* 39 (Spring 1970): 219–41, and (Summer 1970): 395–424.

Carlisle, Robert B. "The Birth of Technocracy: Science, Society, and Saint-Simonians." *Journal of the History of Ideas* 35 (July-September 1974): 445–64.

————. "Saint-Simonian Radicalism: A Definition and a Direction." *French Historical Studies* 5 (Fall 1968): 430–45.

Carlyle, Thomas. "Signs of the Times." *Edinburgh Review* 49 (June 1829): 439–59.

Cary, Francine C. "Shaping the Future in the Gilded Age: A Study of Utopian Thought, 1888–1900." Ph.D. diss., University of Wisconsin, 1975.

————. "*The World a Department Store*: Bradford Peck and the Utopian Endeavor." *American Quarterly* 29 (Fall 1977): 370–84.

Cazes, Bernard. "Condorcet's True Paradox, or, The Liberal Transformed into Social Engineer." *Daedalus* 105 (Winter 1976): 47–58.

Chambers, Clarke A. "The Belief in Progress in Twentieth-Century America." *Journal of the History of Ideas* 19 (April 1958): 197–224.

————. *Paul U. Kellogg and the Survey: Voices for Social Welfare and Social Justice*. Minneapolis: University of Minnesota Press, 1971.

Chambers, John Whiteclay, II. *The Tyranny of Change: America in the Progressive Era, 1900–1917*. New York: St. Martin, 1980.

Chandler, Alfred D., Jr. "The Beginnings of 'Big Business' in American Industry." *Business History Review* 33 (Spring 1959): 1–31.

————. *Strategy and Structure: Chapters in the History of the American Industrial Enterprise*. Cambridge: MIT Press, 1969.

————. *The Visible Hand: The Managerial Revolution in American Business*. Cambridge: Harvard University Press, 1977.

————, and Louis Galambos. "The Development of Large-scale Economic Organizations in Modern America." *Journal of Economic History* 30 (March 1970): 201–17.

————, and Stephen Salsbury. *Pierre S. DuPont and the Making of the Modern Corporation*. New York: Harper and Row, 1971.

Chase, Stuart. *A New Deal*. Macmillan, 1932.

————. *Technocracy: An Interpretation*. New York: Day, 1933.

Chesneaux, Jean. "Egalitarian and Utopian Traditions in the East." *Diogenes* 62 (Summer 1968): 76–102.

Chudacoff, Howard P. *The Evolution of American Urban Society.* Englewood Cliffs, N.J.: Prentice-Hall, 1975.

Cipolla, Carlo M. *Clocks and Culture, 1300–1700.* London: Collins, 1967.

Clarke, I. F. *The Pattern of Expectation, 1644–2001.* New York: Basic Books, 1979.

Clawson, Marion. *New Deal Planning: The National Resources Planning Board.* Baltimore: Johns Hopkins University Press, 1981.

Clayre, Alaisdair, ed. *Nature and Industrialization: An Anthology.* Oxford: Oxford University Press, 1977.

————. *Work and Play: Ideas and Experience of Work and Leisure.* New York: Harper and Row, 1974.

Coben, Stanley, and Lorman Ratner, eds. *The Development of an American Culture.* Englewood Cliffs, N.J.: Prentice-Hall, 1970.

Cochran, Thomas C. *Business in American Life: A History.* New York: McGraw-Hill, 1972.

Cohen, David K. "Lemontey: An Early Critic of Industrialism." *French Historical Studies* 4 (Spring 1966): 290–303.

————. "The Vicomte de Bonald's Critique of Industrialism." *Journal of Modern History* 41 (December 1969): 475–84.

Cohn, Norman. *The Pursuit of the Millennium: Revolutionary Millenarians and Mystical Anarchists of the Middle Ages.* 3d rev. ed. New York: Oxford Galaxy Books, 1970.

Colburn, David R., and George E. Pozzetta, eds. *Reform and Reformers in the Progressive Era.* Westport, Conn.: Greenwood Press, 1983.

Commager, Henry Steele. "America and the Enlightenment." In *The Development of a Revolutionary Mentality,* 7–29. Washington, D. C.: Library of Congress, 1972.

Comte, Auguste. *Auguste Comte and Positivism: The Essential Writings.* Edited by Gertrud Lenzer. New York: Harper Torchbooks, 1975.

————. *Auguste Comte: The Foundation of Sociology.* Edited by Kenneth Thompson. New York: Wiley, 1975.

————. *System of Positive Polity.* Translated by J. H. Bridges et al. 4 vols. New York: Burt Franklin, 1968.

Condorcet, Marquis de. *Condorcet: Selected Writings.* Edited and translated by Keith Michael Baker. Indianapolis: Bobbs-Merrill, 1976.

————. *Fragment sur l'Atlantide ou efforts combinés de l'espèce humaine pour le progrès des sciences.* In his *Oeuvres,* edited by A. Condorcet-O'Connor and M. F. Arago, 6: 597–606. Paris: Firmin Didot Frères, 1847.

————. *Sketch for a Historical Picture of the Progress of the Human Mind.* 1795. Reprint, translated by June Barraclough. New York: Noonday, 1955.

Conkin, Paul K. *Tomorrow a New World: The New Deal Community Program.* Ithaca, N.Y.: Cornell University Press, 1959.

Cooper, Chester L., ed. *Growth in America.* Westport, Conn.: Greenwood Press, 1976.

Copley, Frank B. *Frederick Winslow Taylor, Father of Scientific Management.* 2 vols. New York: Harper, 1923.

Creese, Walter L. *The Search for Environment: The Garden City: Before and After.* New Haven, Conn.: Yale University Press, 1966.

Crenson, Matthew A. *The Federal Machine: Beginnings of Bureaucracy in Jacksonian America.* Baltimore: Johns Hopkins University Press, 1975.

Crews, Frederick. *Out of My System: Psychoanalysis, Ideology, and Critical Method.* New York: Oxford University Press, 1975.

Cro, Stelio. "The New World in Spanish Utopianism." *Alternative Futures* 2 (Summer 1979): 39–53.

Crook, David H. "Louis Sullivan, the World's Columbian Exposition, and American Life." Ph.D. diss., Harvard University, 1963.

Crunden, Robert M. *Ministers of Reform: The Progressives' Achievement in American Civilization, 1889–1920.* New York: Basic Books, 1982. (See also Buenker, John D.)

Cuff, Robert D. "American Historians and the 'Organizational Factor.' " *Canadian Review of American Studies* 4 (Spring 1973): 19–31.

———. "From Market to Manager." *Canadian Review of American Studies* 10 (Spring 1979): 47–54.

———. *The War Industries Board: Business-Government Relations During World War I.* Baltimore: Johns Hopkins University Press, 1973.

———. "We Band of Brothers—Woodrow Wilson's War Managers." *Canadian Review of American Studies* 5 (Fall 1974): 135–48.

Curti, Merle. "America at World Fairs." *American Historical Review* 55 (July 1950): 833–56.

———. *The Growth of American Thought.* 2d ed. New York: Harper, 1951.

Curtis, Peter H. "Bellamy Nationalism and Later Reform Movements, 1888–1940." Ph.D. diss., Indiana University, 1973.

Cutcliffe, Stephen H. et al. *Technology and Values in American Civilization: A Guide to Information Sources.* Detroit: Gale, 1980.

Dahl, Robert A. "The Behavioral Approach in Political Science: Epitaph for a Monument to a Successful Protest." *American Political Science Review* 55 (December 1961): 763–72.

Dahlberg, Jane S. *The New York Bureau of Municipal Research: Pioneer in Government Administration.* New York: New York University Press, 1966.

Danbom, David B. *The Resisted Revolution: Urban America and the Industrialization of Agriculture, 1900–1930.* Ames: Iowa State University Press, 1979.

Daniels, George H. *American Science in the Age of Jackson.* New York: Columbia University Press, 1968.

———. "The Big Questions in the History of American Technology." *Technology and Culture* 11 (January 1970): 1–21.

Darnton, Robert. "Intellectual and Cultural History." In *The Past Before Us: Contemporary Historical Writing in the United States*, edited by Michael Kammen, 327–54. Ithaca, N.Y.: Cornell University Press, 1980.

———. "Reading, Writing, and Publishing in Eighteenth-Century France: A Case Study in the Sociology of Literature." In *Historical Studies Today*, edited by Felix Gilbert and Stephen R. Graubard, 238–80. New York: Norton, 1972.

Davis, David Brion, ed. *Antebellum American Culture: An Interpretive Anthology.* Lexington, Mass.: Heath, 1979.

Davis, J. C. *Utopia and the Ideal Society: A Study of English Utopian Writing, 1516–1700.* New York: Cambridge University Press, 1981.

Dawley, Alan. *Class and Community: The Industrial Revolution in Lynn.* Cambridge: Harvard University Press, 1976.

Deane, Phyllis. *The First Industrial Revolution.* Cambridge: Cambridge University Press, 1969.

de Grazia, Sebastian. *Of Time, Work, and Leisure.* Garden City, N.Y.: Doubleday Anchor Books, 1964.

Delmage, Rutherford E. "The American Idea of Progress, 1750–1800." *Proceedings of the American Philosophical Society* 91 (October 1947): 307–14.

De Maio, Gerald, and Bernard F. Lynch. "The Recrudescence of Political Philosophy." *Centennial Review* 21 (Spring 1977): 159–75.

Demos, John, ed. *Remarkable Provinces: 1600–1760.* New York: Braziller, 1972.

Derber, Milton. *The American Idea of Industrial Democracy, 1865–1965.* Urbana: University of Illinois Press, 1970.

Deutsch, Karl W. "Mechanism, Organism, and Society: Some Models in Natural and Social Science." *Philosophy of Science* 18 (July 1951): 230–52.

DeWitt, Benjamin Parke. *The Progressive Movement.* New York: Macmillan, 1915.

Diamond, Sigmund. *The Reputation of the American Businessman.* Cambridge: Harvard University Press, 1955.

Donald, David. "An Excess of Democracy: The American Civil War and the Social Process." In his *Lincoln Reconsidered: Essays on the Civil War Era,* 209–35. 2d ed. New York: Vintage Books, 1961.

Dorf, Richard C., and Yvonne L. Hunter, eds. *Appropriate Visions: Technology, the Environment, and the Individual.* San Francisco: Boyd and Fraser, 1978.

Dorfman, Joseph. *Thorstein Veblen and His America.* New York: Viking, 1934.

Dorsett, Lyle W. "The City Boss and the Reformer: A Reappraisal." *Pacific Northwest Quarterly* 63 (October 1972): 150–54.

Douglas, Ann. *The Feminization of American Culture.* New York: Knopf, 1977.

Douglas, Jack, ed. *Freedom and Tyranny: Social Problems in a Technological Society.* New York: Knopf, 1970.

Drucker, Peter F. "The Technological Revolution: Notes on the Relationship of Technology, Science, and Culture." *Technology and Culture* 2 (Fall 1961): 342–51.

Dubos, René. "The Despairing Optimist." *American Scholar* 40 (Winter 1970-1971): 16–20, and 42 (Summer 1973): 378–82.

―――. *The Dreams of Reason: Science and Utopias.* New York: Columbia University Press, 1961.

―――. *So Human an Animal.* New York: Scribner, 1968.

Dumazedier, Joffre. *Toward a Society of Leisure.* Translated by Stewart McClure. New York: Free Press, 1967.

Dupree, A. Hunter. *Science in the Federal Government: A History of Policies and Activities to 1940.* Cambridge: Harvard University Press, 1957.

Eakins, David W. "The Development of Corporate Liberal Policy Research in the United States, 1885–1965." Ph.D. diss., University of Wisconsin, 1966.

Easton, David. "An Approach to the Analysis of Political Systems." *World Politics* 9 (April 1957): 383–400.

Eccli, Sandy et al., eds. *Alternative Sources of Energy: Practical Technology and Philosophy for a Decentralized Society.* New York: Seabury, 1974.

Echeverria, Durand. *Mirage in the West: A History of the French Image of American Society to 1815.* Princeton, N.J.: Princeton University Press, 1957.

Egerton, John. *Visions of Utopia: Nashoba, Rugby, Ruskin, and the "New Communities" in Tennessee's Past.* Knoxville: University of Tennessee Press, 1977.

Ehrmann, Jacques. "Homo Ludens Revisited." *Yale French Studies* 41: 31–57.

Ekirch, Arthur A., Jr. *The Idea of Progress in America, 1815–1860.* New York: Columbia University Press, 1944.

―――. *Man and Nature in America.* New York: Columbia University Press, 1963.

Elkins, Stanley M. *Slavery: A Problem in American Institutional and Intellectual Life.* 2d ed. Chicago: University of Chicago Press, 1968.

Elliott, Robert C. "Literature and the Good Life: A Dilemma." *Yale Review* 65 (Autumn 1975): 24–37.

―――. *The Shape of Utopia: Studies in a Literary Genre.* Chicago: University of Chicago Press, 1970.

Ellul, Jacques. *The Political Illusion.* Translated by Konrad Kellen. New York: Knopf, 1967.

―――. *Propaganda: The Formation of Men's Attitudes.* Translated by Konrad Kellen and Jean Lerner. New York: Knopf, 1966.

―――. *The Technological Society.* Translated by John Wilkinson. New York: Vintage Books, 1964.

Elsner, Henry Jr. "Messianic Scientism: Technocracy, 1919–1960." Ph.D. diss., University of Michigan, 1963.

―――. *The Technocrats: Prophets of Automation.* Syracuse, N.Y.: Syracuse University Press, 1967.

Emerson, Harrington. *Twelve Principles of Efficiency*. New York: Engineering Magazine, 1913.

Emmerich, Herbert. *Federal Organization and Administrative Management*. University: University of Alabama Press, 1971.

The Energy Certificate. Rushland, Pa.: Technocracy Inc., 1938.

Epps, P. H. "The Golden Age." *Classical Journal* 29 (January 1934): 292–96.

Eulau, Heinz. "The Behavioral Movement in Political Science: A Personal Document." *Social Research* 35 (Spring 1968): 1–29.

Eurich, Nell. *Science in Utopia: A Mighty Design*. Cambridge: Harvard University Press, 1967.

Ewen, Stuart. *Captains of Consciousness: Advertising and the Social Roots of the Consumer Culture*. New York: McGraw-Hill, 1976.

Farrington, Benjamin. *Francis Bacon: Pioneer of Planned Science*. New York: Praeger, 1963.

Fay, Sidney B. "The Idea of Progress." *American Historical Review* 52 (January 1947): 231–47.

Feibleman, James K. "Pure Science, Applied Science, Technology, Engineering: An Attempt at Definitions." *Technology and Culture* 2 (Fall 1961): 305–17.

Fein, Albert. "The American City: The Ideal and the Real." In *The Rise of an American Architecture*, edited by Edgar Kaufmann, Jr., 51–112. New York: Praeger, 1970.

———. *Frederick Law Olmsted and the American Environmental Tradition*. New York: Braziller, 1972.

Fellman, Michael. *The Unbounded Frame: Freedom and Community in Nineteenth Century American Utopianism*. Westport, Conn.: Greenwood Press, 1973.

———. "Anachronism, Context, and Progress in Nineteenth-Century American Communitarianism." *Canadian Review of American Studies* 15 (Spring 1984): 35–48.

Ferguson, Eugene S. *Bibliography of the History of Technology*. Cambridge: MIT Press, 1968.

———. "On the Origin and Development of American Mechanical 'Know-How.' " *Midcontinent American Studies Journal* 3 (Fall 1962): 3–16.

———. "Toward a Discipline of the History of Technology." *Technology and Culture* 15 (January 1974): 13–30.

Ferguson, John. *Utopias of the Classical World*. Ithaca, N.Y.: Cornell University Press, 1975.

Ferkiss, Victor C. "Bureaucracy." In *Technology and Change*, edited by John G. Burke and Marshall C. Eakin, 86–91. San Francisco: Boyd and Fraser, 1979.

———. *The Future of Technological Civilization*. New York: Braziller, 1974.

———. *Technological Man: The Myth and the Reality*. New York: Mentor Books, 1970.

Feuer, Lewis S. *Ideology and the Ideologists*. New York: Harper and Row, 1975.

Filene, Peter G. "An Obituary for 'The Progressive Movement.' " *American Quarterly* 22 (Spring 1970): 20–34.

Fine, Sidney. *Laissez-faire and the General-Welfare State: A Study of Conflict in American Thought, 1865–1901*. Ann Arbor: University of Michigan Press, 1956.

Finley, M. I. "Utopianism Ancient and Modern." In *The Critical Spirit: Essays in Honor of Herbert Marcuse*, edited by Kurt H. Wolff and Barrington Moore, Jr., 3–20. Boston: Beacon, 1968.

Fisher, Berenice M. *Industrial Education: American Ideals and Institutions*. Madison: University of Wisconsin Press, 1967.

Fisher, Marvin. *Workshops in the Wilderness: The European Response to American Industrialization, 1830–1860*. New York: Oxford University Press, 1967.

Fisher, Seymour, and Roger P. Greenberg. *The Scientific Credibility of Freud's Theories and Therapy*. New York: Basic Books, 1977.

Fishman, Robert. *Urban Utopias in the Twentieth Century: Ebenezer Howard, Frank Lloyd Wright, and Le Corbusier*. New York: Basic Books, 1977.

Fitch, Charles H. "The Rise of a Mechanical Ideal." *Magazine of American History* 11 (June 1884): 516–27.

Fleming, Donald. "Roots of the New Conservation Movement." *Perspectives in American History* 6 (1972): 7–91.

Florman, Samuel C. *Blaming Technology: The Irrational Search for Scapegoats.* New York: St. Martin, 1981.

———. *The Existential Pleasures of Engineering.* New York: St. Martin, 1976.

Flory, Claude R. *Economic Criticism in American Fiction, 1792–1900.* Philadelphia: University of Pennsylvania Press, 1936.

Fogarty, Robert S. "American Communes, 1865–1914." *Journal of American Studies* 9 (August 1975): 145–62.

———. *Dictionary of American Communal and Utopian History.* Westport, Conn.: Greenwood Press, 1980.

Folsom, Michael Brewster, and Steven D. Lubar, eds. *The Philosophy of Manufactures: Early Debates over Industrialization in the United States.* Cambridge: MIT Press, 1982.

Forbes, Allyn. "The Literary Quest for Utopia, 1880–1900." *Social Forces* 6 (December 1927): 179–89.

Foucault, Michel. *The Archaeology of Knowledge and the Discourse on Language.* Translated by A. M. Sheridan Smith. New York: Harper Colophon Books, 1976.

———. *The Order of Things: An Archaeology of the Human Sciences.* New York: Vintage Books, 1973.

Fourier, Charles. *The Utopian Vision of Charles Fourier: Selected Texts on Work, Love, and Passionate Attraction.* Edited and translated by Jonathan Beecher and Richard Bienvenu. Boston: Beacon, 1971. (See also Bienvenu, Richard.)

Fox, Daniel M. *The Discovery of Abundance: Simon N. Patten and the Transformation of Social Theory.* Ithaca, N.Y.: Cornell University Press, 1967. (See also Patten, Simon N.)

Fredrickson, George M. *The Inner Civil War: Northern Intellectuals and the Crisis of the Union.* New York: Harper and Row, 1965.

Freund, Julian. *The Sociology of Max Weber.* Translated by Mary Ilford. New York: Vintage Books, 1969.

Friedmann, Georges. *Industrial Society: The Emergence of the Human Problems of Automation.* Edited by Harold L. Sheppard. Glencoe, Ill.: Free Press, 1955.

Fuller, Henry B. "An Industrial Utopia: Building Gary, Indiana, to Order." *Harper's Weekly* 51 (October 12, 1907): 1482–83, 1495.

Fuller, R. Buckminster. *Utopia or Oblivion: The Prospects for Humanity.* New York: Bantam Books, 1969.

Gager, John G. *Kingdom and Community: The Social World of Early Christianity.* Englewood Cliffs, N.J.: Prentice-Hall, 1975.

Galambos, Louis. *Competition and Cooperation: The Emergence of a National Trade Association.* Baltimore: Johns Hopkins University Press, 1966.

———. "The Emerging Organizational Synthesis in Modern American History." *Business History Review* 44 (Autumn 1970): 279–90.

———. *The Public Image of Big Business in America, 1880–1940: A Quantitative Study in Social Change.* Baltimore: Johns Hopkins University Press, 1975. (See also Chandler, Alfred D., Jr.)

Gardiner, Helen J. "American Utopian Fiction, 1885–1910: The Influence of Science and Technology." Ph.D. diss., University of Houston, 1978.

Garner, John S. "Leclaire, Illinois: A Model Company Town (1890–1934)." *Journal of the Society of Architectural Historians* 30 (October 1971), 219–27.

Geertz, Clifford. "Ideology as a Cultural System." In *Ideology and Discontent,* edited by David E. Apter, 47–76. New York: Free Press, 1973.

———. "Stir Crazy." *New York Review of Books* 24 (January 26, 1978): 3–6.

Gelbart, Nina R. " 'Science' in Enlightenment Utopias: Power and Purpose in Eighteenth-Century French 'Voyages Imaginaires.' " Ph.D. diss., University of Chicago, 1974.

George, Claude S., Jr. *The History of Management Thought.* Englewood Cliffs, N.J.: Prentice-Hall, 1972.

Gerber, Richard. *Utopian Fantasy: A Study of English Utopian Fiction Since the End of the Nineteenth Century.* London: Routledge and Kegan Paul, 1955.

Giddens, Anthony. "Notes on the Concepts of Play and Leisure." *Sociological Review,* n.s. 12 (March 1964): 73–89.

Giedion, Siegfried. *Mechanization Takes Command: A Contribution to Anonymous History.* New York: Norton, 1969.

Giglierano, Geoffrey. "The City and the System: Developing a Municipal Service, 1800–1915." *Cincinnati Historical Society Bulletin* 35 (Winter 1977): 223–47.

Gilbert, James B. *Designing the Industrial State: The Intellectual Pursuit of Collectivism in America, 1880–1940.* Chicago: Quadrangle, 1972.

———. *Work without Salvation: America's Intellectuals and Industrial Alienation, 1880–1910.* Baltimore: Johns Hopkins University Press, 1977.

Gilchrist, David T., and W. David Lewis, eds. *Economic Change in the Civil War Era.* Greenville, Del.: Eleutherian Mills-Hagley Foundation, 1965.

Gimpel, Jean. *The Medieval Machine: The Industrial Revolution of the Middle Ages.* New York: Holt, Rinehart and Winston, 1976.

Ginsberg, Morris. "Progress in the Modern Era." In *Dictionary of the History of Ideas,* edited by Philip P. Wiener, 3: 633–60. New York: Scribner, 1973.

Glaab, Charles N., and A. Theodore Brown. *A History of Urban America.* 2d ed. New York: Macmillan, 1976.

Gluck, Peter R., and Richard J. Meister. *Cities in Transition: Social Changes and Institutional Changes in Urban Development.* New York: Franklin Watts, 1979.

Goist, Park Dixon. *From Main Street to State Street: Town, City, and Community in America.* Port Washington, N.Y.: Kennikat, 1977.

Goldman, Robert B. *A Work Experiment: Six Americans in a Swedish Plant.* New York: Ford Foundation, 1976.

Gombrich, E. H. "Huizinga and 'Homo Ludens.' " *Times Literary Supplement* 3787 (October 4, 1974): 1083–89.

Goodman, Paul. "Utopian Thinking." In his *Utopian Essays and Practical Proposals,* 3–22. New York: Random House, 1962.

———, and Percival Goodman. *Communitas: Means of Livelihood and Ways of Life.* 2d ed. New York: Vintage Books, 1960.

Goodwin, Barbara. *Social Science and Utopia: Nineteenth-Century Models of Social Harmony.* Sussex, Eng.: Harvester Press, 1978.

Gottmann, Jean. *Megalopolis: The Urbanized Northeastern Seaboard of the United States.* Cambridge: MIT Press, 1961.

Gould, Lewis L., ed. *The Progressive Era.* Syracuse, N.Y.: Syracuse University Press, 1974.

———. *Reform and Regulation: American Politics, 1900–1916.* New York: Wiley, 1978.

Gouldner, Alvin W. "Metaphysical Pathos and the Theory of Bureaucracy." *American Political Science Review* 49 (June 1955): 496–507.

Graham, Otis L., Jr. *The Great Campaigns: Reform and War in America, 1900–1928.* Englewood Cliffs, N.J.: Prentice-Hall, 1971.

———. "The Planning Ideal and American Reality: The 1930's." In *The Hofstadter Aegis: A Memorial,* edited by Stanley Elkins and Eric McKitrick, 257–299. New York: Knopf, 1974.

———. *Toward a Planned Society: From Roosevelt to Nixon.* New York: Oxford University Press, 1976.

Grin, Carolyn. "The Unemployment Conference of 1921: An Experiment in National Cooperative Planning." *Mid-America* 55 (April 1973): 83–107.

Gutman, Herbert G. *Work, Culture, and Society in Industrializing America: Essays in American Working-Class and Social History.* New York: Vintage Books, 1977.

Habakkuk, H. J. *American and British Technology in the Nineteenth Century: The Search for Labour-Saving Inventions.* Cambridge: Cambridge University Press, 1962.

Haber, Samuel. *Efficiency and Uplift: Scientific Management in the Progressive Era.* Chicago: University of Chicago Press, 1964.

Habermas, Jurgen. "Ernst Bloch: A Marxist Romantic." In *The Legacy of the German Refugee Intellectuals,* edited by Robert Boyers, 292–306. New York: Schocken, 1972.

———. *Knowledge and Human Interests.* Translated by Jeremy Shapiro. Boston: Beacon, 1971.

———. *Toward a Rational Society: Student Protest, Science, and Politics.* Translated by Jeremy Shapiro. Boston: Beacon, 1970.

Hackett, Alice P. *Seventy Years of Best Sellers, 1895–1965.* New York: Bowker, 1967.

Hall, David D. "Intellectual History as the History of Mentalities." *Intellectual History Group Newsletter* 1 (Spring 1979): 14–16.

———. "The World of Print and Collective Mentality in Seventeenth-Century New England." In *New Directions in American Intellectual History,* edited by John Higham and Paul K. Conkin, 166–80. Baltimore: Johns Hopkins University Press, 1979.

Hall, Peter Dobkin. *The Organization of American Culture, 1700–1900: Private Institutions, Elites, and the Origins of American Nationality.* New York: New York University Press, 1982.

Hall, Tom G. "Agricultural History and the 'Organizational Synthesis': A Review Essay." *Agricultural History* 48 (April 1974): 313–25.

Haller, Mark H. *Eugenics: Hereditarian Attitudes in American Thought.* New Brunswick, N.J.: Rutgers University Press, 1963.

Hancock, John L. "Planners in the Changing American City, 1900–1940." *Journal of the American Institute of Planners* 33 (September 1967): 290–304.

Hansot, Elisabeth. *Perfection and Progress: Two Modes of Utopian Thought.* Cambridge: MIT Press, 1974.

———. "Reflections on War, Utopias, and Temporary Systems." In *Small Comforts for Hard Times: Humanists on Public Policy,* edited by Michael Mooney and Florian Stuber, 246–59. New York: Columbia University Press, 1977.

Harbeson, Robert W. "Railroads and Regulation, 1877–1916: Conspiracy or Public Interest?" *Journal of Economic History* 27 (June 1967): 230–42.

Hardy, Dennis. *Alternative Communities in Nineteenth-Century England.* London: Longman, 1979.

Harris, Neil, ed. *The Land of Contrasts: 1880–1901.* New York: Braziller, 1970.

———. "Utopian Fiction and Its Discontents." In *Uprooted Americans: Essays to Honor Oscar Handlin,* edited by Richard L. Bushman et al., 211–44. Boston: Little, Brown, 1979.

Harrison, Helen A. et al. *Dawn of a New Day: The New York World's Fair, 1939–40.* New York: The Queens Museum/New York University Press, 1980.

Harrison, J. F. C. *Quest for the New Moral World: Robert Owen and the Owenites in Britain and America.* New York: Scribner, 1969.

Hart, James D. *The Popular Book: A History of America's Literary Taste.* Berkeley and Los Angeles: University of California Press, 1971.

Hawkins, Richmond L. *Auguste Comte and the United States (1816–1853).* Cambridge: Harvard University Press, 1936.

———. *Positivism in the United States (1853–1861)*. Cambridge: Harvard University Press, 1938.

Hawley, Ellis W. *The Great War and the Search for a Modern Order: A History of the American People and Their Institutions, 1917–1933*. New York: St. Martin, 1979.

———, ed. *Herbert Hoover as Secretary of Commerce: Studies in New Deal Thought and Practice*. Iowa City: University of Iowa Press, 1981.

———. "Herbert Hoover, the Commerce Secretariat, and the Vision of an 'Associative State,' 1921–1928." *Journal of American History* 61 (June 1974): 116–40.

———. *The New Deal and the Problem of Monopoly: A Study in Economic Ambivalence*. Princeton, N.J.: Princeton University Press, 1966.

Hayden, Dolores. *Seven American Utopias: The Architecture of Communitarian Socialism, 1790–1975*. Cambridge: MIT Press, 1976.

Hays, Samuel P. *American Political History as Social Analysis*. Knoxville: University of Tennessee Press, 1980.

———.*Conservation and the Gospel of Efficiency: The Progressive Conservation Movement, 1890–1920*. New York: Atheneum, 1969.

———. "Political Parties and the Community–Society Continuum." In *The American Party System: Stages of Political Development*, edited by William Nisbet Chambers and Walter Dean Burnham, 152–81. New York: Oxford University Press, 1967.

———. "The Politics of Reform in Municipal Government in the Progressive Era." *Pacific Northwest Quarterly* 55 (October 1964): 157–69.

———. *The Response to Industrialism, 1885–1914*. Chicago: University of Chicago Press, 1957.

———. "The Social Analysis of American Political History, 1880–1920." *Political Science Quarterly* 80 (September 1965): 373–94.

Hayter, Earl W. *The Troubled Farmer, 1850–1900: Rural Adjustment to Industrialism*. DeKalb: Northern Illinois University Press, 1968.

Heald, Morrell. "Management's Responsibilities to Society: The Growth of an Idea." *Business History Review* 31 (Winter 1957): 375–84.

———. *The Social Responsibilities of Business: Company and Community, 1900–1960*. Cleveland: Case Western Reserve University Press, 1970.

Heilbroner, Robert L. "Economic Problems of a 'Postindustrial' Society." *Dissent* 20 (Spring 1973): 163–77.

Henderson, Hazel. *Creating Alternative Futures: The End of Economics*. New York: Berkley Windhover Books, 1978.

Henretta, James A. "Social History as Lived and Written." *American Historical Review* 84 (December 1979): 1293–1322.

Hermand, Jost. "The Necessity of Utopian Thinking." *Soundings* 58 (Spring 1975): 97–111.

Hertzler, Joyce O. *The History of Utopian Thought*. New York: Macmillan, 1923.

———. "On Golden Ages: Then and Now." *South Atlantic Quarterly* 39 (July 1940): 318–29.

Hexter, J. H. *More's Utopia: The Biography of an Idea*. New York: Harper Torchbooks, 1965.

———. "Thomas More: On the Margins of Modernity." *Journal of British Studies* 1 (November 1961): 20–37.

———. *The Vision of Politics on the Eve of the Reformation: More, Machiavelli, and Seysell*. New York: Basic Books, 1973.

Higham, John. *From Boundlessness to Consolidation: The Transformation of American Culture, 1848–1860*. Ann Arbor, Mich.: Clements Library, 1969.

———. "Hanging Together: Divergent Unities in American History." *Journal of American History* 61 (June 1974): 5–28.

———. *Writing American History: Essays on Modern Scholarship.* Bloomington: Indiana University Press, 1970.

Hillegas, Mark R. *The Future as Nightmare: H. G. Wells and the Anti-Utopians.* New York: Oxford University Press, 1967.

Himmelberg, Robert F. *The Origins of the National Recovery Administration: Business, Government, and the Trade Association Issue, 1921–1933.* New York: Fordham University Press, 1976.

Hindle, Brooke. *Technology in Early America: Needs and Opportunities for Study.* Chapel Hill: University of North Carolina Press, 1966.

Hinds, William A. *American Communities.* 1878. Reprint. New York: Corinth Books, 1961.

Hines, Thomas S. *Burnham of Chicago: Architect and Planner.* New York: Oxford University Press, 1974.

Hobsbawm, E. J. "The Machine Breakers." In his *Labouring Men: Studies in the History of Labour,* 7–26. Garden City, N.J.: Anchor Books, 1967.

Hobson, Wayne K. "Professionals, Progressives, and Bureaucratization: A Reassessment." *Historian* 39 (August 1977): 639–58.

Hoke, Donald. "Conference Report: The Rise of the American System of Manufactures, 1800–1970: Smithsonian Institution, March 1978." *Technology and Culture* 21 (January 1980): 67–80.

Holli, Melvin G. *Reform in Detroit: Hazen S. Pingree and Urban Politics.* New York: Oxford University Press, 1969.

Hollister-Short, G. "The Vocabulary of Technology." In *History of Technology,* edited by A. Rupert Hall and Norman Smith, 2: 125–55. London: Mansell, 1977.

Holloway, Mark. *Heavens on Earth: Utopian Communities in America, 1680–1880.* 2d ed. New York: Dover, 1966.

Holt, Glen E. "The Changing Perception of Urban Pathology: An Essay on the Development of Mass Transit in the United States." In *Cities in American History,* edited by Kenneth T. Jackson and Stanley K. Schultz, 324–43. New York: Knopf, 1972.

Honour, Hugh. *The New Golden Land: European Images of America from the Discoveries to the Present Time.* New York: Pantheon, 1975.

Hoogenboom, Ari. *Outlawing the Spoils: A History of the Civil Service Reform Movement, 1865–1883.* Urbana: University of Illinois Press, 1961.

Horkheimer, Max. "The Social Function of Philosophy." In his *Critical Theory: Selected Essays.* Translated by Matthew J. O'Connell et al., 253–72. New York: Herder and Herder, 1972.

Horowitz, Daniel. "Genteel Observers: New England Economic Writers and Industrialization." *New England Quarterly* 48 (March 1975): 65–83.

———. "An Intellectual History of American Industrialization: Academic Economists, New England Publicists, and Engineers, 1830–1910." Manuscript, in possession of author, n.d.

Hovenkamp, Herbert. *Science and Religion in America, 1800–1860.* Philadelphia: University of Pennsylvania Press, 1978.

Howe, Daniel Walker, ed. *Victorian America.* Philadelphia: University of Pennsylvania Press, 1976.

Howe, Henry. *Memoirs of the Most Eminent American Mechanics.* New York: Blake, 1844.

Hughes, Thomas Parke, ed. *Changing Attitudes toward American Technology.* New York: Harper and Row, 1975.

———. *Elmer Sperry: Inventor and Engineer.* Baltimore: Johns Hopkins University Press, 1971.

———. "The Order of the Technological World." In *History of Technology,* edited by A. Rupert Hall and Norman Smith, 5: 1–16. London: Mansell, 1980.

Huizinga, Johan. *Homo Ludens: A Study of the Play-Element in Culture*. New York: Roy, 1950.

Huntington, Samuel P. "Conservatism as an Ideology." *American Political Science Review* 51 (June 1957): 454–73.

Huth, Hans. *Nature and the American: Three Centuries of Changing Attitudes*. Berkeley and Los Angeles: University of California Press, 1957.

Huxtable, Ada Louise. "You Can't Go Home to Those Fairs [of the 1930s] Again." *New York Times*, Arts and Leisure Section, October 28, 1973.

Iggers, Georg G. *The Cult of Authority: the Political Philosophy of the Saint-Simonians*. 2d ed. The Hague: Nijhoff, 1970.

————, ed. *The Doctrine of Saint-Simon: An Exposition, First Year, 1828–1829*. Translated by Iggers. New York: Schocken, 1972.

————. "The Idea of Progress: A Critical Reassessment." *American Historical Review* 71 (October 1965): 1–17.

Inouye, Arlene, and Charles Susskind. " 'Technological Trends and National Policy,' 1937: The First Modern Technology Assessment." *Technology and Culture* 18 (October 1977): 593–621.

Israel, Jerry, ed. *Building the Organizational Society: Essays on Associational Activities in Modern America*. New York: Free Press, 1972.

Jackson, John Brinckerhoff. *American Space: The Centennial Years, 1865–1876*. New York: Norton, 1972.

Jaher, Frederic Cople, ed. *The Age of Industrialism in America: Essays in Social Structure and Cultural Values*. New York: Free Press, 1968.

————. *Doubters and Dissenters: Cataclysmic Thought in America, 1885–1918*. New York: Free Press, 1964.

Jameson, Fredric. *Marxism and Form: Twentieth-Century Dialectical Theories of Literature*. Princeton, N.J.: Princeton University Press, 1971.

Jardim, Anne. *The First Henry Ford: A Study in Personality and Business Leadership*. Cambridge: MIT Press, 1970.

Jay, Martin. *The Dialectical Imagination: A History of the Frankfurt School and the Institute of Social Research, 1923–1950*. Boston: Little, Brown, 1973.

Jennings, Francis. *The Invasion of America: Indians, Colonialism, and the Cant of Conquest*. New York: Norton, 1976.

Jensen, Gordon M. "The National Civic Federation: American Business in an Age of Social Change and Social Reform, 1900–1910." Ph.D. diss., Princeton University, 1956.

Johnson, Christopher H. *Utopian Communism in France: Cabet and the Icarians, 1839–1851*. Ithaca, N.Y.: Cornell University Press, 1974.

Jones, Howard M. *The Age of Energy: Varieties of American Experience, 1865–1915*. New York: Viking Compass Books, 1973.

————. *O Strange New World: American Culture: The Formative Years*. New York: Viking Compass Books, 1967.

Jones, Richard F. *Ancients and Moderns: A Study of the Rise of the Scientific Movement in Seventeenth-Century England*. 2d ed. St. Louis: Washington University Press, 1961.

Jones, Thomas E. "Daniel Bell's Evolving Vision of the Post-Industrial Society." *World Future Society Bulletin* 13 (January-February 1979): 7–23.

Jordy, William H. *Henry Adams: Scientific Historian*. New Haven, Conn.: Yale University Press, 1952.

Juenger, Friedrich G. *The Failure of Technology: Perfection without Purpose*. Translated by F. D. Wieck. Hinsdale, Ill.: Regnery, 1949.

Kahn, Herman. *The Coming Boom: Economic, Political, and Social*. New York: Simon and Schuster, 1982.

———, et al. *The Next 200 Years. A Scenario for America and the World.* New York: Morrow, 1976.

———, and Anthony J. Wiener. *The Year 2000: A Framework for Speculation on the Next Thirty-Three Years.* New York: Macmillan, 1967.

Kakar, Sudhir. *Frederick Taylor: A Study in Personality and Innovation.* Cambridge: MIT Press, 1970.

Kanter, Rosabeth Moss. *Commitment and Community: Communes and Utopias in Sociological Perspective.* Cambridge: Harvard University Press, 1972.

Karl, Barry D. *Charles E. Merriam and the Study of Politics.* Chicago: University of Chicago Press, 1974.

———. *Executive Reorganization and Reform in the New Deal: The Genesis of Administrative Management, 1900–1939.* Cambridge: Harvard University Press, 1963.

———. "Presidential Planning and Social Science Research: Mr. Hoover's Experts." *Perspectives in American History* 3 (1969): 347–409.

Kasson, John F. *Civilizing the Machine: Technology and Republican Values in America, 1776–1900.* New York: Grossman/Viking, 1976.

Kateb, George, ed. *Utopia.* New York: Atherton, 1971.

———. *Utopia and Its Enemies.* New York: Free Press, 1963.

Kauffman, Moritz. *Utopias or Schemes of Social Improvement.* London: Kegan Paul, 1879.

Kaufman, Burton I. *Efficiency and Expansion: Foreign Trade Organization in the Wilson Administration, 1913–1921.* Westport, Conn.: Greenwood Press, 1974.

Keller, Morton. *Affairs of State: Public Life in Late Nineteenth Century America.* Cambridge: Harvard University Press, 1977.

Kennedy, David M., *Over Here: The First World War and American Society.* New York: Oxford University Press, 1980.

———. "Overview: The Progressive Era." *Historian* 37 (May 1975): 453–68.

———. "The Political Economy of World War I." *Reviews in American History* 2 (March 1974): 102–107.

Kennedy, Emmet. *A Philosophe in the Age of Revolution: Destutt de Tracy and the Origins of "Ideology."* Philadelphia: American Philosophical Society, 1978.

Ketterer, David. *New Worlds for Old: The Apocalyptic Imagination, Science Fiction, and American Literature.* Garden City, N.Y.: Doubleday Anchor Books, 1974.

King, Richard. *The Party of Eros: Radical Social Thought and the Realm of Freedom.* Chapel Hill: University of North Carolina Press, 1972.

Kirkland, Edward D. *Dream and Thought in the Business Community, 1860–1900.* Ithaca, N.Y.: Cornell University Press, 1956.

Klare, Michael T. "The Architecture of Imperial America." *Science and Society* 33 (Summer-Fall 1969): 257–84.

Kline, Paul. *Fact and Fantasy in Freudian Theory.* London: Methuen, 1972.

Koistinen, Paul A. C. "The 'Industrial-Military Complex' in Historical Perspective: The Interwar Years." *Journal of American History* 56 (March 1970): 819–39.

———. "The 'Industrial-Military Complex' in Historical Perspective: World War I." *Business History Review* 41 (Winter 1967): 378–403.

Kolakowski, Leszek, and Stuart Hampshire, eds. *The Socialist Idea: A Reappraisal.* New York: Basic Books, 1965.

Kolko, Gabriel. *Railroads and Regulation, 1877–1916.* Princeton, N.J.: Princeton University Press, 1965.

———. *The Triumph of Conservatism: A Reinterpretation of American History, 1900–1915.* Glencoe, Ill.: Free Press, 1963.

Kostelanetz, Richard. "The Five Careers of René Dubos." *Michigan Quarterly Review* 19 (Spring 1980): 194–202.

Kouwenhoven, John A. *The Arts in Modern American Civilization.* New York: Norton, 1967.

Kranzberg, Melvin, and Carroll W. Pursell, Jr., eds. *Technology in Western Civilization.* 2 vols. New York: Oxford University Press, 1967. (See also Pursell, Carroll W., Jr.)

Kraus, Michael. "America and the Utopian Ideal in the Eighteenth Century." *Mississippi Valley Historical Review* 22 (March 1936): 487–504.

Kretzmann, Edwin M. J. "German Technological Utopias of the Pre-War Period." *Annals of Science* 3 (October 15, 1938): 417–30.

Kroeber, A. L., and Clyde Kluckhohn. *Culture: A Critical Review of Concepts and Definitions.* 1952. Reprint. New York: Vintage Books, n.d.

Kuhns, William. *The Post-Industrial Prophets: Interpretations of Technology.* New York: Weybright and Talley, 1971.

Kurland, Gerald. *Seth Low: The Reformer in an Urban and Industrial Age.* New York: Twayne, 1971.

Lakoff, Sanford A. "The Third Culture: Science in Social Thought." In *Knowledge and Power: Essays on Science and Government,* edited by Lakoff, 1–61. New York: Free Press, 1966.

Landes, David S. *The Unbound Prometheus: Technological Change and Industrial Development in Western Europe from 1750 to the Present.* Cambridge: Cambridge University Press, 1969.

Lane, James B. *"City of the Century": A History of Gary, Indiana.* Bloomington: Indiana University Press, 1978.

Lane, Robert E. "The Decline of Politics and Ideology in a Knowledgeable Society." *American Sociological Review* 31 (October 1966): 649–62.

———. "The Politics of Consensus in an Age of Affluence." *American Political Science Review* 59 (December 1965): 874–95.

Lang, S. "The Ideal City: From Plato to Howard." *Architectural Review* 112 (August 1952): 91–101.

Larrabee, Eric, and Rolf Meyersohn, eds. *Mass Leisure.* Glencoe, Ill.: Free Press, 1958.

Lasch, Christopher. *The World of Nations: Reflections on American History, Politics, and Culture.* New York: Knopf, 1973.

Lasky, Melvin J. *Utopia and Revolution.* Chicago: University of Chicago Press, 1976.

Lawson, R. Alan. *The Failure of Independent Liberalism, 1930–1941.* New York: Capricorn Books, 1972.

Layton, Edwin T., Jr. "Frederick Haynes Newell and the Revolt of the Engineers." *Midcontinent American Studies Journal* 3 (Fall 1962): 17–26.

———. "Mirror-Image Twins: The Communities of Science and Technology in 19th-Century America." *Technology and Culture* 12 (October 1971): 562–80.

———. *The Revolt of the Engineers: Social Responsibility and the American Engineering Profession.* Cleveland: Case Western Reserve University Press, 1971.

———. "Technology as Knowledge." *Technology and Culture* 15 (January 1974): 31–41.

———. "Veblen and the Engineers." *American Quarterly* 14 (Spring 1962): 64–72.

Lears, T. J. Jackson. *No Place of Grace: Antimodernism and the Transformation of American Culture, 1880–1920.* New York: Pantheon, 1981.

Levine, Daniel. *Varieties of Reform Thought.* Madison: State Historical Society of Wisconsin, 1964.

LeWarne, Charles Pierce. *Utopias on Puget Sound, 1885–1915.* Seattle: University of Washington Press, 1975.

Lichtheim, George. "The Concept of Ideology." *History and Theory* 4 (1965): 164–95.

———. "Simone Weil." In his *Collected Essays,* 458–76. New York: Viking, 1973.

Lindsey, David. *Nineteenth Century American Utopias.* St. Louis: Forum Press, 1976.

Lipow, Arthur. *Authoritarian Socialism in America: Edward Bellamy and the*

Nationalist Movement. Berkeley and Los Angeles: University of California Press, 1982.

Livesay, Harold C. *Andrew Carnegie and the Rise of Big Business.* Boston: Little, Brown, 1975.

Lloyd, Craig. *Aggressive Introvert: A Study of Herbert Hoover and Public Relations Management, 1912–1932.* Columbus: Ohio State University Press, 1972.

Lobkowicz, Nicholas. *Theory and Practice: History of a Concept from Aristotle to Marx.* Notre Dame, Ind.: University of Notre Dame Press, 1967.

Lokke, Virgil L. "The American Utopian Anti-Novel." In *Frontiers of American Culture,* edited by Ray B. Browne et al., 123–53. West Lafayette, Ind.: Purdue University Studies, 1968.

Lotchin, Roger W. "The City and the Sword: San Francisco and the Rise of the Metropolitan-Military Complex, 1919–1941." *Journal of American History* 65 (March 1979): 996–1020.

———. "The Metropolitan-Military Complex in Comparative Perspective: San Francisco, Los Angeles, and San Diego, 1919–1941." *Journal of the West* 18 (July 1979): 19–30.

Lubove, Roy. *Community Planning in the 1920's: The Contribution of the Regional Planning Association of America.* Pittsburgh: University of Pittsburgh Press, 1963.

———. *The Progressives and the Slums: Tenement House Reform in New York, 1890–1917.* Pittsburgh: University of Pittsburgh Press, 1962.

———. "Social History and the History of Landscape Architecture." *Journal of Social History* 9 (Winter 1975): 268–75.

———, ed. *The Urban Community: Housing and Planning in the Progressive Era.* Englewood Cliffs, N.J.: Prentice-Hall, 1967.

Luce, A. T. "Kincaid, Illinois: A Model Mining Town." *The American City* 13 (July 1915): 10–13.

Ludmerer, Kenneth M. *Genetics and American Society: A Historical Appraisal.* Baltimore: Johns Hopkins University Press, 1972.

Lyons, Gene M. *The Uneasy Partnership: Social Science and the Federal Government in the Twentieth Century.* New York: Russell Sage Foundation, 1969.

Machlup, Fritz. *The Production and Distribution of Knowledge in the United States.* Princeton, N. J.: Princeton University Press, 1962.

MacKenzie, Findlay, ed. *Planned Society: Yesterday, Today, Tomorrow.* New York: Prentice-Hall, 1937.

MacLeish, Archibald. "Machines and the Future." *Nation* 36 (February 8, 1933): 140–42.

Maier, Charles S. "Between Taylorism and Technocracy: European Ideologies and the Vision of Industrial Productivity in the 1920's." *Journal of Contemporary History* 5 (1970): 27–61.

Mannheim, Karl. *From Karl Mannheim.* Edited by Kurt H. Wolff. New York: Oxford University Press, 1971.

———. *Ideology and Utopia: An Introduction to the Sociology of Knowledge.* Translated by Louis Wirth and Edward Shils. New York: Harcourt, Brace, 1936.

Manuel, Frank E. *Freedom from History and Other Untimely Essays.* New York: New York University Press, 1971.

———. *The New World of Saint-Simon.* Notre Dame, Ind.: University of Notre Dame Press, 1963.

———. *The Prophets of Paris: Turgot, Condorcet, Saint-Simon, Fourier, Comte.* New York: Harper Torchbooks, 1965.

———, ed. *Utopias and Utopian Thought: A Timely Appraisal.* Boston: Beacon, 1967.

———, and Fritzie P. Manuel. "Sketch for a Natural History of Paradise." *Daedalus* 101 (Winter 1972): 83–128.

———. *Utopian Thought in the Western World*. Cambridge: Harvard University Press, 1979.

———, eds. and trans. *French Utopias: An Anthology of Ideal Societies*. New York: Schocken, 1971.

Marcell, David W. *Progress and Pragmatism: James, Dewey, Beard, and the American Idea of Progress*. Westport, Conn.: Greenwood Press, 1974.

Marcuse, Herbert. "The End of Utopia." In his *Five Lectures: Psychoanalysis, Politics, and Utopia*. Translated by Jeremy Shapiro and Shierry M. Weber, 62–82. Boston: Beacon, 1970.

———. *Eros and Civilization: A Philosophical Inquiry into Freud*. New York: Vintage Books, 1962.

———. *Negations: Essays in Critical Theory*. Translated by Jeremy Shapiro. Boston: Beacon, 1968.

———. *One-Dimensional Man: Studies in the Ideology of Advanced Industrial Society*. Boston: Beacon, 1964.

———. *Reason and Revolution: Hegel and the Rise of Social Theory*. Boston: Beacon, 1964.

Marlowe, Donald E. "Public Interest—First Priority in Engineering Design?" In *Technology and Change*, edited by John G. Burke and Marshall C. Eakin, 297–301. San Francisco: Boyd and Fraser, 1979.

Marrus, Michael R., ed. *The Emergence of Leisure*. New York: Harper and Row, 1974.

———. *The Rise of Leisure in Industrial Society*. St. Charles, Mo.: Forum Press, 1974.

Marshall, Lynn L. "The Strange Stillbirth of the Whig Party." *American Historical Review* 72 (January 1967): 445–68.

Martin, Albro. *Enterprise Denied: Origins of the Decline of American Railroads, 1897–1917*. New York: Columbia University Press, 1971.

Martin, Jay. *Harvests of Change: American Literature, 1865–1914*. Englewood Cliffs, N.J.: Prentice-Hall, 1967.

Marx, Karl. *Capital*. Edited by Friedrich Engels. New York: International, 1967.

———. *The Grundrisse*. Edited and translated by David McLellan. New York: Harper and Row, 1971.

———. *The Poverty of Philosophy*. Edited by Friedrich Engels. New York: International, 1963.

———, and Friedrich Engels. *Correspondence, 1846–1895: A Selection with Commentary and Notes*. Edited and translated by Dona Torr. New York: International, 1934.

———. *The Marx-Engels Reader*. Edited by Robert C. Tucker. 2d ed. New York: Norton, 1978. (See also Tucker, Robert C.)

Marx, Leo. *The Machine in the Garden: Technology and the Pastoral Ideal in America*. New York: Oxford University Press, 1964.

Masso, Guido. *Education in Utopias*. New York: Columbia University Press, 1927.

Mayr, Otto, and Robert C. Post, eds. *Yankee Enterprise: The Rise of the American System of Manufactures*. Washington, D.C.: Smithsonian Institution Press, 1981.

Mazlish, Bruce. "The Idea of Progress." *Daedalus* 92 (Summer 1963): 447–61.

McClymer, John F. "The Pittsburgh Survey, 1907–1914: Forging an Ideology in the Steel District." *Pennsylvania History* 41 (April 1974): 169–86.

———. *War and Welfare: Social Engineering in America, 1890–1925*. Westport, Conn.: Greenwood Press, 1980.

McCraw, Thomas K. *Morgan vs. Lilienthal: The Feud within the TVA*. Chicago: Loyola University Press, 1970.

———, ed. *Regulation in Perspective: Historical Essays*. Cambridge: Harvard University Press, 1981.

———. *TVA and the Power Fight, 1933–1939*. Philadelphia: Lippincott, 1971.

McDermott, John. "Knowledge Is Power." *Nation* 208 (April 14, 1969): 458–62.

———. "Technology: The Opiate of the Intellectuals." *New York Review of Books* 13 (July 31, 1969): 25–35.

McKenna, George. "On Hannah Arendt: Politics: As It Is, Was, Might Be." In *The Legacy of the German Refugee Intellectuals*, edited by Robert Boyers, 104–22. New York: Schocken, 1972.

McLellan, David. *The Thought of Karl Marx: An Introduction.* New York: Harper and Row, 1971. (See also Marx, Karl.)

McLeod, Richard A. *Workers and Industrialization in Ante Bellum America.* St. Louis: Forum Press, 1977.

McLuhan, Marshall, and Quentin Fiore. *War and Peace in the Global Village.* New York: Bantam Books, 1968.

McMurtry, John. *The Structure of Marx's World-View.* Princeton, N.J.: Princeton University Press, 1978.

McNaught, Kenneth. "American Progressives and the Great Society." *Journal of American History* 53 (December 1966): 504–20.

McRobie, George. *Small Is Possible.* New York: Harper and Row, 1981.

McShane, Clay. *Technology and Reform: Street Railways and the Growth of Milwaukee, 1887–1900.* Madison: State Historical Society of Wisconsin, 1974. (See also Schultz, Stanley K.)

Meadows, Donella H. et al. *The Limits to Growth.* 2d ed. New York: Signet Books, 1974.

Meier, Hugo A. "American Technology and the Nineteenth-Century World." *American Quarterly* 10 (Summer 1958): 116–30.

———. "The Technological Concept in American Social History, 1750–1860." Ph.D. diss., University of Wisconsin, 1950.

———. "Technology and Democracy, 1800–1860." *Mississippi Valley Historical Review* 43 (March 1957): 618–40.

Meikle, Jeffrey L. *Twentieth Century Limited: Industrial Design in America, 1925–1939.* Philadelphia: Temple University Press, 1979.

Melman, Seymour. "Alternative Criteria for the Design of Means of Production." Paper read at annual meeting, Organization of American Historians, 1978.

Mercier, Louis Sebastian. *Memoirs of the Year Two Thousand Five Hundred.* Translated by W. Hooper. London: Robinson, 1772.

Merkle, Judith A. *Management and Ideology: The Legacy of the International Scientific Management Movement.* Berkeley and Los Angeles: University of California Press, 1980.

Merriam, Charles E. "The National Resources Planning Board: A Chapter in American Planning Experience." *American Political Science Review* (December 1944): 1075–88.

Merritt, Raymond H. *Engineering in American Society, 1850–1875.* Lexington: University Press of Kentucky, 1969.

Mészáros, István. *Marx's Theory of Alienation.* New York: Harper Torchbooks, 1972.

Meynaud, Jean. *Technocracy.* Translated by Paul Barnes. London: Faber and Faber, 1964.

Meyerson, Martin. "Utopian Traditions and the Planning of Cities." In *The Future Metropolis*, edited by Lloyd Rodwin, 233–50. New York: Braziller, 1961.

Miller, George H. "Fairfield, a Town with a Purpose." *The American City* 9 (September 1913): 213–19.

Miller, Jonathan. *Marshall McLuhan.* New York: Viking, 1971.

Miller, Perry. *The Life of the Mind in America: From the Revolution to the Civil War.* New York: Harvest Books, 1965.

Miller, Zane L. "Cincinnati: A Bicentennial Assessment." *Cincinnati Historical Society Bulletin* 34 (Winter 1976): 231–49.

———. "Scarcity, Abundance, and American Urban History." *Journal of Urban History* 4 (February 1978): 131–55. (See also Shapiro, Henry D., and Zane L. Miller.)

Mohl, Raymond A., and Neil Betten. "The Failure of Industrial City Planning: Gary, Indiana, 1906–1910." *Journal of the American Institute of Planners* 38 (July 1972): 203–14.

Mondale, Clarence. "Daniel Webster and Technology." *American Quarterly* 14 (Spring 1962): 37–47.

Montgomery, David. "The Shuttle and the Cross: Weavers and Artisans in the Kensington Riots of 1844." *Journal of Social History* 5 (Summer 1972): 411–46.

———. *Workers' Control in America: Studies in the History of Work, Technology, and Labor Struggles.* New York: Cambridge University Press, 1979.

———. "The Working Classes of the Pre-Industrial American City, 1780–1830." *Labor History* 9 (Winter 1968): 3–22.

Moore, Barrington, Jr. "Totalitarian Elements in Pre-Industrial Societies." In his *Political Power and Social Theory: Seven Studies,* 30–88. Cambridge: Harvard University Press, 1958.

Moore, Wilbert E. "The Utility of Utopias." *American Sociological Review* 31 (December 1966): 765–72.

More, Thomas. *Utopia.* Edited by Edward Surtz. New Haven, Conn.: Yale University Press, 1964.

———. *Utopia: Norton Critical Edition,* ed. Robert M. Adams. New York: Norton, 1975.

Morgan, Arthur. *Edward Bellamy.* New York: Columbia University Press, 1944.

———. *Plagiarism in Utopia.* Yellow Springs, Ohio: Morgan, 1944.

Morgan, Edmund S. "The Puritan Ethic and the American Revolution." *William and Mary Quarterly,* 3d ser., 24 (January 1967): 3–43.

Morison, Elting E. *From Know-How to Nowhere: The Development of American Technology.* New York: Basic Books, 1975.

———. *Men, Machines, and Modern Times.* Cambridge: MIT Press, 1966.

Mott, Frank L. *Golden Multitudes: The Story of Best Sellers in the United States.* New York: Macmillan, 1947.

Mowry, George E. *The Era of Theodore Roosevelt and the Birth of Modern America, 1900–1912.* New York: Harper, 1958.

———, ed. *The Twenties: Fords, Flappers, and Fanatics.* Englewood Cliffs, N.J.: Spectrum Books, 1963.

Multhauf, Robert P. "Some Observations on the State of the History of Technology." *Technology and Culture* 15 (January 1974): 1–12.

Mumford, John K. "This Land of Opportunity: Gary, the City That Rose from a Sandy Waste." *Harper's Weekly* 52 (July 4, 1908): 22–23, 29.

Mumford, Lewis. *Art and Technics.* New York: Columbia University Press, 1952.

———. *The Story of Utopias.* New York: Viking Compass Books, 1962.

———. *Technics and Civilization.* New York: Harcourt, Brace, and World, 1963.

Nadworny, Milton J. *Scientific Management and the Unions, 1900–1932: A Historical Analysis.* Cambridge: Harvard University Press, 1955.

Nash, Gerald D. "Experiments in Industrial Mobilization: WIB and NRA." *Mid-America* 45 (July 1963): 157–74.

———. *The Great Depression and World War II: Organizing America, 1933–1945.* New York: St. Martin, 1979.

Nash, Roderick. *The American Conservation Movement.* St. Charles, Mo.: Forum Press, 1974.

———. *Wilderness and the American Mind.* 2d ed. New Haven, Conn.: Yale University Press, 1973.

Negley, Glenn, and J. Max Patrick. *The Quest for Utopia: An Anthology of Imaginary Societies.* New York: Schuman, 1952.

Nelson, Daniel. *Frederick W. Taylor and the Rise of Scientific Management*. Madison: University of Wisconsin Press, 1980.

———. *Managers and Workers: Origins of the New Factory System in the United States, 1880–1920*. Madison: University of Wisconsin Press, 1975.

Nelson, William, ed. *Twentieth Century Interpretations of Utopia: A Collection of Critical Essays*. Englewood Cliffs, N.J.: Spectrum Books, 1968.

Neufeld, Maurice. "The Contributions of the World's Columbian Exposition of 1893 to the Idea of a Planned Society in the United States." Ph.D. diss., University of Wisconsin, 1935.

———. "The White City: The Beginnings of a Planned Civilization in America." *Journal of the Illinois State Historical Society* 27 (April 1934): 71–93.

Neumann, Franz. *The Democratic and the Authoritarian State: Essays in Political and Legal Theory*. Edited by Herbert Marcuse. New York: Free Press, 1964.

Newton, Norman T. *Design on the Land: The Development of Landscape Architecture*. Cambridge: Harvard University Press, 1971.

Nisbet, Robert A. *History of the Idea of Progress*. New York: Basic Books, 1980.

———. *The Quest for Community*. 2d ed. New York: Oxford University Press, 1969.

———. *The Sociological Tradition*. New York: Basic Books, 1966.

Noble, David F. *America by Design: Science, Technology, and the Rise of Corporate Capitalism*. New York: Knopf, 1977.

———. "Social Choice in Machine Design: The Case of Automatically Controlled Machine Tools, and a Challenge for Labor." *Politics and Society* 8 (1978): 313–47.

Nordhoff, Charles. *The Communistic Societies of the United States*. 1875. Reprint. New York: Schocken, 1965.

Normano, J. F. "Saint Simon and America." *Social Forces* 11 (October 1932): 8–14.

———. "Social Utopias in American Literature." *International Review of Social History* 3 (1938): 287–99.

North, Douglas C. *Growth and Welfare in the American Past: A New Economic History*. Englewood Cliffs, N.J.: Prentice-Hall, 1966.

Norton, Paul F. "World's Fairs in the 1930's." *Journal of the Society of Architectural Historians* 24 (March 1965): 27–30.

Noyes, John H. *History of American Socialisms*. 1870. Reprint. New York: Dover, 1966.

Nye, Russel B. *Society and Culture in America, 1830–1860*. New York: Harper and Row, 1974.

O'Gorman, Edmundo. *The Invention of America: An Inquiry into the Historical Nature of the New World and the Meaning of Its History*. Bloomington: Indiana University Press, 1961.

Oliver, John W. *History of American Technology*. New York: Ronald, 1956.

Ollman, Bertell. *Alienation: Marx's Conception of Man in Capitalist Society*. Cambridge: Cambridge University Press, 1971.

Olson, Theodore. *Millennialism, Utopianism, and Progress*. Toronto: University of Toronto Press, 1982.

O'Neill, Gerard K. *The High Frontier: Human Colonies in Space*. Garden City, N.Y.: Doubleday Anchor Books, 1982.

———. *2081: A Hopeful View of the Human Future*. New York: Simon and Schuster, 1981.

Onosko, Tim. *Wasn't the Future Wonderful? A View of Trends and Technology from the 1930's*. New York: Dutton, 1979.

Owen, Robert. *A New View of Society and Report to the County of Lanark*. Edited by V. A. C. Gatrell. Baltimore: Penguin Books, 1970.

Parker, Stanley. *The Future of Work and Leisure*. London: MacGibbon and Kee, 1971.

Parrington, Vernon L., Jr. *American Dreams: A Study of American Utopias*. Providence, R.I.: Brown University Press, 1947.

Parrini, Carl P. *Heir to Empire: U.S. Economic Diplomacy, 1916–1923.* Pittsburgh: University of Pittsburgh Press, 1969.

Parssinen, T. M. "Bellamy, Morris, and the Image of the Industrial City in Victorian Social Criticism." *Midwest Quarterly* 14 (April 1973): 257–66.

Passmore, John. *Man's Responsibility for Nature: Ecological Problems and Western Traditions.* New York: Scribner, 1974.

Patten, Simon N. *The New Basis of Civilization.* Edited by Daniel M. Fox. Cambridge: Harvard University Press, 1968. (See also Fox, Daniel M.)

Pells, Richard H. *Radical Visions and American Dreams: Culture and Social Thought in the Depression Years.* New York: Harper and Row, 1973.

Pemberton, William E. *Bureaucratic Politics: Executive Reorganization During the Truman Administration.* Columbia: University of Missouri Press, 1979.

Penick, James L., Jr. *Progressive Politics and Conservation: The Ballinger-Pinchot Affair.* Chicago: University of Chicago Press, 1968.

Perrin, Noel. *Giving Up the Gun: Japan's Reversion to the Sword, 1543–1879.* Boulder, Col.: Shambhala, 1980.

Persons, Stow, ed. *Evolutionary Thought in America.* New Haven, Conn.: Yale University Press, 1950.

Pessen, Edward. *Jacksonian America: Society, Personality, and Politics.* 2d ed. Homewood, Ill.: Dorsey, 1978.

Peterson, Jon A. "The City Beautiful Movement: Forgotten Origins and Lost Meanings." *Journal of Urban History* 2 (August 1976): 415–34.

———. "The Origins of the Comprehensive City Planning Ideal in the United States, 1840–1911." Ph.D. diss., Harvard University, 1967.

Pétremont, Simone. *Simone Weil: A Life.* Translated by Raymond Rosenthal. New York: Pantheon, 1976.

Pfaelzer, Mary J. "Utopian Fiction in America, 1880–1900: The Impact of Political Theory on Literary Form." Ph.D. diss., University College, London, 1975.

Pickens, Donald K. *Eugenics and the Progressives.* Nashville: Vanderbilt University Press, 1968.

Pieper, Josef. *Leisure: The Basis of Culture.* Translated by Alexander Dru. New York: Mentor-Omega Books, 1963.

Pierce, Roy. *Contemporary French Political Thought.* New York: Oxford University Press, 1966.

———. "Sociology and Utopia: The Early Writings of Simone Weil." *Political Science Quarterly* 77 (December 1962): 505–25.

Pitzer, Donald E., ed. *Robert Owen's American Legacy: Proceedings of the Robert Owen Bicentennial Conference.* Indianapolis: Indiana Historical Society, 1972.

Plath, David W., ed. *Aware of Utopia.* Urbana: University of Illinois Press, 1971.

Plato. *Republic.* Translated by Paul Shorey. In *The Collected Dialogues of Plato,* edited by Edith Hamilton and Huntington Cairns, 575–844. New York: Pantheon, 1966.

Pochmann, Henry A. *German Culture in America: Philosophical and Literary Influences, 1600–1900.* Madison: University of Wisconsin Press, 1957.

Pocock, J. G. A. "Virtue and Commerce in the Eighteenth Century." *Journal of Interdisciplinary History* 3 (Summer 1972): 119–34.

Poggioli, Renato. "The Oaten Flute." In his *The Oaten Flute: Essays on Pastoral Poetry and the Pastoral Ideal,* 1–41. Cambridge: Harvard University Press, 1975.

Polenberg, Richard. *Reorganizing Roosevelt's Government: The Controversy over Executive Reorganization, 1936–1939.* Cambridge: Harvard University Press, 1966.

Pollard, Sidney. *The Genesis of Modern Management: A Study of the Industrial Revolution in Great Britain.* London: Arnold, 1965.

Porter, Glenn. *The Rise of Big Business, 1860–1910.* Arlington Heights, Ill.: AHM, 1973.

Poster, Mark. *The Utopian Thought of Restif De La Bretonne*. New York: New York University Press, 1971.

Potter, David M. *People of Plenty: Economic Abundance and the American Character*. Chicago: University of Chicago Press, 1954.

Pratter, Frederick E. "The Uses of Utopia: An Analysis of American Speculative Fiction, 1880–1960." Ph.D. diss., University of Iowa, 1973.

Pred, Allan R. *The Spatial Dynamics of U.S. Urban-Industrial Growth, 1800–1914: Interpretive and Theoretical Essays*. Cambridge: MIT Press, 1966.

Price, Derek J. de Solla. "Is Technology Historically Independent of Science? A Study in Statistical Historiography." *Technology and Culture* 6 (Fall 1965): 553–68.

———. "On the Historiographic Revolution in the History of Technology." *Technology and Culture* 15 (January 1974): 42–48.

Pursell, Carroll W., Jr., ed. *The Military-Industrial Complex*. New York: Harper and Row, 1972.

———, ed. *Readings in Technology and American Life*. New York: Oxford University Press, 1969. (See also Kranzberg, Melvin.)

Quandt, Jean B. *From the Small Town to the Great Community: The Social Thought of Progressive Intellectuals*. New Brunswick, N.J.: Rutgers University Press, 1970.

———. "Religion and Social Thought: The Secularization of Postmillennialism." *American Quarterly* 25 (October 1973): 390–409.

Quimby, Ian M. G., ed. *Material Culture and the Study of American Life*. New York: Norton, 1978.

Quissell, Barbara C. "The Sentimental and Utopian Novels of Nineteenth-Century America: Romance and Social Issues." Ph.D. diss., University of Utah, 1973.

Racine, Philip N. "A Progressive Fights Efficiency: The Survival of Willis Sutton, School Superintendent." *South Atlantic Quarterly* 76 (Winter 1977): 103–16.

Radosh, Ronald, and Murray N. Rothbard, eds. *A New History of Leviathan: Essays on the Rise of the American Corporate State*. New York: Dutton, 1972.

Rae, John B. "The 'Know-How' Tradition: Technology in American History." *Technology and Culture* 1 (Spring 1960): 139–50.

Ramirez, Bruno. *When Workers Fight: The Politics of Industrial Relations in the Progressive Era, 1898–1916*. Westport, Conn.: Greenwood Press, 1978.

Ransom, Ellene. *Utopus Discovers America, or Critical Realism in American Utopian Fiction, 1798–1900: A Summary of a Ph.D. Thesis* (Vanderbilt University, 1946). Folcroft, Pa.: Folcroft Press, 1970.

Rasmussen, Wayne D. "The Impact of Technological Change on American Agriculture, 1862–1962." *Journal of Economic History* 22 (December 1962): 578–82.

Raucher, Alan R. *Public Relations and Business, 1900–1929*. Baltimore: Johns Hopkins University Press, 1968.

Reingold, Nathan. "Alexander Dallas Bache: Science and Technology in the American Idiom." *Technology and Culture* 11 (April 1970): 163–77.

———. "American Indifference to Basic Research: A Reappraisal." In *Nineteenth-Century American Science: A Reappraisal*, edited by George H. Daniels, 38–62. Evanston, Ill.: Northwestern University Press, 1972.

———, and Arthur P. Molella. "Theories and Ingenious Mechanics: Joseph Henry Defines Science." *Science Studies* 3 (October 1973): 323–51.

Reps, John W. "Ideal Cities." Master of Regional Planning thesis, Cornell University, 1947.

———. *The Making of Urban America: A History of City Planning in the United States*. Princeton, N.J.: Princeton University Press, 1965.

———. *Town Planning in Frontier America*. Princeton, N.J.: Princeton University Press, 1969.

Rescher, Nicholas. "Technological Progress and Human Happiness." In his *Unpopular Essays on Technological Progress*, 3–22. Pittsburgh: University of Pittsburgh Press, 1980.

Rezneck, Samuel. "The Rise and Early Development of Industrial Consciousness in the United States, 1760–1830." *Journal of Economic and Business History*, Supplement to 4 (August 1932): 784–811.

Riasanovsky, Nicholas V. *The Teaching of Charles Fourier*. Berkeley and Los Angeles: University of California Press, 1969.

Rice, Bradley R. *Progressive Cities: the Commission Government Movement in America, 1901–1920*. Austin: University of Texas Press, 1977. (See also Bernard, Richard M.)

Rideout, Walter B. Introduction. In Ignatius Donnelly, *Caesar's Column*, edited by Rideout, vii–xxxi. Cambridge: Harvard University Press, 1960.

Riesman, David. "Some Observations on Community Plans and Utopia." In his *Individualism Reconsidered and Other Essays*, 70–98. Glencoe, Ill.: Free Press, 1954.

Rodgers, Daniel T. *The Work Ethic in Industrial America, 1850–1920*. Chicago: University of Chicago Press, 1978.

Roemer, Kenneth M., ed. *America as Utopia: Collected Essays*. New York: Burt Franklin, 1981.

———. "American Utopian Literature (1888–1900): An Annotated Bibliography." *American Literary Realism* 4 (Summer 1971): 227–54.

———. *Build Your Own Utopia: An Interdisciplinary Course in Utopian Speculation*. Washington, D.C.: University Press of America, 1981.

———. *The Obsolete Necessity: America in Utopian Writings, 1888–1900*. Kent, Ohio: Kent State University Press, 1976.

———. "Using Utopia to Teach the 80's: A Case for Guided Design." *World Future Society Bulletin* 14 (July-August 1980): 1–5.

———. " 'Utopia Made Practical': Compulsive Realism." *American Literary Realism* 7 (Summer 1974): 273–76.

———. "The Yankee(s) in Noahville." *American Literature* 45 (November 1973): 434–37.

Rooney, Charles J., Jr. "Utopian Literature as a Reflection of Social Forces in America, 1865–1917." Ph.D. diss., George Washington University, 1968.

Roper, Laura W. *FLO: A Biography of Frederick Law Olmsted*. Baltimore: Johns Hopkins University Press, 1973.

Rosen, Fred. "Labour and Liberty: Simone Weil and the Human Condition." *Theoria to Theory* 7 (October 1973): 33–47.

Rosenberg, Nathan, ed. *The American System of Manufactures: The Report of the Committee on the Machinery of the United States, 1855, and the Special Reports of George Wallis and Joseph Whitworth, 1854*. Edinburgh: Edinburgh University Press, 1969.

———. "Marx as a Student of Technology." *Monthly Review* 28 (July-August 1976): 56–77.

Rosenbloom, Richard S. "Some Nineteenth-Century Analyses of Mechanization." *Technology and Culture* 5 (Fall 1964): 489–511.

Rosenkrantz, Barbara Gutmann. *Public Health and the State: Changing Views in Massachusetts, 1842–1936*. Cambridge: Harvard University Press, 1972.

Rosenof, Theodore. "Freedom, Planning, and Totalitarianism: The Reception of F. A. Hayek's *Road to Serfdom*." *Canadian Review of American Studies* 5 (Fall 1974): 149–65.

Rotenstreich, Nathan. "The Idea of Historical Progress and Its Assumption." *History and Theory* 10 (1971): 197–221.

Rothenberg, Marc. *The History of Science and Technology in the United States: A Critical and Selective Bibliography.* New York: Garland, 1982.

Rothman, David J. *The Discovery of the Asylum: Social Order and Disorder in the New Republic.* Boston: Little, Brown, 1971.

Rothschild, Emma. *Paradise Lost: The Decline of the Auto-Industrial Age.* New York: Random House, 1973.

Rürup, Reinhard. "Reflections on the Development and Current Problems of the History of Technology." *Technology and Culture* 15 (April 1974): 161–93.

Russell, Frances T. *Touring Utopias: The Realm of Constructive Humanism.* New York: Dial, 1932.

Russett, Cynthia Eagle. *The Concept of Equilibrium in American Social Thought.* New Haven, Conn.: Yale University Press, 1966.

Sadler, Elizabeth. "One Book's Influence: Edward Bellamy's *Looking Backward.*" *New England Quarterly* 17 (December 1944): 530–55.

Saint-Simon, Henri de. *Henri Saint-Simon (1760–1825): Selected Writings on Science, Industry, and Social Organisation.* Edited and translated by Keith Taylor. London: Croom Helm, 1975.

———. *The Political Thought of Saint-Simon.* Edited by Ghita Ionescu and translated by Valence Ionescu. London: Oxford University Press, 1976.

———. *Social Organization, The Science of Man and Other Writings.* Edited and translated by Felix Markham. New York: Harper Torchbooks, 1964.

Salvadori, Mario G. "The Aesthetics of Technology: In Response to David P. Billington." In *Small Comforts for Hard Times: Humanists on Public Policy*, edited by Michael Mooney and Florian Stuber, 199–203. New York: Columbia University Press, 1977. (See also Billington, David P.)

Sanford, Charles L. "The Intellectual Origins and New-Worldliness of American Industry." *Journal of Economic History* 18 (March 1958): 1–16.

———. *The Quest for Paradise: Europe and the American Moral Imagination.* Urbana: University of Illinois Press, 1961.

Sargent, Lyman Tower. *British and American Utopian Literature, 1516–1975: An Annotated Bibliography.* Boston: G. K. Hall, 1979.

———. "A Note on the Other Side of Human Nature in the Utopian Novel." *Political Theory* 3 (February 1975): 88–97.

———. "Utopianism in Colonial America." *History of Political Thought* 4 (Winter 1983): 483–522.

———. "The 'Utopian Tradition.' " Paper in possession of author, n.d.

———. "Utopia: The Problem of Definition." *Extrapolation: A Journal of Science Fiction and Fantasy* 16 (May 1975): 137–48.

Sawyer, John E. "The Social Basis of the American System of Manufacturing." *Journal of Economic History* 14 (Fall 1954): 361–79.

Saxton, Alexander. "*Caesar's Column*: The Dialogue of Utopia and Catastrophe." *American Quarterly* 19 (Summer 1967): 224–38.

Schaffer, Daniel. *Garden Cities for America: The Radburn Experience.* Philadelphia: Temple University Press, 1982.

Scheiber, Harry N. *Ohio Canal Era: A Case Study of Government and the Economy, 1820–1861.* Athens: Ohio University Press, 1969.

Schiesl, Martin J. *The Politics of Efficiency: Municipal Administration and Reform in America, 1800–1920.* Berkeley and Los Angeles: University of California Press, 1977.

Schlesinger, Arthur M., Jr. *The Age of Roosevelt.* Vol. 1, *The Crisis of the Old Order, 1919–1933.* Boston: Houghton Mifflin, 1957.

———. *The Age of Roosevelt*. Vol. 3, *The Politics of Upheaval*. Boston: Houghton Mifflin, 1960.

Schlesinger, Arthur M., Sr. "Biography of a Nation of Joiners." In his *Paths to the Present*, 23–50. New York: Macmillan, 1949.

Schmitt, Peter J. *Back to Nature: The Arcadian Myth in Urban America*. New York: Oxford University Press, 1969.

Schultz, Stanley K., and Clay McShane. "To Engineer the Metropolis: Sewers, Sanitation, and City Planning in Late Nineteenth-Century America." *Journal of American History* 65 (September 1978): 389–411. (See also McShane, Clay.)

Schumacher, E. F. *Good Work*. New York: Harper and Row, 1979.

———. *Small Is Beautiful: Economics as if People Mattered*. New York: Harper and Row, 1979.

Scott, Howard. *Science versus Chaos*. New York: Technocracy Inc., 1933.

——— et al. *Introduction to Technocracy*. New York: Day, 1933.

Scott, Mel. *American City Planning since 1890*. Berkeley and Los Angeles: University of California Press, 1969.

Segal, Howard P. "The American Jeremiad of Technological Progress: Historical Perspectives." *Alternative Futures* 3 (Spring 1980): 139–52.

———. "American Visions of Technological Utopia, 1883–1933." *Markham Review* 7 (July 1978): 65–76.

———. "Appropriate Visions: In Defense of Utopianism Today." *World Future Society Bulletin* 18 (March-April 1984): 24–29.

———. "Are Fairs Obsolete?" *New York Times* Op-Ed Page, June 3, 1981, 25.

———. "Arthur E. Morgan's Conception of a Humane Technological Society: The Tennessee Valley Authority as a Case Study." Paper read at the annual meeting, Southern Historical Association, 1977.

———. "The Automobile and the Prospect of an American Technological Plateau." *Soundings* 65 (Spring 1982): 78–87.

———. "From Utopian Communities to Utopian Writings: A Change in Form and Purpose." *Communal Societies: Journal of the National Historic Communal Societies Association* 3 (Fall 1983): 93–100.

———. "Leo Marx's 'Middle Landscape': A Critique, a Revision, and an Appreciation." *Reviews in American History* 5 (March 1977): 137–150.

———. "Reconsideration: Harold Loeb's *Life in a Technocracy: What It Might Be Like* (1933)." *New Republic* 175 (October 30, 1976): 42–44.

———. "The Technological Utopians." In *Imagining Tomorrow: History, Technology, and the American Future*, edited by Joseph T. Corn, 119–36. Cambridge: MIT Press, 1986.

———. "The Uniquely American Faith in Utopia through Technology." *World's Fair 2* (Fall 1982): 11–15.

———. "Utopian Fairs." *Chicago History: The Magazine of the Chicago Historical Society* 12 (Fall 1983):7–9.

———. "Vonnegut's *Player Piano*: An Ambiguous Technological Dystopia." In *No Place Else: Explorations in Utopian and Dystopian Fiction*, edited by Eric S. Rabkin et al., 162–81. Carbondale and Edwardsville: Southern Illinois University Press, 1983.

———. "Young West: The Psyche of Technological Utopianism." *Extrapolation: A Journal of Science Fiction and Fantasy* 19 (December 1977): 50–58.

Seligman, Edwin R. A., ed. *The Encyclopaedia of the Social Sciences*. 15 vols. New York: Macmillan, 1930–1935.

Shapiro, David. *Neurotic Styles*. Austin Riggs Center Monograph Series No. 5. New York: Harper Torchbooks, 1973.

Shapiro, Henry D. "From Association to Community: The Organization of the A.A.A.S. and the Transformation of Scientific Society in the United States." Paper read at the annual meeting, American Association for the Advancement of Science, 1976.

———. "The Western Academy of Natural Sciences of Cincinnati and the Structure of Science in the Ohio Valley, 1810–1850." In *The Pursuit of Knowledge in the Early American Republic: American Scientific and Learned Societies from Colonial Times to the Civil War*, edited by Alexandra Oleson and Sanborn C. Brown, 219–47. Baltimore: Johns Hopkins University Press, 1976.

———, and Zane L. Miller, eds. *Physician to the West: Selected Writings of Daniel Drake on Science and Society*. Lexington: University Press of Kentucky, 1970. (See also Miller, Zane L.)

Shaw, William H. " 'The Handmill Gives You the Feudal Lord': Marx's Technological Determinism." *History and Theory* 18 (1979): 155–76.

———. *Marx's Theory of History*. Stanford, Cal.: Stanford University Press, 1978.

Shaw, William P. "The World's Columbian Exposition: Its Revelations and Influences." Master's thesis, Clark University, 1935.

Shklar, Judith N. *After Utopia: The Decline of Political Faith*. Princeton, N.J.: Princeton University Press, 1957.

———. *Men and Citizens: A Study of Rousseau's Social Theory*. Cambridge: Cambridge University Press, 1969.

———. "Rousseau's Two Models: Sparta and the Age of Gold." *Political Science Quarterly* 81 (March 1966): 25–59.

Shorter, Edward. "Industrial Society in Trouble: Some Recent Views." *American Scholar* 40 (Spring 1971): 330–48.

———, ed. *Work and Community in the West*. New York: Harper and Row, 1973.

Shover, John L. *First Majority–Last Minority: The Transforming of Rural Life in America*. DeKalb: Northern Illinois University Press, 1976.

Shurter, Robert L. "The Utopian Novel in America, 1888–1900." *South Atlantic Quarterly* 34 (April 1935): 137–44.

———. "The Utopian Novel in America, 1865–1900." Ph.D. diss., Case Western Reserve University, 1936.

Sibley, Mulford Q. *Technology and Utopian Thought*. Minneapolis: Burgess, 1971.

Sills, David L., ed. *International Encyclopedia of the Social Sciences*. 17 vols. New York: Macmillan, 1968.

Simon, Julian L. *The Ultimate Resource*. Princeton, N.J.: Princeton University Press, 1981.

Simon, W. M. *European Positivism in the Nineteenth Century: An Essay in Intellectual History*. Ithaca, N.Y.: Cornell University Press, 1963.

Sinclair, Bruce. *Philadelphia's Philosopher Mechanics: A History of the Franklin Institute, 1824–1865*. Baltimore: Johns Hopkins University Press, 1974.

Skinner, Quentin. "Meaning and Understanding in the History of Ideas." *History and Theory* 8 (1969): 5–53.

Smith, Henry Nash. *Virgin Land: The American West as Symbol and Myth*. Cambridge: Harvard University Press, 1950.

Smith, Merritt Roe. *Harpers Ferry Armory and the New Technology: The Challenge of Change*. Ithaca, N.Y.: Cornell University Press, 1977.

Snow, C. P. *The Two Cultures: And a Second Look*. New York: Cambridge University Press, 1964.

Somkin, Fred. *Unquiet Eagle: Memory and Desire in the Idea of American Freedom, 1815–1860*. Ithaca, N.Y.: Cornell University Press, 1967.

Soule, George. "NESPA: December 1934–December 1940." *Plan Age* 6 (November–December 1940): 289–94.

———. *A Planned Society*. New York: Macmillan, 1932.

———. *Planning U.S.A.* New York: Viking, 1967.

Stark, Werner. *The Fundamental Forms of Social Thought*. London: Routledge and Kegan Paul, 1962.

Steintrager, James. "Plato and More's *Utopia*." *Social Research* 36 (Autumn 1969): 357–72.

Stillman, Richard J., II. *The Rise of the City Manager: A Public Professional in Local Government*. Albuquerque: University of New Mexico Press, 1974.

Stupple, A. James. "Utopian Humanism in America, 1888–1900." Ph.D. diss., Northwestern University, 1971.

Susman, Warren, ed. *Culture and Commitment: 1929–1945*. New York: Braziller, 1973.

Sussman, Herbert L. *Victorians and the Machine: The Literary Response to Technology*. Cambridge: Harvard University Press, 1968.

Swain, Donald C. *Federal Conservation Policy, 1921–1933*. Berkeley and Los Angeles: University of California Press, 1963.

Symonds, Carolyn. "Technology and Utopia." In *The Future of Work*, edited by Fred Best, 174–79. Englewood Cliffs, N.J.: Prentice-Hall, 1973.

Talbert, Roy. "Arthur E. Morgan's Social Philosophy and the Tennessee Valley Authority." *East Tennessee Historical Society Publications* 4 (1969): 86–99.

———. "Beyond Pragmatism: The Story of Arthur E. Morgan." Ph.D. diss., Vanderbilt University, 1971.

———. "The Human Engineer: Arthur E. Morgan and the Launching of the TVA." Master's thesis, Vanderbilt University, 1967.

Tarbell, Ida M. *New Ideals in Business: An Account of Their Practice and Their Effects upon Men and Profits*. New York: Macmillan, 1917.

Tarr, Joel A. "From City to Suburb: The 'Moral' Influence of Transportation Technology." In *American Urban History: An Interpretive Reader With Commentaries*. 2d ed., edited by Alexander B. Callow, 202–12. New York: Oxford University Press, 1973.

———, ed. *Retrospective Technology Assessment—1976*. San Francisco: San Francisco Press, 1977.

Taylor, George Rogers. *The Transportation Revolution, 1815–1860*. New York: Holt, Rinehart and Winston, 1951.

Taylor, Graham R. *Satellite Cities: A Study of Industrial Suburbs*. New York: Appleton, 1915.

Taylor, Walter F. *The Economic Novel in America*. Chapel Hill: University of North Carolina Press, 1942.

Taylor, William R. "Toward a History of Perception." *Intellectual History Group Newsletter* 1 (Spring 1979): 16–18.

"Technocracy: Boon, Blight, or Bunk?" *Literary Digest* 114 (December 31, 1932): 5–6.

Technocracy Briefs. Seattle: Technocracy Inc., n.d.

Technocracy: Technological Social Design. Savannah, Ohio: Technocracy Inc., 1975.

Temko, Allan. "Which Guide to the Promised Land: Fuller or Mumford?" *Horizon* 10 (Summer 1968): 25–30.

Terkel, Studs. *Working: People Talk about What They Do All Day and How They Feel about What They Do*. New York: Pantheon, 1974.

Thal-Larsen, Margaret. "Political and Economic Ideas in American Utopian Fiction, 1868–1914." Ph.D. diss., University of California at Berkeley, 1941.

Thelen, David P. *The New Citizenship: Origins of Progressivism in Wisconsin, 1885–1900*. Columbia: University of Missouri Press, 1972.

———. "Social Tensions and the Origins of Progressivism." *Journal of American History* 56 (September 1969): 323–41.

Thiem, Jon E. "The Artist in the Ideal State: A Study of the Troubled Relations between Arts and Society in Utopian Fiction." Ph.D. diss., Indiana University, 1975.

Thimm, Alfred L. *Business Ideologies in the Reform-Progressive Era, 1880–1914.* University: University of Alabama Press, 1976.

Thomas, George B. "Blueprint for Tomorrow: American Novels of Future Change." Ph.D. diss., Harvard University, 1970.

Thomas, John L. *Alternative America: Henry George, Edward Bellamy, Henry Demarest Lloyd and the Adversary Tradition.* Cambridge: Harvard University Press, 1983.

———. "Antislavery and Utopia." In *The Antislavery Vanguard: New Essays on the Abolitionists,* edited by Martin Duberman, 240–69. Princeton, N.J.: Princeton University Press, 1965.

———. Introduction. In Edward Bellamy, *Looking Backward: 2000–1887,* edited by Thomas, 1–88. Cambridge: Harvard University Press, 1967.

———. "Romantic Reform in America, 1815–1865." *American Quarterly* 17 (Winter 1965): 656–81.

———. "Utopia for an Urban Age: Henry George, Henry Demarest Lloyd, Edward Bellamy." *Perspectives in American History* 6 (1972): 135–63.

Thomas, Keith. "Work and Leisure in Pre-Industrial Society." *Past and Present* 29 (December 1964): 50–62.

Thomas, Robert D. *The Man Who Would Be Perfect: John Humphrey Noyes and the Utopian Impulse.* Philadelphia: University of Pennsylvania Press, 1977.

Thomis, Malcolm I. *The Luddites: Machine-Breaking in Regency England.* New York: Schocken, 1972.

Thompson, E. P. *The Making of the English Working Class.* New York: Pantheon, 1963.

———. "Time, Work-Discipline, and Industrial Capitalism." *Past and Present* 38 (December 1967): 56–97.

Thrupp, Sylvia, ed. *Millennial Dreams in Action: Essays in Comparative History.* The Hague: Mouton, 1962.

Tichi, Cecelia. *New World, New Earth: Environmental Reform in American Literature from the Puritans through Whitman.* New Haven, Conn.: Yale University Press, 1979.

Tihany, Leslie C. "French Utopian Thought, 1676–1790." Ph.D. diss., University of Chicago, 1943.

———. "Utopia in Modern Western Thought: The Metamorphosis of an Idea." In *Ideas in History: Essays Presented to Louis Gottschalk by His Former Students,* edited by Richard Herr and Harold T. Parker, 20–38. Durham, N.C.: Duke University Press, 1965.

Tobey, Ronald G. *The American Ideology of National Science, 1919–1930.* Pittsburgh: University of Pittsburgh Press, 1971.

Tomlin, E. W. F. *Simone Weil.* New Haven, Conn.: Yale University Press, 1954.

Torbet, William R., and Malcolm P. Rogers. *Being for the Most Part Puppets: Interactions among Men's Labor, Leisure, and Politics.* Cambridge, Mass.: Schenkman, 1973.

Tozer, Lowell. "American Attitudes toward Machine Technology, 1893–1933." Ph.D. diss., University of Minnesota, 1953.

———. "A Century of Progress, 1833–1933: Technology's Triumph over Man." *American Quarterly* 4 (Spring 1952): 78–81.

Trachtenberg, Alan. *Brooklyn Bridge: Fact and Symbol.* New York: Oxford University Press, 1965.

———, ed. *Critics of Culture: Literature and Society in the Early Twentieth Century.* New York: Wiley, 1976.

———. *The Incorporation of America: Culture and Society in the Gilded Age.* New York: Hill and Wang, 1982.

Tucker, Robert C. *The Marxian Revolutionary Idea*. New York: Norton, 1969. (See also Marx, Karl, and Friedrich Engels.)

Tunnard, Christopher. *The City of Man*. 2d ed. New York: Scribner, 1970.

Tuveson, Ernest L. "Millenarianism." In *Dictionary of the History of Ideas*, edited by Philip P. Wiener, 3: 223–25. New York: Scribner, 1973.

———. *Millennium and Utopia: A Study in the Background of the Idea of Progress*. New York: Harper Torchbooks, 1964.

———. *Redeemer Nation: The Idea of America's Millennial Role*. Chicago: University of Chicago Press, 1968.

Tyack, David B. *The One Best System: A History of American Urban Education*. Cambridge: Harvard University Press, 1974.

Urofsky, Melvin I. *Big Steel and the Wilson Administration: A Study in Business-Government Relations*. Columbus: Ohio State University Press, 1969.

Van Dalsem, Newton. *History of the Utopian Society in America: An Authentic Account of its Origin and Development up to 1942*. Los Angeles: Utopian Society, 1942.

Van Riper, Paul. "American Civil Service Reform." In *Bureaucracy in Historical Perspective*, edited by Michael T. Dalby and Michael S. Weithman, 124–37. Glenview, Ill.: Scott, Foresman, 1971.

Vasu, Michael Lee. *Politics and Planning: A National Study of American Planners*. Chapel Hill: University of North Carolina Press, 1979.

Veblen, Thorstein. *The Engineers and the Price System*. New York: Viking, 1933.

Venturi, Franco. *Utopia and Reform in the Enlightenment*. New York: Cambridge University Press, 1971.

Veysey, Laurence. *The Communal Experience: Anarchist and Mystical Communities in Twentieth-Century America*. Chicago: University of Chicago Press, 1978.

———. "A New Record of American Civilization." *Reviews in American History* 1 (September 1973): 318–30.

Wagar, W. Warren. *Good Tidings: The Belief in Progress from Darwin to Marcuse*. Bloomington: Indiana University Press, 1972.

———. "Modern Views of the Origins of the Idea of Progress." *Journal of the History of Ideas* 28 (March 1967): 55–70.

———. "The Steel-Gray Saviour: Technocracy as Utopia and Ideology." *Alternative Futures* 2 (Spring 1979): 38–54.

———. *Terminal Visions: The Literature of Last Things*. Bloomington: Indiana University Press, 1982.

Walden, Daniel. "The Two Faces of Technological Utopianism: Edward Bellamy and Horatio Alger, Jr." *JGE: The Journal of General Education* 33 (Spring 1981): 24–30.

Waldo, Dwight. *The Administrative State: A Study of the Political Theory of American Public Administration*. New York: Ronald, 1948.

Walker, Timothy. "Defense of Mechanical Philosophy." *North American Review* 33 (July 1831): 122–36.

Wallace, Anthony F. C. *Rockdale: The Growth of an American Village in the Early Industrial Revolution*. New York: Knopf, 1978.

———. *The Social Context of Innovation: Bureaucrats, Families, and Heroes in the Early Industrial Revolution, as Foreseen in Bacon's New Atlantis*. Princeton, N.J.: Princeton University Press, 1982.

Walsh, Chad. *From Utopia to Nightmare*. New York: Harper and Row, 1962.

Walters, Ronald G. *American Reformers, 1815–1860*. New York: Hill and Wang, 1978.

Warken, Philip W. "A History of the National Resources Planning Board, 1933–1943." Ph.D. diss., Ohio State University, 1969.

Warner, Sam Bass, Jr. *The Private City: Philadelphia in Three Periods of Its Growth*. Philadelphia: University of Pennsylvania Press, 1968.

———. *Streetcar Suburbs: The Process of Growth in Boston, 1870–1900.* Cambridge: Harvard University Press, 1962.

———. *The Urban Wilderness: A History of the American City.* New York: Harper and Row, 1972.

Warner, W. Lloyd, ed. *The Emergent American Society.* Vol.1, *Large Scale Organizations.* New Haven, Conn.: Yale University Press, 1967.

Watzlawick, Paul, John Weakland, and Richard Fisch. "The Utopia Syndrome." In their *Change: Principles of Problem Formation and Problem Resolution,* 47–61. New York: Norton, 1974.

Waxman, Chaim., ed. *The End of Ideology Debate.* New York: Clarion Books, 1969.

Weber, Max. *The Protestant Ethic and the Spirit of Capitalism.* Translated by Talcott Parsons. New York: Scribner, 1958.

Weil, Simone. *Gravity and Grace.* Translated by Arthur Wills. New York: Putnam, 1952.

———. *The Need for Roots: Prelude to a Declaration of Duties toward Mankind.* Translated by Arthur Wills. New York: Harper Colophon Books, 1971.

———. *Oppression and Liberty.* Translated by Arthur Wills and John Petrie. Amherst: University of Massachusetts Press, 1973.

Weinberger, J. "Science and Rule in Bacon's Utopia: An Introduction to the Reading of the *New Atlantis.*" *American Political Science Review* 70 (September 1976): 865–85.

Weinstein, James. *The Corporate Ideal in the Liberal State, 1900–1918.* Boston: Beacon, 1968.

Weiss, Paul. "A Philosophical Definition of Leisure." In *Leisure in America: Blessing or Curse,* edited by James C. Charlesworth, 21–29. Philadelphia: American Academy of Political and Social Science, 1964.

Welter, Rush A. "The Idea of Progress in America: An Essay in Ideas and Method." *Journal of the History of Ideas* 16 (June 1955): 401–15.

———. *The Mind of America, 1820–1860.* New York: Columbia University Press, 1975.

West, Thomas Reed. *Flesh of Steel: Literature and the Machine in American Culture.* Nashville: Vanderbilt University Press, 1967. (See also Burner, David.)

Whalen, Robert K. "Millenarianism and Millennialism in America, 1790–1880." Ph.D. diss., State University of New York at Stony Brook, 1971.

White, Hayden. "Review Essay: George Armstrong Kelly, *Idealism, Politics and History: Sources of Hegelian Thought.*" *History and Theory* 9 (1970): 343–63.

White, Lynn, Jr. *Machina Ex Deo: Essays in the Dynamism of Western Culture.* Cambridge: MIT Press, 1968.

———. *Medieval Technology and Social Change.* New York: Oxford University Press, 1962.

Wiebe, Robert H. *Businessmen and Reform: A Study of the Progressive Movement.* Cambridge: Harvard University Press, 1962.

———. *The Search for Order, 1877–1920.* New York: Hill and Wang, 1967.

Wiener, Norbert. *Cybernetics; Or Control and Communication in the Animal and the Machine.* Cambridge, Mass.: Technology Press, 1948.

———. *The Human Use of Human Beings: Cybernetics and Society.* Boston: Houghton Mifflin, 1950.

Wiener, Philip P., ed. *Dictionary of the History of Ideas.* 4 vols. New York: Scribner, 1973.

Wilkinson, Norman B. "In Anticipation of Frederick W. Taylor: A Study of Work by Lammot du Pont, 1872." *Technology and Culture* 6 (Spring 1965): 208–21.

———. "Brandywine Borrowings from European Technology." *Technology and Culture* 4 (Winter 1963): 1–13.

Williams, George H. *Wilderness and Paradise in Christian Thought.* New York: Harper, 1962.

Williams, Raymond. *Culture and Society, 1780–1950.* New York: Harper Torchbooks, 1966.

———. *Keywords: A Vocabulary of Culture and Society.* New York: Oxford University Press, 1976.

———. "Utopia and Science Fiction." *Science-Fiction Studies* (November 1978): 203–14.

Wilson, Joan Hoff. *American Business and Foreign Policy, 1920–1933.* Lexington: University Press of Kentucky, 1971.

———. *Herbert Hoover: Forgotten Progressive.* Boston: Little, Brown, 1975.

———. *Ideology and Economics: U.S. Relations with the Soviet Union, 1918–1933.* Columbia: University of Missouri Press, 1974.

Wilson, R. Jackson. "Experience and Utopia: The Making of Edward Bellamy's *Looking Backward.*" *American Studies* 11 (April 1977): 45–60.

———. *In Quest of Community: Social Philosophy in the United States, 1860–1920.* New York: Wiley, 1968.

Winner, Langdon. *Autonomous Technology: Technics-Out-of-Control as a Theme in Political Thought.* Cambridge: MIT Press, 1977.

Wirth, Arthur G. *Education in the Technological Society: The Vocational-Liberal Studies Controversy in the Early Twentieth Century.* Scranton, Pa.: Intext, 1972.

Wise, George. "The Accuracy of Technological Forecasts." *Futures* 8 (October 1976): 411–19.

———. "Technological Prediction, 1890–1940." Ph.D. diss., Boston University, 1976.

Wolin, Sheldon S. "Hannah Arendt and the Ordinance of Time." *Social Research* 44 (Spring 1977): 91–105.

———. *Hobbes and the Epic Tradition of Political Theory.* Los Angeles: Clark Memorial Library, University of California, Los Angeles, 1970.

———. "Paradigms and Political Theories." In *Politics and Experience: Essays Presented to Professor Michael Oakeshott on the Occasion of His Retirement,* edited by Preston King and B. C. Parekh, 125–52. Cambridge: Cambridge University Press, 1968.

———. "Political Theory and Political Commentary." In *Political Theory and Political Education,* edited by Melvin Richter, 190–203. Princeton, N.J.: Princeton University Press, 1980.

———. "Political Theory as a Vocation." In *Machiavelli and the Nature of Political Thought,* edited by Martin Fleisher, 23–75. New York: Atheneum, 1972.

———. *Politics and Vision: Continuity and Innovation in Western Political Thought.* Boston: Little, Brown, 1960.

———. "Max Weber: Legitimation, Method, and the Politics of Theory." *Political Theory* 9 (August 1981): 401–24.

Woll, Peter. *American Bureaucracy.* New York: Norton, 1963.

Wood, Gordon, ed. *The Rising Glory of America, 1760–1820.* New York: Braziller, 1971.

"Work and Leisure in Industrial Society." *Past and Present* 30 (April 1965): 96–103.

Work in America: Report of a Special Task Force to the Secretary of Health, Education, and Welfare. Cambridge: MIT Press, 1973.

Wright, Louis B. *The Dream of Prosperity in Colonial America.* New York: New York University Press, 1965.

Wutke, Eugene R. "Technocracy: It Failed to Save the Nation." Ph.D. diss., University of Missouri at Kansas City, 1964.

Index

293